Global Forest Governance

Global Forest Governance

Legal Concepts and Policy Trends

Rowena Maguire

Queensland University of Technology, Australia

Edward Elgar

Cheltenham, UK • Northampton, MA, USA

Published by
Edward Elgar Publishing Limited
The Lypiatts
15 Lansdown Road
Cheltenham
Glos GL50 2JA
UK

Edward Elgar Publishing, Inc.
William Pratt House
9 Dewey Court
Northampton
Massachusetts 01060
USA

A catalogue record for this book
is available from the British Library

Library of Congress Control Number: 2012940703

FSC www.fsc.org MIX Paper from responsible sources FSC® C018575

ISBN 978 0 85793 606 6

Typeset by Servis Filmsetting Ltd, Stockport, Cheshire
Printed and bound by MPG Books Group, UK

Contents

Cases

Legislation

PART I

Overview

1. Introduction

1. OBJECTIVES OF THE BOOK

The objective of this book is to examine the doctrinal concepts that underpin international forestry regulation, and then trace the application of these concepts to a range of public and non-state international institutional forest policies and measures. By examining these concepts it becomes possible to understand the nature and requirements of the international law that governs sustainable forest management and use. Most of the existing literature on international forest regulation comes from scientific and political science authors. This book builds on existing literature discussing sustainable forest management, international forest regulation and politics, the role of international law, governance arrangements, land tenure, property rights, environmental justice and environmental markets.

Legal analysis of the concept of sustainable forest management has been limited to date. Most of it has been narrow in focus and has tended to focus on a particular instrument associated with sustainable forest management, such as National Forest Policies or, more recently, Reduced Emissions from Deforestation and Degradation (REDD+). There is a distinct lack of legal analysis exploring the nature and requirements of international law concerning sustainable forest management. This is probably explained in part by the lack of clarity surrounding the legal status of sustainable forest management obligations. The analysis in this book will be of relevance to those undertaking an examination of a particular issue or forest regulation challenge. The book identifies and explains the underlying legal issues in forest regulation, analyses the dominant regulatory approaches at the international level, and makes some recommendations on how to improve global forest governance.

This book deals with the concept of 'sustainable forest management' as distinct from the concept of 'forests' and deals with the legal principles associated with the concept of sustainable forest management. The purpose of this is to provide an analytical model for legal forest analysis. The concept and meaning of 'sustainable forest management' is further explored in Chapter 4, though for present purposes it can be stated that sustainable forest management is an ever-evolving concept which attempts to incorporate and recognize all forest values (ecological, social and economic). The concept of sustainable forest management is therefore applicable to all forest ecosystems.

Guidance on the definition of 'forest' can be drawn from the Intergovernmental Panel on Climate Change, 4th Assessment Report. The Annex to this report states that a forest may consist either of closed forest formations, where trees of various storey and undergrowth cover a high proportion of the ground, or of open forest. This book adopts a wide definition of 'forest ecosystem' and includes all areas with substantial tree cover , including all types of forest compositions in any geographical range and with any species structure. This broad definition of forest means that rainforests, tropical, boreal, temperate, wetland, grassland and any other areas with reasonably dense tree coverage are within the scope of the definition.

This book uses the terms 'developed' and 'developing' countries to describe the two main categories of membership of international environmental instruments. These terms are used throughout the book to demonstrate the different interests and perspectives of these two groups. Such terms are used in the absence of the availability of better and more precise terms. It is to be noted that there is great diversity among countries in the 'developed' category as there is in the 'developing country' category and that these labels do not accurately reflect the broad range of economic, social and political conditions of all countries within these categories. Despite these limitations, for convenience, this book will use these well-established but out-dated terms to describe the different international legal obligations and policy interests of these two groups.

2. CURRENT TRENDS IN FOREST REGULATION

There are a number of factors currently influencing the development and direction of international forestry policy. It is important to understand these factors, as they will influence the nature of future global forest governance arrangements. The international climate change regime has started to play a dominant role in all forms of international environmental regulation. This is attributable to the high political priority given to climate change negotiations. The climate change regime has significantly influenced global forest governance through the development of the (REDD+) initiative. A second big driver of international forest regulation is the emergence of a series of global networks that are working on the issue of trade in illegal timber. And the third recent trend is the increasing presence of market-based regulatory models being applied over forest areas. These trends indicate a change in the approaches used to regulate forest use and management. These approaches have evolved in response to the long-standing challenges associated with forest regulation. Such major challenges include: the negotiation of international forest instruments; the associated implementation of and compliance with these instruments; the lack of capacity both financial and technical; and the lack of political will to support the adoption of legally binding international forest commitments.

A. The Influence of Climate Change Regime on Forest Policy

The influence of the international climate change regime can be seen within the policies and measures of most international environmental institutions. Variations in the earth's temperature arising from climate change will impact the earth's ecosystems, having significant flow-on effects for communities and the economy. The climate change regime has been able to capture a higher level of political interest in its policies and goals given the potentially significant environmental, societal and economic impacts of climate change, itself. The climate change regime attempts to address all sources and sinks of global greenhouse gas emissions and is interested in forest regulation as it is estimated that deforestation accounts for approximately 20 per cent of global greenhouse gas emissions. The regime has considerably more international political and financial support when compared to other international forest bodies. This support provides the regime with the ability to channel international funds towards projects that lower forest emissions.

Forest practices can be modified to reduce the 'source value' of their greenhouse gas emissions. This occurs when parties adopt more sustainable harvesting and silvicultural practices. Forests are also regulated by the regime for their 'sink values' – that is, their ability to absorb carbon. Forest activities that act as sinks of carbon include afforestation, reforestation, and avoided deforestation. Chapter 6 of the book discusses the mechanisms used by the climate change regime in regulating forest sink and source activities. Such mechanisms include the Land Use, Land-Use Change and Forestry guidelines (LULUCFG), the afforestation and reforestation guidelines of the Clean Development Mechanism (CDM), and the REDD+ initiative.

REDD+ is an initiative that uses donor funds to improve forest management in developing countries. This particular tool has been developed by the climate change regime to try and reduce the emissions arising from deforestation and forest degradation. The initiative is presently funded by voluntary contributions made by parties to the Climate Change Convention. Requests for REDD+ funding have received considerable support, allowing for the implementation of a range of REDD+ capacity building programmes.[1] The financial and political support for REDD+ means that this policy has the potential to significantly influence global forest governance.

REDD+ is a positive mechanism for two reasons. First, the mechanism provides funding for forest conservation as well as providing funding to improve capacity to manage forest areas (to address issues connected with land tenure, monitoring and resource-based shortages). Secondly, involvement in the REDD+ initiative occurs on a voluntary basis (both for the host and the donor of the project), which has allowed REDD+ to avoid dealing with the contentious issue of unfettered sovereign jurisdiction over natural resource products. The REDD+ mechanism, therefore, is the first international forest initiative to emerge that provides finance and high-level political support. The international forestry community is very much hoping that the REDD+ mechanism can succeed in improving forest conservation and management in developing countries.

The international climate change regime encourages the use and development of renewable and sustainable energy production. Forests are now also being valued for their ability to provide a source of renewable energy through bio-energy production. There is, however, some concern that demand for bio-energy from forest products will become a further driver of deforestation as the world's energy demands increase. The Food and Agricultural Organization predicts that the global demand for wood will increase in the future. The drivers of this demand are linked to bio-fuel policies in the European Union, and an increase in per capita consumption of wood products in Asia (though per capita consumption levels are predicted to remain higher in Europe and North America).[2] Furthermore, it is anticipated that by 2060 there will be a six-fold increase in world demand for fuel wood. This will result in an increase of wood prices, and will mean that forest areas previously used to manufacture sawn wood,

---

1 For example, one of the REDD+ demonstrating institutions, the Forest Carbon Partnership Facility has had $US225.4 million deposited in the REDD Readiness Fund and $US204.3 million invested in the Carbon Fund 'Climate Funds Update' (2012), 'Forest Carbon Partnership Facility', available at http://www.climatefundsupdate.org/listing/forest-carbon-partnership-facility (accessed 27 February 2012).

2 Food and Agricultural Organization of the United Nations, *State of the World's Forest 2009* (2009) ix.

panels and pulp will, instead, be used for energy production.[3] The production of energy from timber products will create even more competition for forest use, and it has been suggested that the high wood prices attributed to forest products for bio-fuel production will place further stress on forest health.[4] Such concerns should be addressed through the introduction of regulations that require bio-fuel production to occur from purpose-created timber stocks, managed in accordance with the concept of sustainable forest management.

The impacts of climate change are predicted to have serious repercussions for forest ecosystem functionality. A global study measured the impact of heat stress and drought on forest areas and found that an increase in the frequency, duration and severity of drought associated with climate change has the potential to fundamentally alter the composition, structure, and biogeography of forests in many regions.[5] Of particular concern are increases in tree mortality associated with climate-related changes, such as insect outbreaks and wildfire. The study found 88 examples of forest mortality driven by climatic water or heat stress since 1970. This suggests that climate change adaptation techniques and methodologies will become necessary in future forest governance regimes. Forest adaptation methodologies are still at the developmental stage, but it is anticipated that research in this area will increase as forest ecosystem vitality starts to deteriorate.[6]

Some sectors of the timber industry have already felt the impacts of climate change. Increases in the forest temperature in British Columbia, Canada, have led to a mountain pine beetle epidemic. This beetle infestation has arisen from the unusually warm temperatures of the forests

---

3 Raunikar, Ronald, Joseph Buongiorno, James A. Turner and Shushuai Zhu (2010), 'Global Outlook for Wood and Forests with the Bioenergy Demand Implied By Scenarios of the Intergovernmental Panel on Climate Change', *Forest Policy and Economics*, **12** (1) 48–56, 55.

4 Ibid.

5 Allen, Craig D., Alison K. Macalady, Haroun Chenchouni, Dominique Bachelet, Nate McDowell, Michel Vennetier, Thomas Kitzberger, Andreas Rigling, David D. Breshears, E.H. (Ted) Hogg, Patrick Gonzalez, Rod Fensham, Zhen Zhang, Jorge Castro, Natalia Demidova, Jong-Hwan Lim, Gillian Allard, Steven W. Running, Akkin Semerci, Neil Cobb (2009), 'A Global Overview of Drought and Heat-Induced Tree Mortality Reveals Emerging Climate Change Risks for Forests', *Forest Ecology and Management*, 259 (4), 660–684.

6 Nitschke, Craig R. and John L. Innes (2008), 'Integrating Climate Change into Forest Management in South-Central British Columbia: An Assessment of Landscape Vulnerability and Development of a Climate-Smart Framework', *Forest Ecology and Management*, 256 (3), 313–327.

during all the winters since 1999. Previously, the low winter temperatures killed the beetle, thus regulating the population. Warmer temperatures have allowed the beetle to thrive and the infestation has detrimentally affected 13 million hectares of pine forest in western Canada.[7] The timber industry and Canadian government have invested significant funds in research trying to control and prevent the spread of the infestation to neighbouring forest regions.

B. Common Goal of Banning Trade in Illegal Timber

There are a number of international processes seeking to prevent trade in illegally harvested timber. The oldest body dealing with trade in tropical timber is the United Nations International Tropical Timber Organization (UNITTO). This body was formed in 1986 to deal with concerns surrounding tropical forest conservation from a trade-based focus. The UNITTO is concerned principally with ensuring the existence of sustainably managed tropical forest timber reserves. Its goal is to protect the longevity and the economic viability of the international timber market. In recent years, alternative transnational and market-based frameworks for the forest trade have emerged. These regimes seek to prevent illegally sourced timber from entering global timber markets. The best-known programme of this sort is the European Union's Forest Law Enforcement and Governance Trade (FLEGT) initiative.

The FLEGT process works through the creation of voluntary partnership agreements entered into between consumer and producer timber countries. The terms of these agreements require that all timber supplied under them are legally sourced and comply with sustainable harvesting criteria. Part of the FLEGT process involves capacity building and assistance in implementing legal harvesting practices. The FLEGT initiative, discussed in greater detail in Chapter 7, is another mechanism predicted to have a significant influence on the development of future global forest governance, as it also provides economic and technical assistance for its implementation.

Individual countries are also starting to introduce legislation that prohibits the importation of illegally sourced timber. The United States of

---

[7] A number of articles document the impact of the mountain pine beetle on the forests of British Columbia, for example, Wulder, Michael, Stephanie M. Ortlepp, Joanne C. White, Nicholas C. Coops and Sam B. Coggins (2009), 'Monitoring the Impacts of Mountain Pine Beetle Mitigation', *Forest Ecology and Management*, **258** (7), 1181–1187, which discusses methods for mitigating the effects of the beetle.

America has enacted a piece of legislation, referred to as the Lacey Act.[8] This act aims to curb trade in illegally sourced timber by making its import or export a crime. The legislation, which came into effect on 15 December 2008, creates an offence for anyone who imports illegally harvested timber into the United States of America, or who exports products containing illegally sourced timber. An objective standard is applied to determine liability. If the party knows, or should have known, that the timber was illegal, then they will be found guilty of a crime. The act imposes heavy sanctions, ranging from five years imprisonment to a $US500 000 fine per violation, as well as forfeiture of the merchandise. The legality requirement is enforced by requiring all wood and wood products imported into a country to be accompanied by a customs declaration that provides the scientific name of the wood, its value and quantity, and the name of the country where it was harvested.[9] The reason for the introduction of this law was the contribution to global greenhouse gas emissions of trade in illegally sourced timber. It may be safe to assume that the Act's provisions will impact positively upon the domestic timber industry in the United States of America by making its prices more competitive with timber that is illegally sourced.

The Australian government has suggested that it would 'encourage sourcing of forest products from sustainable forest practices and seek to ban the sale of illegally logged timber exports', though this is yet to amount to enforceable legislation. Australia has also explored the possibility of introducing a policy to prevent trade in illegally sourced timber.[10] The discussion paper outlines five main goals

1. Build capacity within regional governments to prevent illegal harvesting
2. Develop and support certification schemes for timber products sold in Australia
3. Identify illegally logged timber, and restrict its import into Australia
4. Require disclosure at point-of-sale of species, country of origin and certification status
5. Promote the use of market-based incentives in reducing emissions from deforestation in future international climate change negotiations.

---

[8] 16 USC §§ 3371–3378 (1900).

[9] International Network for Environmental Compliance and Enforcement (2008), 'Recent Amendments to U.S. Lacey Act Should Help Protect Forests', available at http://www.inece.org/climate/ClimateComplianceAlert_LaceyAct.pdf (accessed 16 December 2009).

[10] Centre for International Economics, Canberra and Sydney, 'Proposed New Policy on Illegally Logged Timber: Issues Paper' (2009).

C. The Increasing Role of Non-State Forestry Regulation

'Non-state forest regulation' is a term used in this book to refer to international forest regulation by non-governmental or state bodies.[11] The lack of progress by public international forest institutions in implementing forest policy has allowed a space for non-state regulation to emerge.[12] Non-state forest regulation has developed international forest policy and standards away from the political arena of the United Nations. As such, many of the contentious international forest issues – such as sovereignty, capacity building, and technology transfer have been explored by these bodies and they have been able to avoid the usual political deadlock that arises when these issues are discussed within the United Nations framework. Non-state governance bodies have been able to avoid such deadlock by designing regimes based on voluntary participation. As such, participants engage and comply with the requirements of the body on the basis of the incentive offered by the institution. The incentive may be an eco-label, an economic payment, or the development of green credentials.

Pattberg states that 'rules made by private multi stakeholder partnerships in the field of environmental politics not only contain prescriptions of behaviour directed toward the environment. They also define who accounts for compliance with management standards, codes of conduct, or labels, and under what rules'.[13] Aligned with this, Falkner states 'these structuring effects resemble the 'public' governing function of states and intergovernmental institutions, and for this reason the notion of governance, and indeed authority, has been applied to private actors.'[14] This sug-

---

[11] In support of this see Falkner, Robert (2003), 'Private Environmental Governance and International Relations: Exploring the Links', *Global Environmental Politics*, **3** (2), 72–87.

[12] '. . . the result of the State's failure to produce a binding global forest instrument and inaction on forest products ecolabeling, nonstate forest certification schemes have emerged in the shape of powerful market driven governance and rule making systems.' See: Gulbrandsen, Lars (2004), 'Overlapping public and private governance: Can forest certification fill the gaps in the global forest regime?', *Global Environmental Politics*, **4** (2), 75r99, 75, 76. See also: Chan, Sander, and Philipp Pattberg (2008), 'Private rule-making and the politics of accountability: Analysing global forest governance', *Global Environmental Politics*, **8** (3), 103–121; and Murphy, Craig (2009), 'Privatizing environmental governance', *Global Environmental Politics*, **9** (3), 134–138.

[13] Pattberg, Philipp (2005), 'The Forest Stewardship Council: Risk and Potential of Private Forest Governance' 14 *The Journal of Environment and Development*, 14 (3), 356–374, 359.

[14] Falkner, Robert (2003), 'Private Environmental Governance and

gests that non-state regulation has been granted political legitimacy and support from a wide range of stakeholders.

Such stakeholders include forest departments, central governmental agencies, local governmental agencies, pro-active companies, national non-government organizations, consumers, academics, researchers, development assistance donors, consultants, international NGOs, community-based organizations, the logging industry and non-progressive forest producers, communities, farmers and the spectators, and policy target groups.[15] It is easier for a broad range of stakeholders to participate in private forest regimes, as parties are able to interact directly with the governing institution. This is rather different from the merely representative input that such stakeholder groups get at public international negotiations.

This book explores two non-state regulatory approaches to forest management: forestry certification schemes; and payment for ecosystem service regimes (forest markets). Both of the regimes have the following characteristics common to non-state regulation:

- industry has played a role in creating/ driving the regulation;
- both involve market-based regulation;
- both schemes are largely voluntary; and
- both schemes are transnational in character.

Both forestry certification schemes and markets selling forest ecosystem services have influenced the nature of domestic forest regulation. Forestry certification schemes have significantly contributed to the development of productive domestic forest standards. Areas with forest certification status have been directly influenced by non-state regulation. Likewise forest market mechanisms enforce their requirements by providing an economic incentive once it is established that certain sustainable forest management criteria have been met. A number of schemes have been developed to sell forest ecosystem services and, again, non-state regulation is directly affecting the management and use of these forest areas.

Compliance with non-state sustainable forest management standards is generally considered to be higher than compliance with public international sustainable forest management standards. This is because non-state regulation is voluntary, and parties have elected to participate

---

International Relations: Exploring the Links', *Global Environmental Politics*, **3** (2), 72–87, 72–73.

[15] Mayers, James, and Stephen Bass (eds) (1999), *Policy that Works for Forests and People*, London: International Institute for Environment and Development, Ch. 2.

and comply with the regime. As suggested earlier, it is assumed that parties engage with such processes as a result of the incentive offered by the regime. Forest certification provides participants with an eco-label that they can use to differentiate their product in the market, while forest markets provide participants with an economic incentive to manage and use the forest area according to certain criteria. Additionally, the nature of public international environmental law requires states to take the lead in the implementation of international commitments. States have to weigh many factors when considering the implementation of international forest standards. The politics associated with forest regulation often leads to states finding the middle ground in terms of forest regulation, often resulting in voluntary or non-legally binding policies. These factors combined have resulted in stronger compliance with non-state forestry standards than with public international ones.

3. BRIEF SYNOPSIS OF CHAPTERS

This book is divided into five main parts. Part One includes the present introductory chapter, which has explained the purpose of the research, and a short summary of the key legal issues in global forest governance. Part Two has three chapters which explore the legal doctrinal concepts related to the regulation of sustainable forest management: Chapter 2 examines the rule of law and the justice issues that arise in forest regulation; Chapter 3 examines the concepts of state sovereignty and property as principles creating rights and responsibilities in relation to forest use and management; and Chapter 4 examines the governance arrangements required for sustainable forest management.

Part Three focuses on three public international bodies that have played a role in creating and implementing international forestry regulation. The public international bodies chosen for analysis were selected for their different perspectives on forest management. Each chapter in this part explores, for each body, the institutional structure, the founding instruments and the progress of implementation, providing a basis for comparison across these institutional arrangements. Chapter 5 examines the modalities of the United Nations Forum on Forestry. This public international body is the platform that discusses all issues concerning sustainable forest management. Its major accomplishment is the creation of National Forest Policies in most countries. Chapter 6 looks at the legal rules concerning forestry within the international climate change regime. The forest value or ecosystem service regulated by the climate change regime is the carbon storage value associated with forest areas. Chapter 7 analyses the operation of the World Bank forest policy. The World Bank is primarily interested in improving the health and extent of forest areas as a means of alleviating poverty. A strong correlation exists between those experiencing poverty and people living in forest areas, which, again, highlights the need for mutually sustainable practices.

Part Four focuses on non-state forestry regulation. Chapter 8 examines the conditions leading to the creation of markets for forest ecosystem services. This relatively new form of regulation places an economic value upon a particular forest ecosystem service, such as biodiversity, watershed, or scenic amenity. Markets are then established to trade the forest value for economic benefit. Chapter 9 focuses on the operations of forestry certification schemes. These schemes work by creating an eco-label, which is affixed to forest products verifying that the product has complied with the requirements of the certifying body. These schemes work on the basis that consumers will drive the demand for more sustainable forest practices by buying forest products from sustainably managed areas. There are a number of international forestry certification schemes all competing for authority. The original and largest international forest certification scheme is the Forest Stewardship Council. The operational modalities of this body are examined within this chapter.

Part Four also contains Chapter 10, which discusses the overall conclusions and findings of this research. The conclusion makes recommendations for the reform of each of the regulatory bodies examined, followed by some holistic recommendations concerning global forest governance regulation.

PART II

The doctrinal concepts underpinning forest regulation

The purpose of Part Two is to explain the issues, concepts, doctrines, principles and rules relating to forest regulation. This book identifies a number of legal issues that arise in forest regulation. The rule of law requires the existence of clear and enforceable rights. This paradigm should apply within the forest context at both the international and state regulatory levels. The existence of corruption and inefficient governance regimes undermine the rule of law within international and national forest regulation. From an international legal perspective, the concept of state sovereignty remains a relevant consideration when designing international forest obligations. From a state perspective, the most important considerations in the design of forest regulation are issues connected with forest tenure and use rights. Such rights lie at the heart of forest regulation, as they define who holds the right to use the forest area and how it may be used, depending upon the nature of the right conferred. Forest governance is an important legal consideration both at the global and state levels. Governance arrangements define the process for the recognition and enforcement of rights and effective institutional governance structures and processes allow for the implementation of forest norms and rules. Measurement of effectiveness should be assessed in reference to established 'effectiveness criteria'.

These doctrinal concepts are then used in Parts Three and Four to assess the range of policy responses and approaches to forest regulation. Part Three traces the role of these doctrinal concepts in regulation by public international bodies, while Part Four examines the role of these concepts in non-state forest regulation.

This part separates forest issues concepts, doctrines and principles into three categories, which are broken down by chapter:

Chapter 2 examines rule of law and justice issues in forest regulation.

Chapter 3 evaluates the concepts of state sovereignty and property as creating rights to forest areas and limitations on the use of forest areas.

Chapter 4 analyses the governance arrangements required for the implementation of sustainable forest management.

2. Considerations of justice in forest regulation

1. THE ROLE OF LAW AND THE 'RULE OF LAW' IN FOREST REGULATION

Adequate legal frameworks are essential to ensure the implementation of sustainable forest management. Fisher states that rules of law reflect and implement the values and aspirations of the society, from which they emerge. Legal or paralegal rules are the tools used to implement rule of law values and aspirations.[1] Legal rules are defined as enforceable rules of law, such as the imposition of an obligation, duty or standard to be complied with. Paralegal rules refer to principles, objects, strategies, policies, plans, guidelines and codes of practice. Society does not have one common view of how environmental resources should be used and managed so the law does not necessarily reflect all of society's views on it. Law should, however, evolve in line with changes in the views and values of society, and should also seek to accommodate and reflect as many of the different values and interests associated with the environment as possible.

For advances to be made in implementing sustainable forest management, appropriate legal rules, institutions and processes must exist. Law is a crucial component in such change because it is the law that governs most transactions between people. Similarly, the law also governs the forms of transaction between people and the environment. Law defines the role of the state and the relationship between the state and its citizens. International forest law and national forest laws are often criticized for being ineffective and operating in principle only. It is commonly said that the issue in forest regulation is not a lack of paralegal or legal rules but, in fact, an overabundance of inappropriate legal rules, together with weak

[1] Fisher, Douglas (2008), 'Legal and Paralegal Rules for Biodiversity Conservation: A Sequence of Conceptual, Linguistic and Legal Challenges', in Jeffrey, Michael, Jeremy Firestone and Karen Bubna-Litic (eds), *Biodiversity, Conservation, Law + Livelihoods: Bridging the North-South Divide*, New York: Cambridge University Press, pp. 94–131, 94–8.

implementation and compliance with the laws in existence.[2] The role of law in regulating forest use and management is not questioned but the effectiveness of forest law often is.

Rule of law considerations in international forest law

The rule of law requires that certain conditions be satisfied in the creation and enforcement of legal rules. The International Development Legal Organization (IDLO) lists a number of requirements for rule by law. These are[3]

- A set of legally binding rules, which are known in advance
- That these rules be enforced in practice
- That mechanisms exist to ensure the proper application of the rules by properly functioning institutions, and that mechanisms allow for departure from the rules as needed according to established procedures
- That conflicts in the application of these rules be resolved through the binding decisions of an independent judicial or arbitral body
- That there are known procedures for amending these rules when they no longer serve their purpose.

These requirements for the rule of law are not met by international forest regulation. The lack of a legally binding international instrument on forests is often cited as one of the primary failures of the international community in addressing deforestation and sustainable forest management.[4] There are a range of international laws applicable to forest management such as the Convention on Biological Diversity and the Ramsar Convention, for example[5], however these instruments do not comprehensively address all the issues that are relevant to sustainable forest manage-

---

[2] Lindsay Jon, Ali Mekour and Lawrence Christy, FAO Development Law Service (2002), 'Why Law Matters: Design Principles for Strengthening the Role of Forestry Legislation in Reducing Illegal Activities and Corrupt Practices', FAO Legal Papers Online #27, available at http://www.fao.org/legal/prs-ol/lpo27.pdf (accessed 21 March 2012) 3.

[3] International Development Law Organization (2012), 'Rule of Law Assistance Directory', available at http://www.idlo.int/english/Resources/ROL/Pages/default.aspx (accessed 14 March 2012).

[4] Lipschutz, Ronnie D. (2001), 'Why is There no International Forestry Law? An Examination of International Forestry Regulation, both Public and Private', *Journal of Environmental Law and Policy*, 19 (1), 153–179.

[5] The public international arrangements regulating forest use are discussed in greater detail in Part Three of this book.

ment. Instead, each instrument focuses on a particular value associated with forest areas and the regulation thereof. Independently of considerations concerning the legally binding nature, or otherwise, of international forest instruments, the major issue for international forestry regulation is the lack of compliance, transparency and accountability of state parties in implementing the obligations created under the various international and national legal forest regulatory frameworks.[6]

States often perceive no real benefits from complying with international forest standards. From a cost-benefit perspective, the economic and political cost of implementing such initiatives outweighs the international political benefits and the environmental benefits. The institutional global forest governance arrangements overseeing forest management, measures and policies are not in good functioning order. The United Nations Forum on Forests is the body charged with responsibility for forest regulation but it lacks authority and political support. This has led to a situation where a number of international institutions regulate forest areas, resulting in a regime that is fragmented. No international judicial body has examined the conflict in international forest obligations because the obligations are voluntary in nature and, as such, unlikely to be the subject of judicial determination. All the international forest institutions examined have processes that allow for the development of new forest policy. However, the gap lies in the adequate implementation of existing or new forest policy. The governance arrangements underlying international forest regulation are considered in greater detail in Chapter 4, however, it can be accepted for current purposes that the international legal order governing forest use and management does not currently operate to comply with the IDLO rule of law requirements. Consequently, strengthening international forest institutions and instruments is necessary in order for progress to be made in addressing unsustainable forest use and forest degradation and deforestation.

Rule of law considerations domestic forest regulation

At the state level, many jurisdictions face difficulties in implementing sustainable forest management for a range of legal, economic and political reasons. In order to understand the nature of the legal system in operation and the legal or non-legal impediments preventing the operation of sustainable forest management policies, it is important to understand, at the start, the nature of the formal and informal legal systems operating within

---

6 Global Witness (2011), 'Making the Forest Sector Transparent: Annual Transparency Report 2010', available at <http://www.foresttransparency.info/> (accessed 20 January 2012).

the jurisdiction. The legal system may be based on civil law, common law, customary law, Islamic law or a combination of any of these systems. In some instances plural legal systems may be in operation, with a formal state law and an informal but widely applied customary law governing forest use and management.[7] The nature of the legal system becomes important in forest regulation when seeking to understand land tenure rights and other access or use rights (for example, concessions, permits, covenants) attaching to forest estates. The existence of formal and informal rights, public rights, and the sovereign rights of indigenous groups, alongside a web of complex legislative and other legal policies concerning forest estates makes determination of forest rights and forest duties difficult to comprehend. The uncertainty surrounding forest rights and duties then allows corrupt, non-transparent and non-ethical activities to take place on forest estates.

The World Justice Project has created a list of dimensions that are necessary for the rule of law at the domestic level. These are listed as:

- limited government powers;
- absence of corruption;
- order and security;
- fundamental rights;
- open government;
- effective regulatory enforcement;
- access to civil justice;
- effective criminal justice; and
- informal justice.[8]

It is possible, generally, to categorize rule of law issues for developed countries and for developing countries.[9] Developed countries face a number of rule of law challenges in forest regulation. For example, in Australia, the regulation of productive forest practices is achieved by non-binding codes of conduct or advisory policies.[10] The lack of legally

---

7 Development Law Service: FAO Legal Office (2000), 'Legal Framework Analysis for Rural and Agriculture Investment Projects' available at http://www.fao.org/legal/prs-ol/lpo12.pdf (accessed 20 January 2012).

8 The World Justice Project (2011), 'Dimensions on the Rule of Law', available at http://worldjusticeproject.org/rule-of-law-index/dimensions (accessed 20 January 2012).

9 Another publication discusses this category as income-poor countries and income-rich countries. See Mayers, James, and Stephen Bass (eds) (1999), *Policy that Works for Forests and People*, London: International Institute for Environment and Development.

10 On the regulation of Australian forests see Dovers, Stephen (2003), 'Are

binding rules concerning all forest practices leaves gaps in the regulatory process which allows unsustainable harvesting of forest estates to occur in some instances.[11] Furthermore, the practice of conserving forest estates by placing areas under conservation status fails to create adequate management obligations to ensure a high level of ecological functioning within these areas.[12] There are also institutional inefficiencies in forest regulation due to the large numbers of government departments involved. Each government department is normally pursuing an individual objective in connection with the forest estate including: conservation; community engagement; cultural heritage; resource extraction; and state development and planning, causing fragmentation and overlap in the regulation applicable to forest estates. Another common concern is the lack of adequate stakeholder engagement in planning and development processes and in law reform development processes.[13]

Developing countries face the same issues as developed countries in addition to concerns about non-transparent and corrupt forestry practices. In many instances there is a large gap between the forest laws on paper and actual practices across forest areas.[14] This is related to a number of issues such as weak institutional capacity, corruption within government bodies, and the lack of an impartial judiciary.[15] Forest-dwellers, or communities, do not have a voice and often lack the ability to enforce their

---

Forests Different as a Policy Challenge' in David Lindenmayer and Jerry Franklin (eds), *Towards Forest Sustainability*, Collingwood, Victoria: CSIRO Publishing, pp. 15–30.

[11] On the issue of instrument choice, see Bowers, John (2005), 'Instrument Choice for Sustainable Development: An Application to the Forestry Sector', *Forest Policy and Economics*, **7** (1), 97–107. For a focus on Queensland's regime for managing forest resources, see Kehoe, Jo (2009), 'Environmental Law Making in Queensland: The Vegetation Management Act 1999 (Qld)', *Environmental and Planning Law Journal*, **26** (6), 392–410.

[12] Angelstam, Per (2003), 'Forest Biodiversity Management – the Swedish Model' in David Lindenmayer and Jerry F. Franklin (eds), *Towards Forest Sustainability*, Collingwood, Victoria: CSIRO Publishing, pp. 143–159.

[13] Buchy, M. and S. Hoverman (eds) (1999), *Understanding Public Participation in Forest Planning in Australia: How can we Learn from Each Other?* Canberra, ACT: Department of Forestry; and Jack Siry, F.W. Cubbage and M.R. Ahmed (2005), 'Sustainable Forest Management: Global Trends and Opportunities', *Forest Policy and Economics*, **7** (4), 551–561, 554–5.

[14] Blomley, Tom, and Acacia Consulting, International Institute for Environment and Development (2009), *Evaluation of the Work of the Forest Governance Learning Group 2005–2009*.

[15] Richards, Michael (2000), 'Can Sustainable Tropical Forestry be Made Profitable? The Potential and Limitations of Innovative Incentive Mechanisms', *World Development*, **28** (6), 1001–1016, 1003.

forest rights. Corruption is seen as particularly pervasive in the forestry sector:

> Corruption can be driven by politicians, government officials and their business patrons, who make discretionary timber deals without following due process; by the international donor community, which often drives and bankrolls national forest strategies and individual logging operations, but fails to address adequately the corruption that condemns virtually all these ventures to failure; and by the unrelenting international demand for cheap timber.[16]

The evolution of REDD+ and FLEGT initiatives is leading to a renewed interest in combating non-transparent and corrupt forestry practices. Transparency is seen as involving two elements: open disclosure of information to all members of the public and opportunities for the public to participate constructively in decision-making processes.[17] Increasing the transparency of actions in the forest sector is linked with initiatives seeking to reduce corruption. Corrupt activities in forestry are linked to insecure and uncertain land tenure arrangements combined with weak governance and monitoring systems. Around 80 per cent of the global forest estates are publicly owned and managed resources.[18] Such ownership patterns allow for a relatively small number of politicians and civil servants to be in a position to wield huge discretionary power concerning forest use in return for bribes. The physical location of forests also means that they are beyond the public gaze and forest dwelling groups and communities can be excluded from decision-making processes by lack of access to information, resources and influence.[19] These weaknesses allow for a number of criminal activities to occur such as falsification and forgery of documents, fraudulent dealings, bribery, money laundering and tax evasion.[20]

---

16 Alley, Patrick (2011), 'Corruption A Root Cause of Deforestation and Degradation' in Transparency International (eds), *Global Corruption Report: Climate Change*, London, Washington DC: Earthscan, pp. 299–311, 301.

17 Global Witness (2011), 'Making the Forest Sector Transparent: Annual Transparency Report 2010', available at http://www.foresttransparency.info/ (accessed 20 January 2012) 10.

18 Food and Agriculture Organization (2010), 'FAO Global Forest Assessment 2010', available at http://www.fao.org/forestry/fra/fra2010/en/ (accessed 20 January 2012).

19 Alley, Patrick (2011), 'Corruption A Root Cause of Deforestation and Degradation' in Transparency International (eds), *Global Corruption Report: Climate Change*, London; Washington DC: Earthscan, pp. 299–311, 300.

20 Global Witness (2011), 'Forest Carbon, Cash and Crime' available at http://www.globalwitness.org/sites/default/files/library/Forest%20Carbon,%20Cash%20and%20Crime.pdf (accessed 20 January 2012) 5–6.

Such practices are widespread in many regions where REDD+ investments are likely to be significant.[21] There are concerns that the following corrupt activities will continue during REDD+ investments: illegal logging due to limited local capacity to monitor and manage; illegal land grabs in areas perceived as valuable REDD areas; huge corruption in the ownership, trading and regulation of carbon trading as a result of the intangible nature of carbon credits and lack of land and carbon registries recording transactions; theft and misappropriation of REDD+ donor investments and manipulation of carbon measurements.[22] Improving the integrity of legal systems receiving REDD+ funding is of paramount interest to donor countries and international forest institutions. The level of finance to be generated for forest activities as a result of REDD+ investments is unparalleled in terms of commitments from the international community to fund sustainable forest management initiatives. This increases the pressure on REDD+ institutions and projects to deliver forestry projects with a certain level of integrity in order to ensure continued donor support and interest in sustainable forestry activities.

2. GLOBAL FOREST RESPONSIBILITY AND JUSTICE

A. Determining Responsibility and Liability

The management and use of the global forest estate involves a number of 'justice' related considerations.[23] There is a general global consensus that the reduction of global deforestation and forest degradation should be a priority. However, there is no global consensus on how such reductions should occur and who should ultimately bear responsibility or liability for such change. This book will discuss a number of international insti-

---

[21] See diagram which shows the level of legal versus illegal timber harvesting patterns in REDD+ recipient countries: Global Witness (2011), 'Forest Carbon, Cash and Crime' available at http://www.globalwitness.org/sites/default/files/library/Forest%20Carbon,%20Cash%20and%20Crime.pdf (accessed 20 January 2012) 7.

[22] Global Witness (2011), 'Forest Carbon, Cash and Crime' available at http://www.globalwitness.org/sites/default/files/library/Forest%20Carbon,%20Cash%20and%20Crime.pdf> (accessed 20 January 2012) 6.

[23] On the issue of justice and climate change see Richardson, Benjamin J. (ed.) (2009), *Climate Law and Developing Countries: Legal and Policy Challenges for the World Economy*, Cheltenham, UK and Northampton, MA, USA: Edward Elgar Publishing.

tutional approaches that all attempt to deal with aspects of sustainable forest management. Due to the central role that forests play in a number of economies (in combination with other strongly held anthropocentric forest values), all of these institutions fail to assign liability or responsibility to individual states to address global deforestation and forest degradation.

To create a system assigning liability or responsibility to individual states to address global deforestation and degradation require for moral and ethical factors to be considered. Should those countries with large tracts of old growth forest areas be the parties ultimately responsible for conserving and protecting such areas? Does the location of the forest estate make that state responsible for the implementation of sustainable forest management? Should a system of sustainable forest management liability be based upon simple determinants, such as location and ownership, in assigning international liability and responsibility? Or should other factors, such as per capita consumption levels of forest products and previous levels of forest degradation, define liability? Would such an approach be compatible with other international environmental principles, such as common but differentiated responsibility and the 'polluter pays' principle?

B. Historical Approach to Liability

The international concept of state sovereignty would suggest that states have authority and control over forest resources on their land.[24] So, should international responsibility for deforestation and forest degradation be attributed on similar grounds? This would mean that the holders of forest estate are made liable for the implementation of sustainable forest management practices. On first reading, such a simple analysis seemed justifiable. This type of reasoning is reflected in international forest commitments, which make sovereign states responsible for their sovereign forest resources.

Such an approach has, however, failed to deliver the implementation of sustainable forest management practices. Such an approach is not valid on practical grounds because a significant proportion of the old growth forest is found in developing countries that lack the capacity to implement sustainable forest management. In addition, it is arguable that such an

---

[24] See generally on this point: Schrijver, Nico (ed.) (1997), *Sovereignty Over Natural Resources: Balancing Rights and Duties*, Cambridge; New York: Cambridge University Press.

approach is not valid on ethical or moral grounds. Should states that still have the forest reserves be restricted from using their reserves, just because other states damaged their own forest reserves earlier?

Literature that emerged around the time of the Earth Summit depicts developed countries taking the lead on conservation, protection and enhancement of the global forest estate whilst developing countries are portrayed as the parties unwilling to commit to legally-binding forest initiatives until due compensation and technical assistance has been agreed and formally provided.[25] No particular blame is placed upon either group. The literature describes a process whereby two legitimate arguments exist. For example:

> On one hand, it has been claimed that environmental protection must be given priority over economic development programmes because further neglect of the environment threatens all human life including that in the Third World; and on the other hand, it has been said that in the Third World human life is already threatened, in fact being extinguished, so that rapid development must have priority over environmental protection.[26]

The first argument made on the side of developed countries arises from a new appreciation for the value of old growth forest. Conversely, developing countries, many of which are battling a number of other serious, interrelated challenges (such as increasing population, food shortages, and a high incidence of poverty) wish to use their forest resources to improve the living conditions of their people.[27] The international response to these conflicting views was the adoption of a non-legally binding forest instrument that made neither developed nor developing nations liable for implementing sustainable forest management practices.

C. Ecological Justice, Environmental Justice and Environmental Racism

There are a number of theories in the existing literature that can be used to argue against imposing liability based upon sovereign ownership. Three closely related concepts are: ecological justice; environmental justice; and

---

25 Humphreys, David (ed.) (1996), *Forest Politics: The Evolution of International Cooperation*, London: Earthscan Publications.

26 Hunt, Geoffrey (1992), 'Is There a Conflict Between Environmental Protection and the Development of the Third World?' in Robin Attfield and Barry Wilkins (eds), *International Justice and the Third World*, London, New York: Routledge, pp. 117–151.

27 See ibid. 119–40 for an examination of the various solutions to underdevelopment from a market, democratic, green and socialist perspective.

environmental racism. Bosselmann defines *ecological justice* as involving two distinct components. The first examines justice on the basis of the distribution of the environment among people – that is, the access, quality, and quantity of environmental services and products people obtain. The second component examines the relationship between humans and the rest of the natural world. This component focuses more on eco-centric values.[28] *Environmental justice* is distinguished by only examining the first component – essentially, a distribution-focused analysis.[29]

In the United States of America, the term *environmental racism* is used interchangeably with the term *environmental justice*.[30] Environmental racism can be defined as:

> . . . racial discrimination in environmental policy-making and enforcement of regulations and laws, the deliberate targeting of communities of color for toxic waste facilities, the official sanctioning of the presence of life threatening poisons and pollutants in communities of color, and the history of excluding people of color from leadership of the environment movement.[31]

Environmental justice calls for the equitable utilization of resources, procedural fairness, and a safe and healthy environment.[32] An environmental justice movement arose in America in the 1970s due to the existence of inequalities between rich and poor residents and the corresponding local environmental conditions. The movement identified the inequitable burden that poor or coloured neighbourhoods faced in hosting activities potentially dangerous to human health (for example, nuclear power

---

[28] Bosselmann, Klaus (ed.) (2008), *The Principle of Sustainability: Transforming Law and Governance*, Aldershot, England; Burlington, VT: Ashgate, 79.

[29] Often in the environmental justice discussion, the issue of a human right to a healthy environment emerges. While human rights and environment are of course interrelated, this is not the focus of this discussion. See, generally, Doelle, Meinhard (2004), 'Climate Change and Human Rights: The Role of International Human Rights in Motivating States to Take Climate Change Seriously', *Macquarie Journal of International and Comparative Environmental Law*, **1** (2), 179–216, Ch 4.

[30] Ali, Asghar (2006), 'A Conceptual Framework for Environmental Justice Based on Shared but Differentiated Responsibilities' in Tony Shallcross and John Robinson (eds), *Global Citizenship and Environmental Justice*, Amsterdam; New York: Rodopi, pp. 41–50.

[31] Ibid.

[32] Shelton, Dinah (1999), 'Environmental Justice in the Postmodern World' in Klaus Bosselmann and Benjamin Richardson (eds), *Environmental Justice and Market Mechanisms: Key Challenges for Environmental Law and Policy*, London: Kluwer Law International, pp. 21–26.

plants), which released high levels of pollutants, or which affected the surrounding local environmental conditions in a significantly negative way (for example, processing and industrial centres). Specifically, the movement was concerned with addressing the issue of outsourcing dirty, unsustainable and potentially dangerous activities into areas of poor or coloured communities.[33] It has been suggested that justice in this sense equates to fair treatment, and that other issues such as past injustices, existing inequalities and the imbalance of wealth are omitted by such an analysis. [34] The concerns raised by the domestic environmental justice movements have application to the inequalities that exist at the international level concerning forest benefits and burdens.[35]

In the postmodern world, there are serious challenges to environmental justice. On the positive side, it is apparent that more people are living in democratic regimes than in any other time in history, and free-market economies are becoming the norm. Faxes, phones and the Internet have made communication immediate and global. At the same time, climate change, deforestation, pollution and other environmental stresses are the consequences of the enormous pressure placed on resources by an ever-increasing population, industrialization, land conversion, urbanization and rising consumption.[36]

Schlosberg defines the global environmental justice movement as having four distinct components:

- justice as distribution;
- justice as recognition;
- justice as participation;
- procedural justice and justice as capabilities and functioning.[37]

---

[33] See generally Cutter, Susan (1995), 'Race, Class and Environmental Justice', *Progress in Human Geography*, 19 (1), 111–122; Manuel Pastor, Jim Sadd and John Hipp (2001), 'Which Came First? Toxic Facilities, Minority Move-In, and Environmental Justice', *Journal of Urban Affairs*, 23 (1), 1–21; Taylor, Dorceta (2000), 'The Rise of the Environmental Justice Paradigm: Injustice Framing and the Social Construction of the Environmental Discourses', *American Behavioral Scientist*, 43 (4), 508–580.

[34] Bosselmann, Klaus (1999), 'Justice and the Environment: Building Blocks for a Theory on Ecological Justice' in Klaus Bosselmann and Benjamin Richardson (eds), *Environmental Justice and Market Mechanisms: Key Challenges for Environmental Law and Policy*, London: Kluwer Law International, pp. 30–31.

[35] Schlosberg, David (ed.) (2007), *Defining Environmental Justice: Theories, Movements, and Nature*, Oxford; New York: Oxford University Press, 79.

[36] Shelton, above n 32, 23.

[37] Schlosberg, above n 35.

D. Justice as Distribution

The most common analysis of 'justice' is work concerning the concept of distributive justice.[38] The underlying theory supporting the concept of distributive justice is Rawls's 'theory of justice'.[39] Justice, in this sense, is associated with the distribution of goods within society. Distributional analysis can be undertaken from a number of perspectives, as identified by Weiss: environmental equity between countries; environmental equity within countries; and the ability of the international regime to deliver environmental equity.[40] In addition to these perspectives, the issue of distributional justice between present and future generations should also be considered.

Environmental equity among countries

Weiss discusses environmental equity between countries in relation to: the allocation of resources between nations; responsibility and liability for pollution; and the international responsibility to ensure that activities under a state's control do not cause harm to areas outside that state's control.[41] The allocation of resources between nations can be considered from two perspectives. First, consideration could be given to the global geographical spread of forests: such an approach would see those states with high forest coverage as having more benefits then states with low forest coverage. A second perspective would focus on where forest goods and services were enjoyed. This approach would view states that consume high levels of forest goods and services as being the ones that enjoy forests' benefits. It is suggested that the second approach is more appropriate from an international perspective as it reflects the true distribution of the benefits and burdens associated with forest areas on a global scale.

Benefits associated with forest estates could be viewed as encompassing ecological, economic and social benefits for local communities and stakeholders but also all the benefits deriving from the enjoyment of forest products such as wood and paper products, foods and medicines.

---

38 Ibid. 3. Regarding the definition of distribution, see Nozick, Robert (2000), 'Distributive Justice', in Richard Epstein (ed.), *Modern Understanding of Liberty and Property*, New York: Garland Pub, pp. 317–495.

39 Rawls, John (ed.) (1972), *A Theory of Justice*, Oxford: Clarendon Press.

40 Weiss, Edith Brown (1995), 'Environmental Equity: The Imperative for the Twenty-First Century' in Winfried Lang (ed.), *Sustainable Development and International Law*, London; Boston: Graham & Trotman/M. Nijhoff, pp.17–23, 19–23.

41 Ibid. 19.

Forest products are harvested and then exported for local, regional or international export to consumers who are able to afford and enjoy forest products and services. This creates a scenario whereby local stakeholders lose the benefits associated with a forest estate (ecological and social) in return for only a small economic gain. In many instances, the economic benefits associated with timber harvesting are not shared with local or indigenous forest communities due to corrupt practices and insecure rights to land. The benefit, from this perspective, is enjoyed by countries that are far removed from any of the impacts of deforestation. This pattern of distributional impacts occurs in all natural resources extractions and exportations and is not unique to forest products. The unique component, from a forest perspective, is the wide range of values that attach to forest areas, making the distribution of benefits and burdens complicated. Consideration should also be given to the burdens associated with implementing sustainable forest management. Environmental equity between countries requires a more nuanced approach to imposing liability for implementing sustainable forest management that takes into account both the location of the forest product service and the place in which the forest good or service is enjoyed. A potential model is outlined further, below.

This form of equity requires that the forest activities of one state do not cause harm outside the state's borders. Depending on the definition of harm, this could impose quite heavy restrictions on a state's forest practices. A recent decision by the International Court of Justice (ICJ), however, suggests a more cautious approach to imposing liability based on the 'no harm' principle. The *Pulp Mills on the River Uruguay*[42] dispute arose from a complaint by Argentina that construction of a pulp mill in Uruguay would adversely affect the shared watercourses and ecological barriers between the two countries. The court accepted evidence presented on behalf of Argentina that Uruguay had not conformed to its procedural obligations under the Statue on the River Uruguay,[43] however the court decided to not impose any penalty in this instance given the stage of development of the project site and the economic significance of the project to Uruguay.[44] The court could have used this opportunity to further

---

42 *Pulp Mills on the River Uruguay* (AR VuRU) Judgement 67 (April 20, 2010) available at http://www.icj-cij.org/docket/index.php?p1=3&p2=3&case=135&code=au&p3=4 (accessed 26 March 2012).

43 Statute on the River Uruguay (1975) 1295 UNTS 340.

44 McDermott, Michael (2011), 'International Environmental Disputes and the Need for Court-Commissioned Independent Experts', *Boston College International and Comparative Law Review*, 34, 67–80.

examine the 'no harm' principle; however it chose to limit its considerations to adherence with the statue in question, suggesting that the court is unwilling to restrict sovereign rights to harvest forest products even where there is evidence of harm being suffered to shared watercourses.

Environmental equity within countries

Environmental equity within countries examines the equity that exists at the domestic level between poor people and environmental burdens.[45] In the forest context this would involve consideration of the impacts of deforestation on local communities and stakeholders. The values attached to the importance of sustainable forest management, however, vary according to the economic position of local stakeholders. Weiss states

> Some economic studies have suggested that the empirical relationship between environment and development is an inverted U-shaped curve, in which countries pollute the environment up to the point at which they become sufficiently industrialised to indulge their concern for protecting the environment.[46]

This theory is known as the 'EKC theory' (Environmental Kuznets Curve). Kuznets hypothesized that income inequality first rises and then falls as sustainable economic development proceeds.[47] More recently, this theory has come under scrutiny, with many studies demonstrating that economic growth alone does not necessarily lead to sustainable development.[48] Improvements in environmental quality may, in reality, be indicators of the increased ability of consumers in wealthy nations to distance themselves from the environmental degradation associated with their consumption.[49] The livelihood considerations of local forest stakeholders are therefore extremely relevant in designing and implementing measures and policies aimed at improving sustainable forest management outcomes.

---

45 Ibid. 22.

46 Ibid.

47 Stern, David (2004), 'The Rise and Fall of the Environmental Kuznets Curve', *World Development*, **32** (8), 1419–1439. Also see an earlier articles which applies the theory in the forest context: Stern, David, Michael Common and Edward Barbier (1996), 'Economic Growth and Environmental Degradation: The Environmental Kuznets Curve and Sustainable Development', *World Development*, **24** (7), 1151–1160.

48 Cavigila-Harris, Jill, Dustin Chambers and James Kahn (2009), 'Taking the "U" out of Kuznets: A Comprehensive Analysis of the EKC and Environmental Degradation', *Ecological Economics*, **68** (4), 1149–1159.

49 Rothman, Dale (1998), 'Environmental Kuznets Curves – Real Progress or Passing the Buck? A Case for Consumption-based Approaches', *Ecological Economics*, **25** (2), 177–194.

Third World Approaches to International Law (TWAIL) challenges western perceptions of history and the content of international law, by highlighting the inequitable nature of the body of rules transferred from the past.[50] The impacts of colonization both past and present are examined to demonstrate that a system of continued western superiority operates at the international level, with particular relevance to economic development and advancement. In the environmental context, attention is directed towards the impacts of industrialization and the accompanying undesirable environmental effects, such as pollution and land and water degradation. These issues have been framed as issues and challenges of common concern, without due regard to those responsible for generating them.[51] Developing countries, perceive themselves as being asked to manage their resources for the benefit of western interests. A statement from the Prime Minister of Malaysia at the Earth Summit negotiations states the position eloquently:

> The poor are not asking for charity. When the rich chopped down their own forests, built their own poison-belching factories and scoured the world for cheap resources, the poor said nothing. Indeed they paid for the development of the rich. Now the rich regulate the development of the poor countries. And yet any suggestion that the rich compensate the poor adequately is regarded as outrageous. As colonies we were exploited. Now as independent nations are we to be equally exploited.[52]

TWAIL perspectives have influenced the development of international environmental law. Evidence of such perspectives can be found in the operation of the principle of common but differentiated responsibility and in provisions within environmental treaties calling for finance and technology transfer. The principle of common but differentiated responsibility has not been firmly embedded into international forest instruments and it is suggested that inclusion of this principle is essential in order for international forest-related laws to emerge in a fashion that can be considered

---

50 Mickelson, Karen (1998), 'Rhetoric and Rage: Third World Voices in International Legal Discourse', *Wisconsin International Law Journal*, **16** (2), 353–419, 354.

51 A foundational resource on the topic of international environmental law and developing countries is Anand, Ruchi (1980), 'Development and the Environment: The Case of Developing Countries', *Indian Journal of International Law*, **20**, 1–19.

52 United Nations Conference on Environment and Development, Non-Legally Binding Authoritative Statement of Principles for a Global Consensus on the Management, Conservation and Sustainable Development of all Types of Forests, UN Doc A/Conf 151/126 (14 August 1992) 23.

equitable. The REDD+ mechanism is innovative in many ways, but the most important aspect of this initiative is its ability to transfer funds for sustainable forest management. The REDD+ mechanism will therefore be the first international instrument to provide finance and capacity building assistance for sustainable forest management initiatives and as such go some way to addressing the concerns of the TWAIL movement.

The efficiency of international environmental regimes

Weiss discusses the efficiency of the international environmental legal regimes as an element of equity. She suggests that inefficient instrument negotiation processes and reporting regimes impact negatively on developing countries' abilities to participate in such processes. The concept of 'treaty congestion' is used to describe the overabundance of international environmental instruments and processes. Weiss argues that many countries, especially those with limited resources, complain about the demands that these negotiations place on them for staffing and funding in order to participate in negotiations.[53] In the forestry context, there is evidence of treaty congestion and regime inefficiency, which significantly undermine the efficiency of the regime.[54]

Equity for future generations

Weiss has written extensively on the principle of intergenerational equity.[55] According to Weiss, this principle implies that humans are 'stewards' of the earth and its resources. It imposes obligations upon us to care for the planet, and gives us certain rights to use it.[56] The principle of intergenerational equity involves three components: the conservation of options;

---

[53] Weiss, above n40, 23–4.

[54] Dimitrov, Radoslave (2005), 'Hostage to Norms: States, Institutions and Global Forest Politics', *Global Environmental Politics*, 5 (4), 1–24.

[55] Weiss, Edith Brown (ed.) (1989), *In Fairness to Future Generations: International Law, Common Patrimony and Intergenerational Equity*, Tokyo, Japan: United Nations University; Dobbs Ferry N.Y.: Transnational Publishers; Weiss, Edith Brown (1984), 'The Planetary Trust: Conservation and Intergenerational Equity', *Ecology Law Quarterly*, 11 (4), 495–582.

[56] The principle of intergenerational equity locates its authority by reference to the concept of future interests in the environment in a number of international instruments. The following instruments recognize the interests of future generations: the Stockholm Declaration on the Human Environment; the 1972 London Ocean Dumping Convention; the 1973 Convention on International Trade in Endangered Species; the 1972 Convention Concerning the Protection of World Cultural and Natural Heritage; the 1992 Convention on Biological Diversity; the 2007 non-legally binding instrument on all types of forests; and the United Nations Framework Convention on Climate Change.

the conservation of quality; and the conservation of access.[57] Problems of equity between generations arise from: the depletion of non-renewable resources and the elimination of renewable resources; degradation in quality (of environmental resources such as air, water, soils); the loss of environmental services performed by natural resources; the loss of cultural resources; the lack of effective access to natural and cultural resources.

The stewardship concept has been in existence for a long time. References to the stewardship ethic and its responsibilities are evident in both religious and philosophical texts.[58] From a legal perspective, the ethic of stewardship creates responsibilities for individuals, governments and private institutions. Looking at existing analysis on stewardship it can be concluded that:

- humans have a responsibility to take care of the environment;[59]
- humans have a responsibility to pass on to their descendants a natural environment that is more fruitful as a result of their efforts;[60]
- stewardship involves a move from narrow individual interests towards wider common interests (the wider common interest being a healthy environment).[61]

The stewardship ethic emphasizes that humans are only temporary inhabitants of the earth. As temporary owners and users of the earth, they have a responsibility to take care of it for future generations. Stewardship requires that the current generation fulfil a caretaker position that is considerate and acts with thought for the future. Even though a stewardship ethic is not prevalent in environmental regulation as yet, there is great recognition and support for the principle within society.

> Fortunately, the notion that each generation holds the earth as a trustee or steward for its descendants strikes a deep chord with all cultures, religions and nationalities. Nearly all human traditions recognise that we, the living are sojourners on earth and temporary stewards of our resources.[62]

---

57 Ibid.

58 Passmore, John (ed.) (1974), *Man's Responsibility for Nature: Ecological Problems and Western Traditions*, London: Duckworth, 28–32.

59 Fisher, Douglas (ed.) (2003), *Australian Environmental Law*, Pyrmont, NSW: Lawbook Co., 7.

60 Morgan, G. (1993), 'The Dominion of Nature: Can Law Embody a New Attitude?', *Bulletin of the Australian Society of Legal Philosophy*, **18** (60) 43–76, 59.

61 Evans, Michael (ed.) (2000), *Principles of Environmental and Heritage Law*, St Leonards, NSW: Prospect Media, 84.

62 Weiss, above n 40, 20.

The traditional approach to environmental regulation is an anthropocentric approach. This approach places humans at the focus of the universe and views and interprets everything in terms of human experience and values. Using forests as an example, forest goods and services have generally only been attributed value when humans attach value to each individual good or service from the forest. Generally speaking, forests were previously valued for their economic benefits only. This approach to forest valuation is changing, as the list of forest values and services recognized by humans grows longer. Humans have started to realize that they are dependent upon the earth and its natural resources for survival, and that the supply of natural resources is finite. As such, society is beginning to accept that reform needs to take place in order not only to preserve the environment but also to preserve humanity.[63] The legal system needs to evolve so as to create responsibilities and duties to protect the earth and its natural resources for their intrinsic value.[64]

Proposed new system of responsibility based on distributive justice

Rawls's 'difference principle' requires the law to take into account the different circumstances of members of society when creating and implementing rights and obligations. The difference principle recognizes 'disadvantage' within society. The theory adopts a substantive approach to justice by advocating for different treatment for different individuals or groups. The implementation of the difference principle requires recognition of legal advantages and disadvantages between individuals and the implementation of strategies to reduce such inequalities. It is proposed that distributive justice in the global environmental area concerns the distribution of benefits and burdens of environmental resources among nations. It implies that relevant dissimilarities between the subjects of the law warrant special attention or special treatment.[65] Therefore, nations suffering disadvantage – which may be in the form of economic, social,

---

[63] McGregor, Joan L. (1999), 'Property Rights and Environmental Protection: Is This Land Made for You and Me?', *Arizona State Law Journal*, 31 (2), 391–437, 391–2.

[64] Many examples of intrinsic protection legislation or legal instruments exist at the international level – for example, the Convention on Biological Diversity. However, like all international legal instruments, it is debatable whether these international instruments create responsibilities and duties.

[65] Chandani, Achala (2007), 'Distributive Justice and Sustainability as a Viable Foundation for the Future Climate Regime', *Carbon and Climate Law Review*, (2), 152–163, 156.

or environmental disadvantage – should be assisted by international law principles to overcome the practices leading to such disadvantage.

The application of distributive justice principles in imposing liability for the implementation of sustainable forest management practices would result in a fundamentally different regime. Distributive justice principles would require international forest law to respond to the disadvantage that exists among different nations in relation to their capacity to implement sustainable forest management practices. According to such principles, different standards and accompanying liability would be created for states (determined by the existence of disadvantage). The international climate change regime has created a system that is reflective of distributive justice principles and the international legal principle of 'common but differentiated responsibility'. This has been achieved by increasing the liability of industrialized nations to reduce emissions, and requires industrialized countries to assist developing countries in implementing emission reduction clean energy development.[66]

It is proposed that, in the international forestry context, different standards should be created for developed and developing countries. This would be an example of applying the international environmental legal principle of 'common but differentiated responsibility'.[67] Disadvantage in this instance would be based upon a country's OECD status, along with other relevant considerations. Developed countries would be expected to implement sustainable forest management commitments. Such a system would allow for the development of an internationally agreed set of sustainable forest management criteria and practices. Developing countries would, similarly, negotiate a series of targets and goals in relation to the implementation of sustainable forest management. The development of an international forest regime that creates standards based upon capacity may prove to be a useful tool in moving forward the international forest agenda.[68] In addition to two sets of globally agreed standards for sustainable forest management, a new system of liability should also be imposed.

This new system of liability would need to recognize the role that developed countries play in the promotion of unsustainable forest and

---

[66] Ibid. 152.

[67] Ibid. 154. Common but differentiated responsibility recognizes: the common responsibility of countries to protect the environment; the differing contributions of countries to environmental issues; and the differences in the abilities of countries to prevent, reduce and control the threat of environmental degradation.

[68] For further information on the politics and difficulties of the international forest regime in creating enforceable targets and standards see Chapter 4.

land practices. Such an approach is consistent with the international environmental principle that the 'polluter pays'.[69] The international timber market is one of the significant drivers of global deforestation;[70] the main consumers of timber products being the citizens of developed countries.[71] It therefore seems appropriate to impose liability for the implementation of sustainable forest management on not only those countries that produce the timber, but also on those countries that are major consumers of forest products.

Furthermore, a system of liability for the implementation of sustainable forest management practices should also recognize that countries with high forest cover will bear a disproportionate responsibility for the implementation of agreed practices. Many of the benefits of sustainable forest management practices may be enjoyed at the local level (for example, improved soil and water quality), but a number of other benefits are enjoyed by the international community at large (for example, biodiversity and air quality). The current system of imposing liability for sustainable forest management does not take into account these additional justice-related perspectives. There is a need to create a system of liability to make all countries responsible for the implementation of sustainable forest management principles.

In summary, the international sustainable forest management regime could be improved if it adopted a distributive justice approach and:

- created two sets of sustainable forest management standards, reflecting the capacity of developed and developing countries to implement such practices;
- created a system of liability which acknowledges that the timber and agriculture product consumption of developed countries contributes significantly to unsustainable timber practices in developing countries; and
- created a system that shares liability for the implementation of sustainable forest management practices among the global community,

---

[69] This principle requires that the polluter bears the costs of remediation. See Chandani, above n 65 at 154. Pollution in this regard is deforestation, and the cost of remediation could be a range of activities such as the cost of restoring forest land, implementing sustainable forest management, and investing in the conservation of forest land.

[70] Bryant, Dirk, Daniel Nielsen and Laura Tangley (eds) (1997), *The Last Frontier Forests: Ecosystems and Economies on the Edge*, Washington DC: World Resources Institute, Forest Frontiers Initiative.

[71] International Tropical Timber Organization, *Annual Review and Assessment of the World Timber Situation* (2008), 10.

reflecting the benefits that the global community obtains as a result of sustainable forest management practices.

E. Justice as Recognition

Some authors argue that the concept of justice involves more than merely a distributional analysis.[72] Young argues that while theories of distributive justice offer models and procedures by which distribution may be improved, none of them thoroughly examine the social, cultural, symbolic and institutional conditions underlying poor distribution in the first place.[73] In support of this, Fraser suggests 'justice today requires both redistribution and recognition, as neither alone is sufficient'.[74] We ought to resist the presentation of the politics of redistribution and the politics of recognition as mutually exclusive alternatives. Instead, we should concentrate our efforts on searching for an alternative framework that can accommodate both types of demand. Recognition of rights is of crucial importance in the environmental context and such rights are often created as property-based rights.

In the international forestry context, there needs to be recognition of the events and circumstances that have landed our global forest estate in its current condition. It is estimated that around a half of the earth's original forest cover is gone, much destroyed within the past three decades.[75] The *Last Frontier Forests* report found that frontier forests (that is, large tracts of relatively undisturbed forest areas) are at threat from a number of human-induced activities. Logging presents the biggest threat to frontier forests, affecting more than 70 per cent of the world's threatened frontiers.[76] Energy, mining and new infrastructure development affect around 40 per cent of all frontier forest. One-fifth of the world's threatened frontiers are directly endangered by farmers who clear forests for cropland and pasture, while excessive vegetation removal, overhunting, and conversion

---

72 Schlosberg, above n 35.

73 Young, Iris (ed.) (1990), *Justice and the Politics of Difference*, Princeton, N.J.: Princeton University Press.

74 Fraser, Nancy (ed.) (1997), *Justice Interruptus: Critical Reflections on the 'Postsocialist' Condition,* New York: Routledge; and Robeyns, Ingrid (2003), 'Is Nancy Fraser's Critique of Theories of Distributive Justice Justified', *Constellations*, **10** (4), 538–554.

75 Bryant, Dirk, Daniel Nielsen and Laura Tangley (eds) (1997), 'The Last Frontier Forests: Ecosystems and Economies on the Edge' available at http://www.globalforestwatch.org/english/pdfs/Last_Frontier_Forests.pdf (accessed 26 March 2012).

76 Ibid. 15.

to plantation (with the introduction of non-native species), also significantly affect the health of frontier forests.[77] It is again proposed that the majority of these damaging activities occur for the benefit of developed or rich countries and rich individuals in developing countries that can afford to consume the products generated from such activities.[78] Aligned with theories of justice that focus on justice as recognition, liability for the implementation of sustainable forest management principles should be attributed on the basis of recognizing those who are ultimately responsible for the current patterns of unsustainable forest use (that, the consumer of the forest and agriculture goods and services).

Forests raise a number of other justice issues as well. One of these should be the recognition of the high correlation between those living in densely forested areas and the high levels of poverty and marginalization of these groups. The World Bank has carried out a number of studies into this relationship, and has found that people living in such forest areas are usually fringe dwellers of society – the poorest of the poor being forced to live hand to mouth from forest resources.[79] People in such circumstances should practically and ethically not be attributed responsibility for the implementation of sustainable forest management practices.

Recognition of indigenous rights in forest areas

Another significant issue in forestry regulation is the pressing need for the introduction of reforms to recognize indigenous and customary rights in forest areas. The United Nations Declaration on the Rights of Indigenous People contains a number of provisions relevant to forest dwelling indigenous communities. Rights enshrined in the document include: the right not be forcefully removed from lands;[80] rights to maintain and strengthen

---

[77] Ibid. 16.

[78] In general, for support of the argument that the majority of the world's consumption of natural resources is enjoyed in rich nations see Myers, Norman (1997), 'Consumption in Relation to Population, Environment and Development', *The Environmentalist*, 17 (1), 33–44 and de Sherbinin, Alex, David Carr, Susan Cassels and Leiwen Jiang (2007), 'Population and Environment', *The Annual Review of Environmental Resources*, 32 (1), 345–373.

[79] Angelsen, Arild and Sven Wunder (eds) (2003), *Exploring the Forest-Poverty Link: Key Concepts, Issues and Research Implications*, Bogor, Indonesia: Centre for International Forestry Research; and Sunderlin, William D., Sonya Dewi and Atie Puntodewo, 'Poverty and Forests: Multi-country Analysis of Spatial Association and Proposed Policy Solutions' (Occasional Paper No 47, Centre for International Forestry Research, 2007).

[80] United Nations Declaration on the Rights of Indigenous People, GA Res 61/295, 107th plen mtg, UN Doc A/61/L.67/ (13 September 2007) art. 10.

a relationship with lands and waters;[81] rights to ensure that the environment is protected;[82] and rights to compensation when land is taken, used or damaged.[83] Article 26 is the key provision from a land rights perspective, providing for: a right to land; rights to use and control resources on the land; and obligations for states to give legal recognition to such rights. The instruments emphasize the importance of obtaining 'free, prior and informed consent from indigenous groups before dealing with land under their control'. The language of 'free, prior and informed consent' has influenced the development of REDD+ policies and measures.

The 'UN Special Rapporteur on the Situation of Human Rights and Fundamental Freedoms of Indigenous People' considered the plight of forest-dwelling indigenous communities.[84] Of particular concern to the special rapporteur, was the lack of recognition by formal tenure systems of customary ownership rights over forest areas and the granting of timber concession rights to large forestry corporation over areas of indigenous or customary land. The special rapporteur found that the forests of indigenous people have been systematically affected by the activities of large forestry corporations resulting in the progressive destruction of their traditional means of subsistence. Such experiences have affected the living conditions of indigenous communities in the equatorial forests of Central Africa, the Amazon basin, the boreal forests of Siberia and America, the Andean range and South-East Asia, as well as the Pacific islands. The special rapporteur found that one of the key reasons underlying the high poverty levels of indigenous forest dwellers was corrupt practices in relation to the granting and recognition of rights over indigenous land.

> Some 60 million indigenous people in the world depend almost entirely on the forests for their survival. Hiding behind forest legislation, the authorities tend to sacrifice the rights of local communities to the interests of commercial firms, and resources are often utilized for illegal activities protected by corrupt officials and entrepreneurs. In many countries, eviction of indigenous people from their traditional forests as a result of such activities is one of the essential causes of their impoverishment.[85]

---

81 Ibid. art. 25.

82 Ibid. art. 29.

83 Ibid. art. 28.

84 United Nations, Commission on Human Rights, *Report of the Special Rapporteur on the Situation of Human Rights and Fundamental Freedoms of Indigenous People*, Rodolfo Stavenhagen, 61st sess., Agenda Item 15, UN Doc A/HRC/4/32 (27 February 2007), paras 25–6.

85 United Nations, Commission on Human Rights, *Report of the Special Rapporteur on the situation of human rights and fundamental freedoms of indigenous*

The case of *the Saramaka People v. Suriname*[86] in the Inter-American Court of Human Rights discussed a number of safeguards that must be satisfied prior to granting concessions or exploration resource licences on indigenous lands. The three safeguards were described as effective participation, benefit-sharing, and prior environmental and social impact assessments.[87] The court was required to decided if the Saramaka people has a traditional relationship with the forest estate and if, in finding that such relationship existed, there has been breach by the state in granting timber concessions. When determining if there was a traditional relationship between the Samaka people and the forest estate, the court took into consideration the following evidence

- Knowledge of the forests including the location and use of a variety of trees within the area
- The use of certain trees for specific purposes, such as the building of boats and canoes, roofs, fruits for consumption and other subsistence purposes
- The manner in which the Saramaka people respected and cared for the forest. Evidence was tendered to establish that the group was able to enter the forest and obtain wood without destroying the environment
- Evidence that the Saramaka people depended upon the extraction of wood as part of their economic structure and for subsistence purposes.[88]

These factors led to a finding by the court that the Saramaka people have traditionally harvested, used, traded and sold timber and non-timber forest products, and have continued to do so until the present day. The state should therefore not have granted logging concessions over Saramaka lands without fulfilling the three safeguards of effective participation, benefit-sharing, and prior environmental and social impact assessments. This decision therefore established the requirements

---

people, Rodolfo Stavenhagen, 61st sess., Agenda Item 15, UN Doc A/HRC/4/32 (27 February 2007) para 37.

[86] *Case of the Saramaka People v. Suriname. Preliminary Objections, Merits, Reparations and Costs*, Judgment of 28 November 2007, 575 I/A Court H.R., Series C No. 172 available at http://www.forestpeoples.org/sites/fpp/files/publication/2010/09/surinameiachrsaramakajudgmentnov07eng.pdf, (accessed 26 March 2012).

[87] Ibid. para 46.

[88] Ibid. paras 44–5.

that must be fulfilled by the state in future when granting forest-logging concessions.

In summary, it can be proposed that 'justice as recognition' in the international forestry context could be used as a basis to argue that:

- consumption patterns of forest products should bear some relation to the liability imposed for the implementation of sustainable forest management practices;
- the circumstances that have pushed certain 'forest dwellers' into their current position should be recognized by the creation of a system to assist them with the implementation of sustainable forest management practices;
- because justice as recognition requires the law to recognize the legitimate claims of indigenous or customary owners of forest areas, it follows that recognition, tenure and land rights should be secured for such groups and that the granting of concessions, licences, and permits over indigenous land should not be granted without the free, prior and informed consent of indigenous groups.

F. Justice as Participation and Procedural Justice

Weiss and Schlosberg agree that an element of justice relates to opportunities for meaningful participation, administrative justice and capacity for participation in the creation of international legal regimes.[89] This form of justice could be classified as justice within the governance process. While there are a number of ways in which governance-related considerations could be improved in the international forest regime, it is proposed that this form of justice is already recognized as being important within international forest regimes. For example, stakeholder participation and consultations are already part of all international regulatory processes. Issues concerning governance and stakeholder engagement are discussed in more detail in Chapter 5. However, as discussions concerning TWAIL perspectives and the examination of indigenous groups' forest rights demonstrate, issues of procedural justice require significant attention at the domestic level.

---

[89] Weiss, above n 40, 23 and Schlosberg above n 35 at 89–91.

3. CONCLUSION

This chapter has argued that rule of law and justice-based criteria are of paramount importance in forest regulation. At the international level, reform should take place so as to meet rule of law criteria with particular attention being directed to implementing and enforcing international forestry standards. One of the biggest challenges in meeting rule of law criteria at the domestic level is the elimination of corruption within the forestry sectors, which should be achieved by the gradual introductions of reforms aimed at increasing the transparency and integrity of forestry practices. This chapter has also argued that justice considerations are significant in forest regulation through analysis of distribution, recognition and participation models. At the global level distributive justice theories have been advanced to support this chapter's argument that responsibility for the implementation of sustainable forest management should be changed from a system based on sovereignty-imposed liability to a system based on the consumption and enjoyment of forest products as the means of assessing liability. In addition, the concept of common but differentiated responsibilities needs to be more fully integrated into international forest regulation, allowing for the development of individual forest targets for all states. Justice as recognition requires the international legal order to recognize the role that industrialized nations have played in global forest degradation and deforestation and to create liabilities for those countries to contribute financially and technically in conserving and improving the remaining forest lands. Recognition of indigenous rights to forest lands, and rights to the use and management of forest resources, should also be a key feature of any future international forest instruments. The FLEGT and REDD+ initiatives demonstrate an understanding by the global community of the justice and rule of law considerations that are related to forest regulation. The challenge for all international forest initiatives, however, remains the implementation of policies and measures that effectively address these concerns.

3. Legal concepts creating forest rights and limitations

1. INTRODUCTION

This chapter explores two legal concepts that have greatly influenced the development of international and national forest laws and policies. These concepts are that of state sovereignty and the legal concept of property. At the international level, states have been unwilling to accept binding legal obligations concerning sustainable forest management due to concerns that such an agreement would limit absolute sovereign rights and discretion about how forest estates were managed. Sovereignty provides rights to states – both externally within international negotiations, and also internally to self-govern. At the national level, the concept of property underpins all natural resources and management by providing rights and interests in land and resources. Property rights in forest estates can take a number of forms including access rights, use rights, usufruct rights and occupation rights. While both of these concepts provide rights, they also inherently involve restrictions on how these rights can be used and enjoyed. The evolution of international and national environmental regulation also places limitations upon how these rights can be exercised. This chapter will examine the following four issues

- Nature and rights of sovereignty in international forest law
- Limitations of sovereignty in international forest law
- Nature and role of property in forest regulation
- Property rights and forest regulation

2. NATURE AND RIGHTS OF SOVEREIGNTY IN INTERNATIONAL FOREST LAW

A. The Meaning of 'Sovereignty'

Sands states that the rules of international environmental law have evolved from two fundamental objectives which have pulled in different directions: first, that states have sovereign rights over natural resources and, secondly, that states must not cause damage to the environment.[1] Principle 21 of the Stockholm Declaration provides:

> States have, in accordance with the Charter of the United Nations and the principles of international law, the sovereign right to exploit their own resources pursuant to their own environmental policies, and the responsibility to ensure that activities within their jurisdiction or control do not cause damage to the environment of other States or of areas beyond the limits of national jurisdiction.

The principle of sovereignty is dual natured, in that it creates both rights and responsibilities. Schrijver states that the principle of sovereignty over natural resources was introduced into United Nations instruments in order to

> underscore the claim of colonial people and developing countries the right to enjoy benefits of resource exploitation, and in order to allow inequitable legal arrangements under which foreign investors had obtained title to exploit resources in the past, to be altered or even annulled because these conflicted with the concept of permanent sovereignty.[2]

The concept of absolute and permanent sovereignty over natural resources has been limited by the introduction of international environmental laws, which seek to address global concerns to ensure the preservation and restoration of a clean and healthy environment for present and future generations.[3] The principle of sovereignty, while an essential component of the international legal order, raises difficulties when one is seeking to determine the scope, definition and application of the concept.[4]

Litfin[5] discusses 'unbundling' the concept of 'sovereignty', which involves dismantling and studying the various terms associated with defin-

---

[1] Sands, Philippe (2003), *Principles of International Environmental Law* (2nd edition), UK: Cambridge University Press, 235.

[2] Schrijver, Nico (1997), *Sovereignty over Natural Resources: Balancing Rights and Duties*, UK: Cambridge University Press, 1.

[3] Van der Vyver, Johan (2009), 'The Environment: State Sovereignty, Human Rights and Armed Conflict', *Emory International Law Review*, **23** (1), 85–112, 88.

[4] On the different disciplines' understanding of the concept (international law, international relations, philosophy and economics) see Lewis, Melea, Charles Sampford and Ramesh Thakur (2008), 'Introduction' in Jacobsen, Trudy, Charles Sampford and Ramesh Thakur (eds), *Re-Envisioning Sovereignty: The End of Westphalia,* UK: Ashgate Publishers, pp. 1–8.

[5] Kuehls, T. (1998), 'Between Sovereignty and Environment: An Exploration of the Discourse of Government' in Liftin, K. (ed.) *The Greening of Sovereignty in World Politics,* Cambridge, Mass: MIT Press, pp. 31–53.

ing it. This unbundling exercise identifies a list of terms linked with the meaning of 'sovereignty'. They include: 'territory'; 'autonomy'; 'authority'; 'control'; and 'population'.[6] In addition to these terms, any analysis of sovereignty also needs to consider rights, capacities and responsibilities in three spheres: those under a state's jurisdiction; those under other states' jurisdictions; and those under a common jurisdiction.[7]

B. Rights Associated with Sovereignty

The terms associated with 'sovereignty' can be brought together to enrich its meaning. Sovereignty must be exercised over a clearly defined area and identifiable population, and the powers being exercised must be independent, must be within the limits of authority to act and and there must be the requisite control to carry out the necessary management of the state. Sovereignty, therefore, implies a plenary of powers to organize and control activities within state boundaries over state citizens.

The exercise of sovereign rights within the state's jurisdiction fits with the explanation of sovereignty offered above. That is, the freedom to exercise power and control over state citizens within state boundaries. A more complex exercise of sovereign powers involves states exercising sovereignty within other states. This could be in the form of recognition from other jurisdictions – acknowledging and respecting the sovereignty of the state – or this could occur by way of agreement: State A may accept State B's sovereignty in State A as part of an international agreement.

Finally, the operation of state sovereignty within common jurisdictions is of relevance to international law and obligations. The manner in which state sovereignty is exercised in international proceedings can be distinguished from the sovereign powers exercised within state boundaries. Under the common jurisdiction (a good example being conferences held at the international level), each state is interested in promoting its own national position on the issue in question.[8] Within the international forestry arena, each state comes together to protect its sovereign rights in relation to forestry, whilst accepting that some common standards might be created between all state parties concerning forests. This acceptance of

---

6 Terms that are synonymous with the requirements of statehood under the Montevideo Convention are: territory; government; population; and capacity to enter into legal relations.

7 Ibid.

8 For further information on the three realms of sovereignty, see Van der Vyver, Johan (2009), 'The Environment: State Sovereignty, Human Rights and Armed Conflict', *Emory International Law Review*, **23** (1), 85–112.

common standards could result in common responsibilities being shared by all states party to the international agreement. These new responsibilities might either maintain state sovereignty in cases where the international obligations are similar to existing domestic legislation, or weaken it where states have to adjust existing domestic laws.

C. Preservation of Sovereign Rights within International Forest Instruments

In 1992, during the Earth Summit negotiations, states were concerned about their sovereign rights over forest areas being eroded and were unwilling to commit to a legally binding agreement concerning forest use and management.[9] Such concerns led to the adoption of the Forest Principles 1992 (the Rio Forest Principles), a set of non-legally binding principles governing sustainable forest management. In 2007, a second non-legally binding instrument on forests[10] (Forest Principles 2007) was created in order to try and strengthen global political commitment to implementing sustainable forest management. States, in seeking to protect rights associated with sovereignty, have refused to commit to legally binding obligations concerning forest use and management. The statement from Mickelson below suggests why a legally binding instrument has been opposed:

> An essential difficulty that arises in the context of developing a forests convention, is that of avoiding the perception that such an agreement constitutes an infringement of sovereignty. In fact, if an instrument on forests is qualitatively different from previous international instruments in the environmental area it is precisely because its potential impact on sovereignty over resources appears to be much more direct.[11]

The Rio Forest Principles illustrate the significance of states intentions to protect sovereign forests within national control. The first two principles of the Rio Forest Principles reiterate the importance of sovereign rights.

---

[9] See generally on the tension between state sovereignty and environmental governance, Bosselmann, Klaus (2004), 'Environmental Governance: A New Approach to Territorial Sovereignty' in Goldstein, Robert J. (ed.), *Environmental Ethics and Law*, Aldershot: Ashgate, pp. 293–313.

[10] United Nations Forum on Forestry, *Non-Legally Binding Instrument on All Types of Forests*, 2nd Comm, 62nd sess, Agenda Item 54, UN Doc A/C.2/62/L.5 (22 October 2007) annex.

[11] Mickelson, Karin (1996), 'Seeing the Forest, the Trees and the People: coming to terms with developing perspectives on the proposed global forest regime', in Johnson, S. (ed.), *Global Forests and International Environmental Law*, London; Boston: Kluwer Law International, pp. 239–264, 248.

Principle 1(a) of the Rio Forest Principles, recalling earlier international concepts provides:

> States have, in accordance with the Charter of the United Nations and the principles of international law, the sovereign right to exploit their own resources pursuant to their own environmental policies and have the responsibility to ensure that activities within their jurisdiction or control do not cause damage to the environment of other States or of areas beyond the limits of national jurisdiction.[12]

Principle 2(a) of the Rio Forest Principles extends sovereign rights over forests further by providing:

> States have the sovereign and inalienable right to utilize, manage and develop their forests in accordance with their development needs and level of socio-economic development on the basis of national policies consistent with sustainable development and legislation, including the conversion of such areas for other uses within the overall socio-economic development plan based on rational land-use policies.

This statement on sovereignty actually provides for and allows for further deforestation. It confirms that states have the inalienable right to utilize, manage and develop their forest areas. This inalienability aspect suggests that states are not able to delegate or transfer power to regulate forest use and management outside the state – so, standards and rules concerning forest use and management prescribed by international institutions would be inconsistent with this statement. Of additional concern is the statement of approval for further development in forest areas, with no qualifying statement as to the nature of this development. This statement clearly supports views expressed by pro-development states at Rio de Janeiro.

The Forest Principles 2007[13] reaffirm the message about sovereignty of the Forest Principles 1992 within the preamble to the agreement. However, states did not insist on recognizing sovereign rights in the articles of this agreement. This might be explained by the voluntary and non-legally binding nature of the agreement, and perhaps a small shift in states' willingness to accept some small limitation upon domestic forest use and management. This, however, could not be construed as suggesting that

---

[12] This provision is very similar to the wording of Article 29 of the Convention on Biological Diversity, opened for signature on 5 June 1992, UNTS 30619 (entered into force 29 December 1995).

[13] United Nations Forum on Forestry, *Non-Legally Binding Instrument on All Types of Forests*, 2nd Comm, 62nd sess., Agenda Item 54, UN Doc A/C.2/62/L.5 (22 October 2007) annex.

states are willing to create a legally binding instrument on forest use and management. Any attempt to curtail sovereign rights over forest estates remains controversial, given the economic contribution of forest resources to many economies. The rights associated with sovereignty within the realm of a state's jurisdiction have hence been preserved and protected by international forest instruments.

3. LIMITATIONS ON THE RIGHTS OF SOVEREIGNTY

A. Restrictions upon State Sovereignty

When states sign international agreements (such as the Forest Principles 1992 and 2007), they are agreeing (at least in principle) to approve and attempt adoption of the responsibilities created by the instrument. The adoption of the agreement may require changes to domestic policy and action. As such, it removes some of the state's autonomy in the creation of new policies and measures. In theory, states are not meant to implement policy that is inconsistent with international obligations, even when the international environmental obligations impact negatively upon a state's economic interests. This means that the implementation of international environmental obligations has the ability to undermine most of the rights commonly associated with the exercise of state sovereignty.[14]

Looking more specifically at the second component of sovereignty found in principle 1(a) of the Rio Forest Principles, it requires states to think beyond the consequences within state boundaries and consider all possible consequences of their actions. The principle requires that states 'do not cause damage to the environment of other states or of areas beyond the limits of national jurisdiction'.[15] This indicates that international duties exist in relation to both the individual states and to the international community as a whole.[16]

---

[14] Fisher, Douglas (ed.) (2009), *The Law and Governance of Water Resources: The Challenge of Sustainability*, Cheltenham, UK and Northampton, MA, USA: Edward Elgar Publishing, 192.

[15] United Nations Conference on Environment and Development, *Non-legally Binding Authoritative Statement of Principles for a Global Consensus on the Management, Conservation and Sustainable Development of All Types of Forests*, UN Doc A/Conf 151/126 (14 August 1992), principle 1(a).

[16] Tarlock, Dan (1997), 'Exclusive Sovereignty versus Sustainable Development of a Shared Resource: The Dilemma of Latin American Rainforest Management', *Texas International Law Journal*, **32** (1), 37–66, 49.

Environmental damage is defined in the Convention on Civil Liability for Damage Resulting from Activities Dangerous to the Environment (Convention on Civil Liability)[17]. The objective of the European instrument is to provide adequate compensation for damage resulting from activities dangerous to the environment; and also to provide means of prevention and reinstatement. Under this convention, it appears that 'damage to the environment' means loss or damage by impairment to the environment. Article 7 defines damage to mean

- Loss of life or personal injury: 7(a)
- Loss or damage to property: 7(b)
- Loss or damage by impairment of the environment, this form of damage being limited to a remedy for the costs of reinstatement: 7(c)
- The cost of preventative measures and any loss or damage caused by preventive measures: 7(d).

Article 7(a) of the Convention on Civil Liability could be broadly construed to include loss of human, animal and plant life. This broad construction is prevented by principle 7(c), because it specifically mentions the environment. The term 'environment' is defined in Principle 10 to include natural resources, including abiotic and biotic specimens, property which forms part of the cultural heritage, and the characteristic aspects of the landscape. This implies that 'loss of life' is only intended to cover the loss of human life. In terms of the natural environment, principle 7(c) is the most relevant. However, it is worth noting that this type of damage is limited to a remedy for the costs of reinstatement only, meaning that other restitutive remedies are specifically excluded for cases of environmental damage.

The emergence of markets for environmental resources has led to the commoditization of environmental values. Markets require the existence of rights in the commodity being traded. The creation of such rights limits the sovereign powers of the state if the interest associated with the rights is transferred by the state to an individual. Foreign investment in natural resources interests through the operation of environmental markets also leads to a reduction in a sovereign state's rights or authority. International environmental markets strip away sovereign rights by imposing their own criteria concerning the management and use of the resource. This, therefore, limits the application of state imposed regulation. The emergence

---

[17] Lugano Convention on Civil Liability for Damage Resulting from Activities Dangerous to the Environment (1993) ILM 1228.

of REDD+ investments and demonstration activities has the potential to raise concern about sovereign rights being eroded. There has, however been more willingness to accept this erosion of sovereign rights because a financial incentive is provided and involvement in the REDD+ program occurs on a voluntary basis.

B. Limitations on Sovereignty in International Forest Governance

Arguably, all international instruments whose objective is to conserve and protect the natural environment affect individual states' rights to deal with their natural resources. The concept of sustainable forest management, like all other accepted environmental norms or standards, impinges upon states' sovereign rights.[18] One eminent writer on international forest regulation has suggested that:

> Sovereignty is a legal fiction which has always been compromised by transnational economic and social forces with assertions of sovereignty serving to insulate the state from the international environmental effects of its policies.[19]

Mickelson[20] explores the importance of sovereignty to developing nations. Following World War II, sovereignty was a concern for many states. Sovereignty over natural resources became linked with concerns about human rights and the right to self-determination.[21] In the UN's *Progressive Development of the Principles and Norms of International Law Relating to the New International Economic Order*[22] the importance of sovereignty to self-determination was highlighted by a Chilean proposal:

> The right of peoples to self determination should also include permanent sovereignty over their natural resources and that in no case might a people

[18] Weiss, Edith Brown (1993), 'International Environmental Law: Contemporary Issues and the Emergence of a New World Order', *Georgetown Law Journal*, **81** (3), 675–710, 703–5.

[19] Humphreys, David (ed.) (1996), *Forest Politics: The Evolution of International Cooperation*, London: Earthscan Publications, 171.

[20] Mickelson, Karin (1996), 'Seeing the Forest, the Trees and the People: Coming to Terms with Developing Perspectives on the Proposed Global Forest Regime', in Johnson, S. (ed.), *Global Forests and International Environmental Law*, London; Boston: Kluwer Law International, pp. 239–264.

[21] Ibid.

[22] United Nations Institute for Training and Research (UNITAR), Progressive Development of the Principles and Norms of International Law Relating to the New International Economic Order, UNITAR/DS/5 (15 August 1982).

be deprived of its own means of subsistence on the grounds of any rights that might be claimed by other states.[23]

International forestry law is yet to be regulated by a legally binding instrument so the issues connected with sovereignty have yet to be addressed.[24] Any instrument seeking to limit sovereign rights over forests would need to be drafted in such a way as to provide clarity on the relationship between sovereign rights and the obligations prescribed by the text.[25] This could, for example, involve a draft that recognizes state sovereignty as being important to states, whilst also recognizing that forest areas are of critical importance to states for many reasons. States may then agree to erode some of the rights associated with sovereignty. This will require a paradigm shift, from forests been valued as a natural resource, which provides resources, capital and jobs, to a view in which forests are recognized as having multiple values that are best managed under an international approach.[26]

State sovereignty involves concepts such as control, autonomy and authority to deal with all transactions occurring within a state's borders.[27] International environmental agreements, on the other hand, prescribe rights and duties for states in relation to the natural environment. The concept of sustainable development was created to balance out the incompatibility between state sovereignty and the obligations created in international agreements.[28] In theory, the concept of sustainable forest management operates to ensure that states balance the competing ecological, social and economic values associated with forestry equally, resulting in management that could be considered sustainable. In practice, economic interests still prevail as the leading concern, and any developments

---

[23] Ibid.

[24] See generally on the tension between state sovereignty and environmental governance Bosselmann, Klaus (2004), 'Environmental Governance: A New Approach to Territorial Sovereignty' in Robert J. Goldstein (ed.), *Environmental Ethics and Law*, Aldershot: Ashgate, pp. 293–313.

[25] Weiss, above n 18, 703–5.

[26] For a discussion of classifying forests as global commons see Tarlock, above n 16, 44–47.

[27] Stuart, Jill (2009), 'Unbundling Sovereignty, Territory and the State in Outer Space: Two Approaches' in Bormann, Natalie and Michael Sheehan (eds), *Securing Outer Space: International Relations Theory and the Politics of Space*, Oxford, UK: Routledge, pp. 8–23.

[28] Kuehls, T (1998), 'Between Sovereignty and Environment: An Exploration of the Discourse of Government' in Liftin, K. (ed.) *The Greening of Sovereignty in World Politics,* Cambridge, Mass: MIT Press, pp. 31–53, 48 and Tarlock above n 16, 42.

in forestry law that impact adversely upon economic interests are fraught with political tension.

It has been suggested that an alternative to a legally binding forest instrument could involve the adoption of the approaches outlined below which side-step sovereignty concerns[29]

- Restoration of forests damaged by development agendas. This process would involve rebuilding or restoring forests to their original biological make-up
- Debt reduction programs for developing countries to prevent further deforestation taking place. This may involve forgiving the debt
- International funding for environmental protection
- Debt for nature swaps; that is, swaps that create financial incentives for rainforest nations to preserve biodiversity.[30]

All of these policy options would impact positively upon forest cover and forest health but the challenge with such measures is the provision of political and economic support to implement them. The Food and Agricultural Organization's (FAO) *State of the World's Forests* of 2007[31] highlighted the urgent need for increased funding for international forestry projects. The FAO had funding to support ten new forest programs annually, however the demand for these programs significantly exceeds the capacity of the FAO.[32] REDD + has risen to this challenge with the potential to deliver between $US 17–33 billion every year (which is the equivalent to a quarter of OECD aid flows in 2010) for sustainable forest management initiatives.[33] Investments made under REDD+ avoid dealing with sovereignty-based concerns directly. Such measures, however, will have a large impact on sovereign rights, which might in the future result in states being more willing to compromise once trust and experience is

[29] Eshbach, R. (1990), 'A Global Approach to the Protection of the Environment: Balancing State Sovereignty and Global Interests', *Temple International and Comparative Law Journal*, **4** (2), 271–305, 282.

[30] Tarlock, above n 16, 64.

[31] Food and Agricultural Organization of the United Nations, *State of the World's Forests 2007* (2007).

[32] Ibid. 70.

[33] Global Witness (2011), 'Forest Carbon, Cash and Crime', available at http://www.globalwitness.org/sites/default/files/library/Forest%20Carbon,%20Cash%20and%20Crime.pdf (accessed 20 January 2012).

gained with the implementation of international forest initiatives through REDD+.[34]

4. THE NATURE AND ROLE OF PROPERTY IN FOREST REGULATION

A. Legal Concept of Property

The rights and duties associated with property law underpin all environmental regulation in non-customary legal systems. This is because property law defines use, access and other necessary rights related to the environment. The legal concept of property is used as an instrument to recognize and allow for control. Fisher proposes the following legal definition of property:[35]

> . . . the relationship between the owner of the object or the asset, and the object or asset itself. The owners – the holders of the rights associated with ownership – are generally acknowledged to have available to them a number of rights: to use, not to use, to permit someone else to use and to alienate.

Property law hence significantly influences the use and management of forest areas. Property rights in relation to interests in the environment are the ultimate right that a forest user can obtain. This is because property rights are generally considered as being secure in form. Additionally, property rights give the holders of such rights a broad spectrum of rights that are not as limited as rights given in other forms (such as under statute or contract). Some benefits associated with property rights can include constitutional protection, the ability to prevent trespass, conversion or nuisance under the common law, a right to obtain a mortgage over a property in question, a right to alienate or divide the property between present and future interests, rights to receive special treatment under federal or state tax laws, and rights to impose or avoid trade constraints.[36]

---

[34] See, on the evolution of international regimes, Bodansky, Daniel and Elliot Diringer (2010), 'The Evolution of Multilateral Regimes: Implications for Climate Change', available at http://www.environmentportal.in/files/evolution-multilateral-regimes-implications-climate-change.pdf (accessed 26 January 2012).

[35] Fisher, Douglas (ed.) (2009), *The Law and Governance of Water Resources: The Challenge of Sustainability*, Cheltenham, UK and Northampton, MA, USA: Edward Elgar Publishing, 67.

[36] Zellmer, Sandra and Jessica Harder (2008), 'Unbundling Property in Water', *Alabama Law Review*, **59** (3), 679–745, 682.

On a more philosophical level, it has been suggested that private property plays a significant role in society.[37] For example, Hegel believed that ownership is extremely important to individuals, since it is only through owning and controlling property that one can 'embody' one's will in external objects and, thereby, start the process of transcending the subjectivity of one's immediate existence. An individual develops their personality by controlling, using or working on external objects.[38] According to Kant, it is necessary for individuals to have control over external objects, as this respects their autonomy.[39] Rawls has suggested that the existence of personal property allows for a sufficient material basis for a sense of personal independence and self-respect, both of which are essential for the development and exercise of the moral powers.[40] McGregor suggests that private ownership is necessary to protect and enhance individual self-determination and autonomy, including the ability to accomplish future projects, to autonomously develop one's occupation, to promote independence in a democratic nation and to foster human happiness.[41]

The existence of property rights over natural resources is often justified in reference to Hardin's tragedy of the commons theory. This theory suggests that when things are left open to the public they are wasted by overuse or underuse. No one wishes to invest in something that may be taken from him tomorrow, and no one knows whom to approach to make exchanges. All resort to snatching up what is available for 'capture' today, leaving behind a wasteland.[42] Therefore, it is believed that private property rights provide an incentive for better management of natural resources areas. The incentive comes in the form of long-term secure and transferable rights in forest property. Where rights to the environment are secure and transferable, they provide holders of such rights with the certainty that their rights will be valuable in the future. Such certainty

---

37 McGregor, Joan L. (1999), 'Property Rights and Environmental Protection: Is This Land Made for You and Me?', *Arizona State Law Journal*, **31** (2), 391–437.

38 Hegel, Georg Wilhelm Friedrich, *Hegel's Philosophy of Right* (T.M. Knox trans, Oxford University Press, 1967) [trans. of *Grundlinien der Philosophie des Rechts* (first published 1821)] 37–41.

39 Kant, Immanuel, *The Metaphysical Elements of Justice – Part 1 of the Metaphysical Morals* (John Ladd trans., Hackett Publishing, 1965) [trans. of *Die Metaphysiche Anfangsgründe der Rechtslehre*, Part 1 of the *Metaphysik der Sitten* (first published 1797)] 76–7.

40 Rawls, John (ed.) (1972), *A Theory of Justice*, Oxford: Clarendon Press, 289.

41 McGregor, above n 38, 422.

42 Rose, Carol (1986), 'The Comedy of the Commons: Custom, Commerce and Inherently Public Property', *The University of Chicago Law Review*, **53**(3), 711–781, 712.

provides an incentive to rights holders to manage rights so as to increase their economic value.

All international institutions and many international forest instruments recognize the important role that secure tenure and property rights play in achieving sustainable forest management.[43] The terms 'property' and 'tenure' are used interchangeably in international institutions. The legal definitions of 'property' and 'tenure' differ from what other disciplines recognize as property- or tenure-based rights,[44] and the legal definition of 'property' is commonly misunderstood.[45] The term is often used in non-legal disciplines to correspond with the concept of ownership.[46] This analysis, therefore, examines the nature of the item or thing that is owned. By way of example, the 'thing' in the forest context may be land upon which the forest exists, or it may be the individual trees. Under this interpretation of property, ownership is seen as the ultimate right.

The legal definition of property, however, recognizes that a number of rights may arise as a result of the property relationship. As such, the legal definition tends to focus on the nature of the rights that attach to the property interest. The traditional approach to describing the legal definition of property is the 'bundle of rights' approach.[47] The bundle of rights analysis merely examines the individual rights associated with the property holding. It has been suggested that the bundle of rights metaphor fails to assess either the character of the thing in question or the nature of the human relationship with it.[48] Arnold has suggested a new metaphor to

---

43 Sunderlin, William, Jeffrey Hatcher and Megan Liddle, Rights and Resource Initiative, *From Exclusion to Ownership? Challenges and Opportunities in Advancing Forest Tenure Reform* (2008).

44 For example, in an economist's language, the phrase 'property right' is typically little more than a synonym for 'ownership' or perhaps possession. See Scott, Anthony (ed.) (2008), *The Evolution of Resource Property Rights*, Oxford; New York: Oxford University Press, 4.

45 See Fisher, D. E., 'Rights of Property in Water: Confusion or Clarity' (2004) 21 *Environmental and Planning Law Journal* **21**(3), pp. 200–226, where it is stated: 'It is axiomatic that linguistic precision is essential for an effective and enforceable regime of law. Yet words are often as much a source of confusion as of clarity. This is particularly true of the word 'property'. The word has assumed political, economic and cultural as well as legal perspective over the years. These distinctive perspectives may well conflict'.

46 A search of the major online dictionaries confirms that property is generally defined by its ability to be owned.

47 This approach is not necessarily ideal, but is suggested in order to fill a 'vacuum' in the description legal property relations. See: Penner, J.E. (1996), 'The Bundle of Rights Picture of Property', *UCLA Law Review*, **43** (3), 711–820.

48 Zellmer, above n 37, 684.

describe property: the 'web of interests' paradigm. This approach places property at the centre of the web and examines relationships with the things and incidents of private ownership as well as public and communal rights from the internal strands of the web and surrounding 'webframe'.[49] In the forestry context, 'the bundle of rights approach' only examines the nature of an individual's rights (the right to harvest, right of entry, and other relevant rights). This is clearly not an adequate method of analysis because it is possible to have a number of competing rights over forest areas – and the bundle of rights paradigm does not recognize the existence of competing or compatible interests in forest land.[50] The 'web of interests' paradigm is more useful in the forest context because it places the forest at the centre of the web and then examines the interests and rights of all holders over the forest area.

Zellmer and Harder take this web approach one step further and suggest that the starting point should be consideration of whether property exists in the first place.[51] In relation to recognizing property in natural resources, it is necessary to examine whether the resource is capable of being classified as property. This type of analysis requires us to identify and re-examine the values that society associates with property interests.[52]

B. Legal Property Characteristics

As Zellmer and Harder suggest, the starting point for legal analysis concerning property interests should start with an examination of whether the thing in question is capable of being recognized as property?[53] If it is, legal recognition and legal rights will accrue to the holder of the property interest. If it is not, then alternative legal protection may need to be sought (for example, protection under contract law or under statute). There is no simple test to apply in order to determine if the item or thing in question is of a proprietary nature. This has led to some confusion in the forest

---

49 Arnold, Craig Anthony (2002), 'The Reconstitution of Property: Property as a Web of Interests', *Harvard Environmental Law Review*, **26** (2), 281–295, 291–5.

50 See Sheehan, John, and Garrick Small (2005), 'Biota and the Problem of Property', *Environmental and Planning Law Journal*, **22** (2), 158–163, which discusses some of the issues with the bundle of rights approach and new biota property challenges.

51 Zellmer and Harder's analysis examines the nature of property rights in water.

52 The scarcer the resource, the tighter the enforcement. As long as resources exist in abundance, societies find no need to invest attention in them. See Kant, above n 40.

53 Zellmer, above n 37, 684.

context as the legal system attempts to articulate an interest in particular forest items (for example, the carbon stored in trees or the biodiversity of the forest), thereby classifying each of these forest services as independent property-based interests.[54]

There are a number of characteristics that are associated with property-based rights. It should be noted, though, that the existence or absence of one of these characteristics does not definitively resolve the existence or non-existence of property interests.[55] However, these characteristics are useful for identifying the legal nature of the interests attached to the thing in question. In *National Provincial Bank Ltd v Ainsworth*[56] it was stated:

> Before a right or an interest can be admitted into the category of property, or of a right affecting property, it must be definable, identifiable by third parties, capable in its nature of assumption by third parties, and have some degree of permanence or stability.

Fisher states that property rights can broadly be identified by examining:

- whether the right can be enforced within the legal system;
- whether the substance of the right can be clearly defined and identified;
- whether the right is exclusive to the holder of the right;
- whether the right is sufficiently permanent and stable to attract a sufficient degree of security; and
- whether the right can be transferred.[57]

---

54 Two states in Australia (Victoria and South Australia) allow the creation of Forest Property Agreements which include rights to plant, maintain and harvest forest property (which includes carbon sequestered by trees). These agreements can then be registered on the land register, which allows the owner of the forestry right to enforce contractual obligations related to the right. The registration of this interest in the land also allows the owner of the right to enforce these contractual obligations against future owners of the land. For further information see: Streck, Charlotte (2008), 'Forests, Carbon Markets, and Avoided Deforestation: Legal Implications', *Carbon and Climate Law Review*, (3), 239–247, 246; and Passero, Michelle (2008), 'The Nature of the Right of Interest Created by a Market for Forest Carbon', *Carbon and Climate Law Review*, (3), 248–253, 249, 252.

55 Fisher, above n46, 211.

56 [1965] AC 1175, 1247–8.

57 Fisher, above n46, 218. These criteria were derived from the following cases: *Bank of New South Wales v Commonwealth* (1948) 76 CLR 1, 349–50; *National Provincial Bank v Ainsworth* [1965] AC 1175, 1247–8; *R v Toohey ex parte Meneling Station Pty Ltd* (1982) 158 CLR 342 and *Western Mining Corporation Ltd v Commonwealth* (1994) 121 ALR 661, 682–4.

In addition to these, Scott identifies a further three characteristics:

- the duration of the right;
- the flexibility of the right; and
- the divisibility of the right.[58]

Enforcement

To determine if the right can be enforced by the legal system, it will be necessary to see if the legal system recognizes its existence. Thomas Jefferson argued that property is strictly a social creation; and because society creates property rights, it should continue to control them.[59] On this view, property exists within a social and political context, and is dependent on this context for its form and existence.[60] This sounds somewhat abstract but an example in the forestry context should help to demonstrate its significance. It is increasingly common in domestic legal frameworks to recognize the existence of a right to forest carbon. However, this recognition is only relatively new, and many legal systems are still determining the correct mechanism for recognizing such forest interests. This suggests that when novel interests arise, the legal system may be slow in recognizing such rights. If the legal system recognizes the new interest, the rights associated with the interest are capable of being enforced. Therefore, recognition and enforcement of legal rights are two key features of a legal right. In many developing countries, the lack of recognition of customary, indigenous, or local people's interests in forests areas prevents such groups being able to enforce their rights.[61]

Clear definition and identification

For the legal system to recognize property, there must be clarity as to the definition of the item, as well as a formal process that identifies the property interest. The definition and identification of forest-related interests in certain instances may be relatively straightforward, but in others it may be much more complicated. It is common for a number of forest interests to exist over the one parcel of land. First, there are interests attaching to the land that the forest stands upon. In the majority of developed countries, various kinds of land boundary, as well as remote mapping technology have made land identification a formal and accountable process. In

---

58 Scott, above n 45, 6–10.
59 McGregor, above n 38, 394.
60 Ibid. 395.
61 Sunderlin, above n 44.

many developing countries, land boundaries are often unclear and remote mapping technologies are limited.[62] Secondly, the vegetation of the forest needs to be capable of definition and identification. This process involves the assistance of scientific, indigenous, and other relevant stakeholders to ensure that genetic and other forest values are clearly understood. Identifying and defining forest ecosystem values should involve a process that is both flexible and dynamic, and which evolves as new knowledge is acquired concerning forest values. Thirdly, other interests in the forest need to be capable of definition and identification. These other interests may be harvesting rights, access rights, carbon, biodiversity or ecosystem rights, or some other form of right. These other forest rights may involve linguistic challenges concerning definitions and similarly complex identification procedures.[63]

Exclusivity

Scott suggests that exclusivity involves two distinct theories. First, it can refer to the reduction or avoidance of physical interference with the right holder's use of the resource. Interference, in this sense, refers to the sharing of the resource with other owners. Secondly, exclusivity can also refer to the right holder's degree of independence or freedom from government regulations that operate to restrict the use of a resource in order to promote the public good and the government's objectives.[64] In relation to the first definition of exclusivity within the forestry context, it may be possible to have compatible interests within the one parcel of forestland. In such instances, there is no need for exclusivity. For example, one party may have rights to harvest the fruit from the trees, while another has rights to gather honey from the trees. In other instances, the exclusivity of the forest right will determine the economic, environmental and social value of the right. Often, in the forestry context, individual interests in forests are incompatible with one another. For example, it is difficult to recognize harvesting rights and biodiversity forest rights on one piece of forestland at the same time. In relation to the second theory of exclusivity, it is suggested that modern environmental regulation has impinged upon absolute rights to natural resource use and management, and that it would be rare to find examples of absolute exclusivity from government intervention.

---

62 Sayer, Jeffrey, Jeffrey McNeely, Stewart Maginnis, Into Boedhihartono, Gill Shepherd and Bob Fisher, Rights and Resource Initiative, (2008) *Local Rights and Tenure for Forests: Opportunity or Threat for Conservation*?

63 See Chapter 6, which examines the complexity of defining and identifying forest carbon interests.

64 Scott, above n 45, 6.

It has been stated that the right to exclude is the most characteristic right of private property.[65] It is suggested that this right makes:

> . . . private property fruitful by enabling owners to capture the full value of their individual investments, thus encouraging everyone to put time and labour into the development of resources . . . Thus exclusive private property is thought to foster the well-being of the community, giving its members a medium in which resources are used, conserved and exchanged to their greatest advantage.[66]

In the forestry context, there is compelling evidence to suggest that private property rights, including the right to exclude, are not sufficient alone to ensure sustainable forest management practices.[67] Conversely, it is frequently recognized that exclusivity and security of tenure contribute to improving the implementation of sustainable forest management practices.[68] It is suggested that there must be a robust legal and policy framework to support private ownership rights while ensuring that sustainable forest management practices are carried out.

Permanency and security

Fisher suggests that for certain rights to be recognized as property, they must endure for a sufficiently long period without any prospect of substantial change or variation, so as to create a degree of security.[69] Evidence of a permanent and secure right to forest areas is a prerequisite for investment in the land. This is particularly so in the case of forestry investment, as the fauna and flora within forests takes a long time to mature.

Transferability

The ability to transfer ownership or rights in an item is a valid indication of the existence of property and an essential requirement for the operation of modern market economies. The transfer of property interests usually takes place in exchange for monetary compensation. This process allows individuals, groups, governments and businesses to buy and sell property to suit their interests. Therefore, proponents of market regulation support the creation of regimes that create property rights over a wide range of interests. For an interest to be transferred, other property criteria such as recognition, enforcement, identification and definition must be met.

---

65 Rose, above n 43.

66 Ibid. 711.

67 White, Andy, and Alejandra Martin, Forest Trends, (2002) *Who Owns the World's Forests? Forest Tenure and Public Forests in Transition* 15–17.

68 Sunderlin, above n 44.

69 Fisher, above n 46, 211.

In the forest context, the ability to transfer interests in forest property increases the attractiveness of investments in it since people generally feel more confident investing in a process that allows them to transfer their interest when it no longer meets their needs. Most forest property, once recognized, will meet transferability requirements.

Duration and flexibility

Duration refers to the timeframe for which the interest exists.[70] The duration of forest property interests will vary according to the interest. The harvest of forest property may require the interest to exist during the propagation and subsequent harvesting period. Rights arising from forest carbon or biodiversity may be required in perpetuity, while other rights (for example, to process non-timber forest products) may only be required for temporary periods. Flexibility is the extent to which the powers and obligations associated with the rights can be adjusted without weakening title.[71] Therefore, flexibility allows for re-negotiation of the terms or conditions during the duration of the holder's occupation. This is a particularly useful characteristic, because it allows the property interest to evolve in accordance with changes in society.

Divisibility

Scott identifies three types of divisibility: horizontal, vertical, and multiple use.[72] All of them are useful in the context of forest property: horizontal divisibility allows the holder of a lot to transfer rights to forest property to third parties; vertical divisibility allows for multiple interests in forest products to be created; and multiple use divisibility allows for the recognition of a number of forest property interests on the one lot.

Summary

The characteristics described as enforcement, definition, identification, exclusivity, permanency and security, transferability, duration, flexibility and divisibility are all common features of property rights. To determine if new interests in forests are capable of being recognized as property, these kinds of criteria should be taken into account. Most legislatures have decided to avoid the issue of determining if interests in forests are property-based and, instead, have referred to such interests as 'rights'.[73]

---

70 Scott, above n 45, 7.
71 Ibid. 7.
72 Ibid. 10–12.
73 Streck, above n 56, 246 and Passero, above n 56, 252.

This, however, does not ultimately determine whether the nature of the right is property-based. In common law systems, the judiciary may, in due course, be called upon to decide the precise nature of interests in forest resources.[74]

5. PROPERTY RIGHTS AND FOREST REGULATION

A. Land Tenure and Forest Management

The majority of the world's forests are held under public arrangements.[75] Globally, there is a shift towards an increase in private ownership,[76] with 2009 figures showing the following break-down in forest ownership: 65 per cent administered by government; 4 per cent reserved for community and indigenous groups; 18 per cent held under private community indigenous arrangements; and 13 per cent held under individual or private holdings.[77] It has been hypothesized that this shift is occurring for two main reasons: first, there is increased recognition that official forest tenure systems discriminate against the rights and claims of indigenous people and local communities; and secondly, there is growing evidence to suggest that governments have not been good stewards for public forest areas.[78]

Tenure reform

There is a growing realization that insecure property rights are a key underlying cause of forest degradation. Property rights to forest areas are often contested, overlapping or unenforceable, and insecure tenure forest

---

[74] In Australia, the High Court was called upon to determine if 'water rights' were a property interest and whether the removal of such rights allowed a claim for compensation under the Federal constitutions. See Maguire, Rowena and Angela Springer (2011), 'The Role of Property Law in Environmental Management: An Examination of Environmental Markets', *Environmental and Planning Law Journal,* **28** (4), 215–242 for a discussion of the *ICM Agriculture Pty Ltd v Commonwealth* (2009) 240 CLR 140.

[75] Ibid. See also studies conducted by Sunderlin, William, Jeffrey Hatcher and Megan Liddle, Rights and Resource Initiative (2008), *From Exclusion to Ownership? Challenges and Opportunities in Advancing Forest Tenure Reform*, and Hatcher, Jeffrey, and Luke Bailey, Rights and Resource Initiative, (2009), *Tropical Forest Tenure Assessment: Trends, Challenges and Opportunities.*

[76] Sunderlin above n 44 and Hatcher and Bailey, above n 77.

[77] Hatcher and Bailey, above n 77.

[78] White and Martin, above n 69, 7–8.

rights affects a significant proportion of the global forest estate. This insecurity acts to undermine sustainable forest management imperatives since, without secure rights, forest holders have few incentives and often lack the legal status to invest in managing and protecting their forest resources.[79] While secure property rights alone cannot ensure sustainable protection and investment, they are a necessary component of sustainable management frameworks.[80]

All forest tenure reform processes pass through three stages: the granting of statutory rights; the implementation of rights in practice; and the development of specific measures, which accompany reforms and enable communities to derive benefits from the new rights.[81] The FAO has considerable experience in implementing policies and measures to improve forest tenure and has identified four main types of forest tenure reform: strengthening customary or traditional tenure rights; privatization initiatives; granting concessions and contracts for productive forest purposes; and community forestry ventures.[82]

Strengthening customary or traditional tenure requires recognition of the rights as the first step. It is common to find overlapping rights arising at the intersection between formal and informal tenure systems. The FAO notes that 'Not only can there be different and sometimes conflicting views of ownership and tenure between local people and government agencies, but traditional systems themselves also frequently accommodate overlapping use rights'.[83] Customary rights often involve a combination of primary rights (for example harvesting) and secondary rights (for example grazing rights), which may also be seasonal in nature.

The disconnection between statutory and customary tenure systems is an issue that is receiving increased attention as a result of REDD+ investment. Past research and experience demonstrates, however, that harmonizing the two systems is not easy and involves more then merely granting customary holders legal title within statutory schemes.[84] Doherty and

---

79 White and Martin, above n 69.

80 Sayer, above n 64.

81 Larson, Anne, (2011), 'Forest Tenure Reform in the Age of Climate Change: Lessons for REDD+', *Global Environmental Change*, **21** (2), 540–549, 544.

82 Food and Agriculture Organization (2011), 'Reforming Forest Tenure: Issues, Principles and Process', available at http://www.fao.org/docrep/014/i2185e/i2185e00.pdf (accessed 27 January 2012) 23.

83 Food and Agriculture Organization (2011), 'Reforming Forest Tenure: Issues, Principles and Process', available at http://www.fao.org/docrep/014/i2185e/i2185e00.pdf (accessed 27 January 2012) 24.

84 Larson, Anne, (2011), 'Forest Tenure Reform in the Age of Climate Change: Lessons for REDD+', *Global Environmental Change*, **21** (2), 540–549.

Schroeder state: 'In most cases, customary systems will predate new laws emanating from the state and will be viewed as more legitimate.'[85] As such, local communities may not value the statutory recognition and the local community may ignore new laws applied by the statutory regime.

Indigenous communities are identified as a group particularly vulnerable to insecure or informal lands rights. A study conducted in five African countries found that Indigenous people have lost forest resources and land to colonists, commercial timber enterprises, and conservation initiatives.[86] The effects of colonization, in particular, have had serious ramifications for indigenous groups. Colonization led to indigenous and customary rights being overlooked, the state proclaiming all rights to land, waters and territories and, hence, disempowering indigenous groups and fundamentally changing the relationship between the indigenous people and their ancestral lands.

The second type of tenure reform involves privatization initiatives. Privatization is a common type of forest reform and refers to processes that create new property rights on forestland, effectively transferring rights from the state to individuals.[87] Such measures are based on the 'tragedy of the commons' theory discussed above and have been applied in countries reforming their land systems from systems based on communal rights to systems based on individual rights (for example China and Central and Eastern Europe). Privatization initiatives are complex and potentially contentious processes due to the large number of potential rights holders. Such initiatives can involve selling, allocating or providing licences for forestland or individual forest rights (carbon, hunting, harvesting), which may or may not provide for exclusive use.

The third form of tenure reform involves initiatives that examine forest concession arrangements. Concessions are a contract between a forest owner and another party for the harvesting or management of specified resources from a specified forest estate. The FAO suggests that concessions can help to achieve sustainable forest management if steps are taken

---

[85] Doherty, Emma and Heike Schroeder (2011), 'Forest Tenure and Multilateral Governance in Avoiding Deforestation under REDD+', *Global Environmental Politics*, **11** (4), 66–88, 78.

[86] Couillard, V., J. Gilbert, J. Kenrick and C. Kidd, , Forest Peoples Program (2009), 'Land Rights and the Forest Peoples of Africa, Historical, Legal and Anthropological Perspectives – Overview: Analysis and Context', available at http://www.forestpeoples.org/sites/fpp/files/publication/2010/05/overviewlandrightsstudy09eng.pdf (accessed 27 January 2012).

[87] Food and Agriculture Organisation (2011), 'Reforming Forest Tenure: Issues, Principles and Process', available at http://www.fao.org/docrep/014/i2185e/i2185e00.pdf (accessed 27 January 2012) 27.

to strengthen compliance with contract terms, supervision and monitoring. Measures, which the FAO recommends, include

- Granting concessions on a competitive basis, which encourage forest enterprises to improve their efficiency and sustainability
- Establishing an independent observer of both the concession allocation process and field operations to improve accountability and compliance with the rule of law
- Strengthening legal frameworks and governance arrangements at the local level.[88]

The fourth type of forest reform initiative involves measures aimed at devolution of responsibility for forest governance to local communities. Such initiatives have been widely applied in Nepal and India. 'Community forestry' is a generic term used to describe a range of participatory approaches including: participatory forest management, community-based forest management and joint forest management. The degree to which access, usufruct and management rights are devolved varies from program to program.[89] The success or failure of community forestry initiatives is dependent upon a number of factors aside from the provision of tenure rights. Communities often receive ownership or management rights for low-quality or degraded forests, which require large investments of resources and time to make them profitable. So, additional capacity building and technology and financial assistance are also necessary for the implementation of such measures.

Investments made under REDD+, or payment for ecosystem services, will require further forest tenure reforms in order to recognize carbon or biodiversity interests in forest estates. Market-based regulation requires the development of secure and clearly defined property rights.[90] Due diligence in forest carbon projects involves consideration of the following

- Clarification of the property being transferred: sequestered carbon; carbon sinks (rights to land above ground, below ground and trees); carbon sequestration potential (right to manage sink to maximize carbon storage potential); and carbon credits
- Clarification of other usufruct property rights on the forestland such

---

[88] Ibid. 31.

[89] Ibid. 33.

[90] See Eckersley, Robyn (1995), 'Markets, the State and the Environment: An Overview' in Robyn Eckersley (ed.) *Markets, the State and the Environment: Towards Integration*, South Melbourne: Macmillan Education Australia, pp. 8–9 and Dovers, Stephen, and Su Wild River (eds) (2003), *Managing Australia's Environment*, Annandale, NSW: Federation Press, 504–6.

as easements, leases, concessions, *profit à prendre* that may undermine carbon sequestration activities

- Clarification of ownership of property: laws may exist preventing foreign ownership of sovereign resources; such rights would need to be protected in other ways (for example contract)
- Clarification of the beneficiaries of the forest carbon project and specification of the conditions to be met and how distribution of profits is to occur
- Clarification of who bears liability should the contracted forest carbon not materialize and a specification of how disputes will be adjudicated.[91]

B. Land Rights Associated with Forest Holdings

The law has developed to recognize a number of rights over land. Some of these rights are compatible with other rights, while some rights require exclusivity in order for the right to be meaningfully enjoyed. Interests which the law may recognize include

- *Freehold interest in land* – a right which accords most easily with the notion of ownership. Freehold title generally provides holders with the greatest amount of freedom in deciding how to use and manage their land. This is limited by government intervention, which may prescribe legislative provisions, such as codes of productive forest conduct or protective forestry standards
- *Customary/native title interests in land* – rights generally held as group or as communal rights. Each legal jurisdiction has its own process for recognizing customary interests in land. In Australia, native title rights are seen a bundle of rights and each individual right must be established before a court will recognize their existence. Rights that would be relevant in the forest context might include a right to carry out indigenous forest farm silvicultural practices, a right to harvest and use forest products, and a right to exclude others from certain forest areas
- *Leasehold interests in land* – these confer a right of general occupation of the land for a specified period of time. Land may be leased

[91] Takacs, David (2009), 'Forest Carbon and Property Rights', Conservation International, available at http://www.conservation.org/Documents/CI_Climate_Forest-Carbon_Law-Property-Rights_Takacs_Nov09.pdf (accessed 27 January 2012) 2.

from the state or from a private owner of the freehold estate. Leases generally include a number of covenants concerning the use and management of land. These covenants, in addition to the defined duration of the agreement, and government environmental legislation, limit the leaseholder's use of the land

- *Easement interests in land* – these interests in land create rights and responsibilities for landholders. Holders may obtain an easement right to use another's land for a specific purpose, which another landholder must then recognize. This type of interest is often given to provide a right of way across another's land, a right to enjoy an area of another's land as a garden, or a right to park a vehicle on another's land[92]
- *Covenant* – a restrictive covenant is a promise made between parties where one party agrees to restrict the use of the land.[93] Covenants have been used to bind the current holder of the land and all future holders of it to carry out particular environmental management activities[94]
- *Profit à prendre* – a right to enter upon the land of another and take away something that forms part of the land or its natural produce.[95] These interests in land have been granted to allow holders to remove timber and other natural resources[96]
- *Forest property rights* – there are conceptual difficulties associated with using existing interests in land as a basis for recognizing rights to natural resources or their processes. Some states in Australia have dealt with the issue by creating instead a 'new interest' in land that can be included on the register.[97] Australia and New Zealand are leading the way in developing legal interests in rights arising from carbon storage in forests.[98]

---

[92] Law Book Company, *Laws of Australia*, vol 28 (at 24 July 2009) 28 Real Property, '28.13.1 Easements'.

[93] Law Book Company, *Laws of Australia*, vol 28 (at 24 July 2009) 28 Real Property, '28.14.1 Restrictive Covenants'.

[94] Queensland Department of Natural Resources and Water (2007), 'Statutory Covenants' available at http://www.nrw.qld.gov.au/factsheets/pdf/land/l98.pdf (accessed 8 March 2012).

[95] Law Book Company, *Laws of Australia*, vol 28 (at 24 July 2009) 28 Real Property, '28.10.550 Profit à prendre'.

[96] *Reid v Moreland Timber Co Pty Ltd* (1946) 73 CLR 1, 16.

[97] Eckert, Sandra and Richard McKellar (2008), 'Securing Rights to Carbon Sequestration: The Western Australian Experience', *Sustainable Development Law and Policy*, **8** (2), 30–33.

[98] Streck, above n 56, 246 and Passero, above n 56, 252.

C. Limitations on Land Tenure

Property rights can be limited in a number of ways. Rights over natural resources may be limited by legislative intervention prescribing how the environmental resource is to be used and managed in accordance with the concept of sustainable development. Similarly, the legislature may make statutory reservations over certain environmental values on private land. Property rights can also be restricted by the operation of environmental markets. As with legislative intervention, markets require that certain criteria and practices are complied with. Both statutory and market prescriptions operate to restrict the way in which property may be used. For this reason, some commentators have suggested that the operation of environmental regulation undermines private property interests.[99] This argument is advanced on the basis that environmental regulation removes a number of rights attaching to land that might otherwise have remained in existence. However, the doctrine of internal state sovereignty and the legal definition of property mean that land ownership had never been absolute, and has always been subject to some form of limitation.

The concept of sustainable forest management is yet to obtain legally binding status and so the broader concept of sustainable development is the concept that limits environmental property holders' rights. Sustainability concepts introduce a number of issues that states must consider in the design of their regulatory schemes:

- the needs of present and future generations must be taken into account (intergenerational equity);
- the needs of the world's poor must receive priority and abject poverty must be eliminated (intra-generational equity);
- the environment needs to be preserved, at least to a significant degree;
- economic, social and environmental policies must be integrated.[100]

The concept of sustainable development introduces a number of responsibilities for governments. It requires them to examine environmental, economic, and social considerations before allowing action to take place.

---

[99] Gardner, Royal (1997), 'Taking the Principle of Just Compensation Abroad: Private Property Rights, National Sovereignty, and the Cost of Environmental Protection', *University of Cincinnati Law Review*, **65** (2), 539–593.

[100] Magraw, D. and D. Hawke (2007), 'Sustainable Development' in Bodansky, Daniel, Ellen Brunée, Ellen Hay (eds), *Oxford Handbook of International Environmental Law*, Oxford: Oxford University Press, pp. 613–638, 619.

This requires the government to gather data on all values associated with the environment, and to consult relevant stakeholder groups prior to approving plans for the relevant area of land. The concept of sustainable development therefore imposes new responsibilities on governments and restrictions on property holders' rights.

6. CONCLUSION

The legal concepts of sovereignty and property both heavily influence the regulation of forests at the international and national level. These are difficult concepts, appearing to be contradictory in terms of the rights associated with them. On one hand, they provide either the state or an individual with certain rights, inherent in which are restrictions on how these rights may be exercised. Restrictions on sovereign and property rights arise as a result of public values associated with the environment. This broader public interest in the environment takes precedence over the individual interests in the environment. The legal system provides recognition both for broader public interests and individual interests by granting rights encumbered with restrictions. Those in charge of regulating forest areas therefore have to grapple with granting certain rights to individuals while, at the same time, ensuring that broader public interests in forests are also maintained.

4. The sustainable governance of forest resources

1. INTRODUCTION

There are no legally binding obligations to ensure that forests are managed sustainably. Negotiations at the international level concerning forest regulation have proven to be politically charged. The competing values associated with forest areas – ecological, economic, and social – have prevented parties from reaching agreement on international standards concerning forest use and management. The United Nations Forum on Forestry (the key public international body charged with regulating forest use and management) has been unable to deal effectively with the political issues surrounding forest regulation. Lack of progress by the United Nations Forum on Forestry has led to the development of a number of alternative international regulatory approaches to forests.

Currently, global forest governance is patched together, with different international bodies regulating individual forest values. By way of example, the international climate change regime regulates forest carbon, the Secretariat of the Convention on Biological Diversity regulates ecological forest values, and the international tropical timber institution regulates the productive or economic value of forests. This means that current global forest governance arrangements result in duplication, overlap, and confusion as to the applicable international standards, rules, or objectives of international forest regulation.

Despite these fragmented governance arrangements, it is possible to identify a common concept permeating all international forest policy. This is the concept of sustainable forest management. There appears to be global consensus that the concept of sustainable forest management is a desirable outcome. The means of achieving and implementing this concept remain open ended. Sustainable forest management, in essence, is the management of forest areas taking into consideration the competing anthropocentric economic, ecological and social values assigned to forest areas. It is a concept that gives recognition to all forest values. This recognition of various forest services, therefore, requires the management of forest estates to ensure that all forest values and services continue to flourish.

Presently, there is no one authoritative definition of sustainable forest management, no one list prescribing the criteria for sustainable forest management, nor one globally agreed form of implementing sustainable forest management. This concept of sustainable forest management is therefore somewhat abstract due to the number of competing definitions, obligations and requirements associated with it. Despite this obvious disadvantage, it is, nevertheless, useful to have at least an agreement in principle that the overall objective for global forestry regulation should be the achievement of sustainable forest management practices. This chapter explores the legal meaning and requirements of sustainable forest management.

2. THE FUNCTION OF GOVERNANCE

The concept of governance has made its way into the writings of most academic disciplines. In some ways, the concept is similar to the concept of sustainable development, in that it is widely used without having a standard meaning, or there being a common understanding of it. There is no one definition of governance. In part, this may be explained by the use of the concept across a number of disciplines. For the concept of governance within forest regulation, it is suggested that environmental governance includes the various institutions and structures of the authority engaged in the protection of the natural environment.[1] Underpinning environmental governance is the concept of sustainability, which acknowledges the fundamental importance of the preservation of earth's ecological integrity.

Fisher explores the meaning and evolution of the concept of governance.[2] He suggests that, traditionally, the term was used to refer to the act or process of governing by regulation or sanction by the government. However, the term has now evolved to include other means of regulating – such as market, self- and voluntary regulation – and to include other bodies performing regulatory duties (for example, industry, the community, and the private sector).[3] The evolution of new kinds of mechanisms and instruments to govern sustainable forest management is evident within the international forestry regime. Similarly, a number of non-state

---

1 Bosselmann, Klaus (ed.) (2008), *The Principle of Sustainability: Transforming Law and Governance*, Aldershot, England; Burlington, VT: Ashgate, 175.

2 Fisher, Douglas (ed.) (2009), *The Law and Governance of Water Resources: The Challenge of Sustainability*, Cheltenham, UK and Northampton, MA, USA: Edward Elgar Publishing, 44.

3 Ibid.

bodies have engaged in the international regulation of forestry. Such developments are discussed in further detail in Part Four of this book, which deals with non-state forest institutions.

Writings on the concept of governance tend to include discussions of how to improve it. Generally, achieving good governance, or creating a system which meets the requirements of good governance, is encouraged and supported across the literature. The requirements of good governance, however, remain elusive.[4] Common themes within the literature discussing good governance are: integration; stakeholder participation; accountability; enforceability; and transparency. Bosselmann found that the term is used to describe democracy, transparency, effectiveness, procedural rights and the rule of law.[5] Of particular relevance to international forest regulation is the concept of integration. This book suggests that a major impediment to the effectiveness of current international forest regulation is the lack of integration within the bodies creating forest standards, rules and processes.

3. THE ORIGIN AND MEANING OF SUSTAINABLE FOREST MANAGEMENT

Sustainable forest management has been adopted as the overall goal for global forest management.[6] The concept is an ever-evolving concept that attempts to incorporate and recognize all values associated with forests and, further, attempts to give equal weighting to all of these varying and potentially conflicting forest values. Common values identified in forest areas include: ecological and environmental values; social and cultural values; and trade and development values. Providing recognition for all of these within a legal framework has proved difficult for policy makers at the global level and, similarly, at the domestic level. The numerous values associated with forests present a unique legal challenge.

The management of forest areas (like other natural resources) is inherently political. However, forests, unlike other elements of the natural environment (for example, water, air and fauna), have for some time allowed for ownership rights. Existing individual rights in environmental resources

---

4 Bosselmann, above n 1.

5 Bosselmann, above n 1, 61.

6 Mendoza, Guillermo and Ravi Prabhu (2000), 'Development of a Methodology for Criteria and Indicators of Sustainable Forest Management: A Case Study on Participatory Assessment', *Environmental Management*, **26** (6), 659–673.

presents a challenge for regulators when introducing policy that restricts such rights. The ability of public and private bodies to own forest areas provides these owners with certain property rights and responsibilities. These rights and responsibilities in relation to forest areas contribute to the complexity of creating a mutually acceptable regime for managing the world's forests. The concept of sustainable forest management is internationally accepted as recognizing and promoting the equal consideration of all rights and interests in forest areas.

The concept of sustainable forest management emerged from private transnational regulation. In Strasbourg in 1990, ministers from European countries came together to discuss the need for greater protection and conservation of forest areas. This meeting was known as the Ministerial Conference on the Protection of Forests in Europe and it represented around 40 European countries. In 1993, a second meeting took place, known as the Helsinki Process.[7] This involved European countries meeting to develop and define sustainable forest management. Importantly, at the Helsinki Process, a workable definition of sustainable forest management was created. This is to be found in Resolution H1: General Guidelines for the Sustainable Management of Forests in Europe, paragraph D.[8] The parties agreed that 'sustainable management' of forests means

> The stewardship and use of forests and forest lands in a way, and at a rate, that maintains their biodiversity, productivity, regeneration capacity, vitality and their potential to fulfil, now and in the future, relevant ecological, economic and social functions, at local, national, and global levels, and that does not cause damage to other ecosystems.

At the follow-up meeting held in 1994, criteria and indicators were created to further define the elements of sustainable forest management.[9] These

---

[7] Officially entitled The European Process on Criteria and Indicators for Sustainable Forest Management.

[8] 'General Guidelines for the Sustainable Management of Forests in Europe', available at http://www.mcpfe.org/resolutions/helsinki/resolution_h1.pdf (accessed 30 November 2006).

[9] The criteria and indicator approach was originally pioneered by the International Tropical Timber Organization in 1992. This report was called 'Criteria for the Measurement of Sustainable Tropical Forest Management' and was published in March 1992. The International Tropical Timber Organization updated its criteria and indicators in 1998 and 2005. An electronic copy of the updated criteria and indicators is in International Tropical Timber Organization (1992), 'Criteria for the Measurement of Sustainable Tropical Forest Management', available at http://www.itto.or.jp/live/PageDisplayHandler?pageId=201 (accessed 30 November 2006).

criteria and indicators were based upon Resolution H1 and H2 from the Helsinki Process.[10] A criterion in this context is a category of conditions or processes by which sustainable forest management may be assessed. Similarly, an indicator is a quantitative or qualitative variable that can be measured or described. The criteria and indicators were designed to identify national trends and patterns in forest conditions and management at the individual European country levels.

The Helsinki Process created six criteria for sustainable forest management[11]

1. Maintenance and appropriate enhancement of forest resources and their contribution to global carbon cycles
2. Maintenance of forest ecosystems health and vitality
3. Maintenance and encouragement of productive functions of forests (wood and non-wood)
4. Maintenance, conservation and appropriate enhancement of biological diversity in forest ecosystems
5. Maintenance and appropriate enhancement of protective functions in forest management (notably soil and water)
6. Maintenance of other socio-economic functions and conditions.[12]

Criterion 6 deals with the legal issues associated with sustainable forest management. The indicators relevant for legal analysis include

> 6.28 – Existence of a legal/regulatory framework, and the extent to which it provides for programs and management guidelines, which recognise cultural heritage in relation to forestry.
> 6.29 – Existence and capacity of an institutional framework to: develop and maintain programs to conserve culturally valuable sites and landscapes.
> 6.30 – Existence of economic policy framework and financial instruments, and the extent to which it: provides for sufficient financial incentives for acknowledgement of cultural values in forest management planning.
> 6.31 – Existence of informational means to implement the policy framework,

[10] 'General Guidelines for the Sustainable Management of Forests in Europe', available at http://www.mcpfe.org/resolutions/helsinki/resolution_h1.pdf (accessed 30 November 2006).

[11] These are the original six criteria but they have undergone changes. An electronic copy is available at http://www.mcpfe.org/mcpfe/resolutions/lisbon/resolution_l2a2.pdf (accessed on 30 November 2006).

[12] Most criteria and indicator processes now have seven criteria and the original six criteria have been modified. See Chapter 3 for more detailed discussion of the current criteria and indicators.

and the capacity to: conduct studies on proportion of culturally valuable sites and sites with special visual value.

The criteria and indicator approach to sustainable forest management does not carry with it international normative power. This is because it was created by transnational negotiation and, furthermore, no commitments were pledged for its implementation. Rather, the criteria and indicator approach serves as a definitional tool by outlining the requirements for sustainable forest management. These criteria are designed for use by states at the individual state level, and each is responsible for ensuring that their forest regulation addresses the criteria outlined in the definition. The process provides states with a framework for assessing their progress towards sustainable forest management. The approach to sustainable forest management of defining criteria and indicators should be adopted as the universal measurement of sustainable forest management. The adoption of such an approach would go some way to resolving the confusion over global forest norms, standards and processes.[13]

4. INTERNATIONAL GOVERNANCE OF SUSTAINABLE FOREST MANAGEMENT

The concept of sustainable forest management permeates in one form or another all global instruments relating to forest use and management. Despite this, a clear definition of sustainable forest management has not emerged. The lack of clarity at the international level as to the concept's meaning and purpose has, thus, given states considerable flexibility in interpreting and implementing the concept. Recurring themes that surround the concept of sustainable forest management include: balancing economic and ecological interests associated with forest areas; representation and recognition of all forest interests and values; and the promotion and certification of sustainable timber harvesting practices.[14]

13 Following on from the original concept of using a criteria and indicators approach to monitor progress towards implementing sustainable forest management, a number of additional programs emerged also using the criteria and indicator approach to defining sustainable forest management. The table at the end of this chapter demonstrates that practically all countries have engaged in a criteria and indicator approach to defining sustainable forest management.

14 Gluck, P., R. Tarasofsky, N. Byron and I. Tikkanen, European Commission, European Forest Institute, The World Conservation Union and the Centre for International Forestry Research (1997), *Options for Strengthening the International Legal Regime for Forests*, 26.

It has been suggested that, in essence, the concept is generally understood to refer to the ways and processes of managing forest resources to meet society's varied needs, today and tomorrow, without compromising the ecological capacity and the renewal potential of the forest resource base.[15]

International forestry regulation is extremely fragmented in form.[16] In addition, sustainable forest management requirements remain largely voluntary in nature. Many developing countries require significant capacity-building assistance[17] and direct foreign investment to be able to implement sustainable forest management practices.[18] A number of competing public international regimes – such the United Nations Forum on Forestry, the United Nations Framework Convention on Climate Change, and the World Bank (among others) – all create individual sustainable forest management standards and requirements. In addition, private forestry international institutions – such as the criteria and indicator processes, forestry certification schemes and forestry markets – have emerged. They create new processes and mechanisms for achieving sustainable forest management.[19] It is widely recognized within all of these institutions that there is duplication, overlap, and need for a streamlining of sustainable forest management obligations.[20]

This fragmentation has resulted in a situation in which there is no one global body responsible for steering the creation of global obligations associated with sustainable forest management. Instead, a fragmented system exists in which a number of public and private forestry institutions

---

15 Wang, Sen (2004), 'One Hundred Faces of Sustainable Forest Management', *Forest Policy and Economics* **6** (3–4), 205–213, 206.

16 See the following report, which summarizes all the international bodies involved in international forest law: Feje Lesniewska (2005), 'Laws for forests: an Introductory Guide to International Forest and Forest Related Materials that Shape Forest Ethics and Practice', available at http://www.policy-powertools.org/Tools/Engaging/docs/law%20for%20forests2.pdf (accessed 1 February 2012).

17 Szaro, Robert, David Langor and Atse Yapi (2000), 'Sustainable Forest Management in the Developing World: Science Challenges and Contributions', *Landscape and Urban Planning*, **47** (3–4), 135–142.

18 Laaksonen-Craig, Susanna (2004), 'Foreign Direct Investment in the Forest Sector: Implications for Sustainable Forest Management in Developed and Developing Countries', *Forest Policy and Economics*, **6** (3–4), 359–370.

19 All of these processes are discussed in greater detail throughout this book.

20 The Collaborative Partnership on Forests has created a report and website dedicated to streamlining sustainable forest management requirements. For further information, see Collaborative Partnership on Forestry (1998), 'International Forestry Directory', available at http://www.fao.org/forestry/site/1662/en (accessed 16 June 2007).

operate.[21] States are, therefore, able to 'forum shop' among these competing international institutions. The participation of states with international forest regimes is dependent upon them seeing some form of benefit or incentive from engaging with that particular institution. A benefit or incentive may come in the form of political respect at the international or domestic level for taking action on an issue, or may come in the form of a financial incentive (for example, new business opportunities). The current international public institution that focuses solely on forest issues (the United Nations Framework on Forests) offers no significant incentives to governments and, furthermore, provides no determinative guidance on issues related to sustainable forest management. The role of the United Nations Forum on Forestry has been questioned, given its inadequacy in addressing global forest issues.[22]

Current operational deficiencies in the international regulation of forest use and management must be addressed. The current arrangement does not provide clear rules and measurable standards, appropriate financing and positive incentives, technology transfer, consensus on a bottom-up and holistic approach, coordination between institutions and policies at all levels, a permanent form for dialogue, nor mechanisms for equitable resolution of conflicts.[23] The live issues requiring attention at the international level include increasing the amount of protected forest estate in line with specified targets, managing productive forests in a sustainable manner and in building capacity within developing countries for forest conservation and management.

5. THE LEGAL REQUIREMENTS OF SUSTAINABLE FOREST MANAGEMENT

The definition of sustainable forest management involves a number of concepts that need to be further analysed to understand the role of law in implementing the sustainable forest management concept. The criteria

---

[21] For a discussion of public and private forest regulation at the state level see Cubbage, Frederick, and David Newman (2006), 'Forest Policy Reformed: A United States Perspective', *Forest Policy and Economics*, **9** (3), 261–270.

[22] Humphreys, David (ed.) (2006), *Logjam: Deforestation and the Crisis of Global Governance*, London: Earthscan Publications.

[23] Tarasofsky, Richard (1999), 'Assessing the International Forest Regime: Gaps, Overlaps, Uncertainties and Opportunities' in Tarasofsky, Richard (ed), *Assessing the International Forest Regime*, Gland, Switzerland: IUCN – The World Conservation Union, pp. 1–3.

and indicator definition of sustainable forest management will be examined to identify and analyse all the forest values, interests and services recognized by the concept

- *Biodiversity values*: this will require law that seeks to protect forest areas of high biological wealth
- *Productive values*: this will require law that defines certain forest areas as productive, and prescribes how forest operations are to be conducted within these defined areas
- *Regeneration capacity*: this will require the law to define regeneration and set out guidelines that require a certain standard
- *Vitality*: again, the law will have to define vitality and set out guidelines that require a certain standard
- *Ecological values*: the law must give recognition to all of the ecological services that forest areas provide. Once these services are recognized, the law must ensure that these services continue by creating regulation that prohibits interference with these services
- *Economic values*: the law must give recognition to all forest economic interests. This must include emerging economic forest values such as PES (Payment for Ecosystem Services) and traditional timber and non-timber economic values
- *Social values*: the law must identify all social values associated with forest areas. This will include stakeholder engagement with indigenous groups, community groups, landholders and other interested parties. Once social values are identified, the law must recognize and, where necessary, create protection for forest areas with strong social values
- *Local interests*: the law must identify local interests in forest areas. This will require active stakeholder engagement. Once local interests are identified, the law must give recognition and, where necessary, protection to local forest interests
- *National interests*: the law must identify national interests in forest areas; this will require research into all forest uses from a national level. Once a report is completed, the law may require amendment to recognize and protect all national forest interests
- *Global interests*: the law must identify global interests associated with forest areas. Such data can be retrieved from global forest reports, such as *The State of the World's Forests*. Relevant international environmental multilateral agreements will also indicate the international communities' interests in relation to forest areas.
- *Future interests*: the law must also recognize that future generations will have interests associated with forest areas. Forest regulation

must be created that ensures all current forest values are available for future generations.

There are a number of regulatory approaches to forest management. Previously, the predominant approach was to deem forest areas to be either a 'protective' forest estate or a 'productive' forest estate. Increasingly, there is a preference for managing forest areas to meet a number of ecological, economic and social criteria. An example of this is the development of forest certification schemes, primarily concerned with productive forest values, which now require that certain ecological criteria be maintained during timber growth and harvesting practices. Similarly, forest areas previously managed for protective values – such as world heritage areas, and national or conservation forest estates – have now begun to be managed to meet economic criteria with the advent of payments for ecosystem regulation.

A. Areas of Law Related to SFM

Implementing sustainable forest management involves many areas of the law. A single legal instrument providing for all aspects of sustainable forest management is unlikely. A combination of legal instruments – including legislation, regulation, policy, standards, and codes of conduct – will need to be created. Additionally, in common law systems, judicial decisions will contribute to the implementation of sustainable forest management. It is crucial that all forest legal instruments are consistent with one another and that duplication or lack of clarity is avoided. Governing departments must be clear on their roles in forest regulation; easy-to-follow flow charts explaining the roles of private and public institutions in forest regulation should be made available to the public. The following areas of law are related to the implementation of sustainable forest management.

Environmental law

Bates' work suggests that the main function of environmental law is to ensure the management of natural resources, protection of biodiversity, management of pollution, waste and contaminated sites, and the management of energy production and climate change.[24] Governments seek to ensure that natural resource use and management is carried out sustainably, howeverthe legal status of the concept of sustainable development remains unclear. Bosselmann explores the status of the principal

[24] Bates, Gerry (ed.) (2010), *Environmental Law in Australia* (7th edition), Chatswood, NSW: LexisNexis Butterworths.

citing authors such as Lowe, who describes it as a 'meta' principle, and Sands, who describes it as a principle of customary international law.[25] Ultimately, it is found that the principle of sustainability is best defined as the duty to protect and restore the integrity of the earth's ecological systems.[26] Fisher expands upon the three limbs of sustainable development and provides the following explanation of each limb:

- economic – the perspective of resource development;
- ecological – the perspective of environmental protection;
- social – the perspective of conservation for present and future human needs.[27]

Forest regulation at the state level can be categorized as being either a protective forest regime or a productive forest regime. Both protective and productive forest regimes seek to incorporate the three pillars or limbs of sustainable development, though the objective of the instrument may favour one of the pillars.

Planning and development law

The largest cause of deforestation worldwide is the clearing of forest areas for urban and agricultural growth.[28] The area of law which regulates urban and agricultural development is planning and development law. Such laws can prevent or minimize the impact of development on forest estates. Mechanisms used within this area of law include environmental impact assessment (EIA), strategic environmental assessment (SEA) and the zoning of areas for conservation purposes. The purpose of an environmental impact assessment is to inform decision-makers and, usually, the general public about the predicted outcomes of a proposal, and of the plans to manage the environmental effects of such development.[29] One

---

25 Bosselmann, above n 1, 50.

26 Ibid. 53.

27 Fisher, above n 2, 23.

28 Lambin, Eric F., B.L. Turner, Helmut J. Geist, Samuel B. Agbola, Arild Angelsen, John W. Bruce, Oliver T. Coomes, Rodolfo Dirzo, Gunther Fischer, Carl Folke, P.S. George, Katherine Homewood, Jacques Imbernon, Rik Leemans, Xiubin Li, Emilio F. Moran, Michael Mortimore, P.S. Ramakrishnan, John F. Richards, Helle Skanes, Will Steffen, Glenn D. Stone, Uno Svedin, Tom A. Veldkamp, Coleen Vogel and Jianchu Xu (2001), 'The Causes of Land-Use And Land-Cover Change: Moving Beyond the Myths', *Global Environmental Change*, **11** (4), 261–269.

29 Bates, Gerry (ed.) (2006), *Environmental Law in Australia*, (6th edition), Chatswood, NSW: LexisNexis Butterworths, 313.

way for planning and development law to embrace the concept of sustainable forest management is to require developers to replace all forest values lost during their development. This would normally require reforestation or afforestation of a nearby area.[30]

Property law

Property law underlies most forest regulation. Fisher suggests that rights of ownership and control are the fundamental values that underpin the legal arrangements by which natural resources are managed.[31] As discussed in Chapter 3, property law creates rights and responsibilities for landholders, which significantly impact upon their use and management of the land. In some jurisdictions, the existence of property rights in environmental resources may help give individuals a *locus standi* in legal proceedings.[32]

Constitutional law

> A constitution carries with it great weight and influence simply because of its status. In this sense it represents a set of principles and even ideologies, whether stated directly or implied, that are seen to sustain and permeate the rest of the legal system to the extent that they are relevant to the issue under consideration. It represents, in other words, a number of the fundamental values of a legal system.[33]

As the above quotation suggests, constitutions have great influence over every area of law. In federal countries, a constitution may broadly define the role of federal and state government in environmental management and protection.[34] Some constitutions contain direct statements concerning forest use and management, while others may impact indirectly upon forest regulation – for example, statements about land and property use. Constitutional norms that oblige governments to protect the environment, or give citizens enforceable rights to a clean and healthy environmental,

---

[30] See on general concept of environmental offsets and banking: Scanlon, Jane (2007), 'An Appraisal of the NSW Biobanking Scheme to Promote the Goal of Sustainable Development in NSW', *Macquarie Journal of International and Environment Law*, **4** (1), 71–133.

[31] Fisher, above n 2, 79.

[32] Bates, above n 29, 150.

[33] Fisher, Douglas (ed.) (2010), *Australian Environmental Law: Norms, Principles and Rules* (2nd edition), Rozelle, NSW: Thomson Reuters, 82.

[34] For a summary and analysis of the role of the Australian federal government's role in environmental regulation see ibid. 94 and Bates above n 24, 105.

may be more secure than statutory rights, which are more vulnerable to change or repeal by the government of the day.[35] However, the enforceability of constitutional provisions is dependent upon the willingness of the courts to interpret and elaborate on their application.

Indigenous law

Indigenous law is based upon the customary laws and traditions of first-nation people. A central feature of most indigenous cultures is a close connection with nature. As such, indigenous law can and has been incorporated into a number of environmental regulatory regimes.[36] Indigenous groups often have a strong connection with the natural environment (including forest areas). This area of law has the ability to change the manner in which the forest is managed through inclusion of indigenous practices, which may be adopted by the government and private landholders. In Australia, the forest fire-burning practices of the Aboriginal population have informed national park management practices. These practices have been adopted to improve biological diversity in areas that previously evolved in response to periodic burning of the *Spinifex* landscape.[37] The United Nations Declaration on the Rights of Indigenous Peoples creates a number of principles aimed at increasing the legal recognition and legal rights of indigenous people. These principles should influence the development of environmental and natural resource laws in countries where indigenous populations exist.

International law

> Environmental protection is an international or global issue due to the biophysical reality that supersedes the political division of the globe into sovereign states and areas beyond state jurisdiction. Ecosystems are interrelated in profound and complex ways. Due to these interrelationships, environmental

---

[35] Richardson, Benjamin, and Jona Razzaque (2006), 'Public Participation in Environmental Decision-Making' in Richardson, Benjamin J. and Stepan Wood (eds), *Environmental Law for Sustainability*, Oxford; Portland, Oregon: Hart Publishing, pp. 166–177.

[36] Richardson, Benjamin, and Donna Craig (2006), 'Indigenous Peoples, Law and the Environment' in Richardson, Benjamin J. and Stepan Wood (eds), *Environmental Law for Sustainability*, Oxford; Portland, Oregon: Hart Publishing, pp. 210–224. This work examines the incorporation of indigenous practices across many regions, such as Latin America, East and South Asia, Scandinavia, Canada, Australia, New Zealand, and the United States of America.

[37] Kimber, Richard (1983), 'Black Lightning Aborigines and Fire in Central Australia and the Western Desert', *Archaeology in Oceania*, **18** (1), 38–45.

> impacts can have widespread repercussions across vast distances and over long periods of time. As a result, no state acting along can hope either to protect the environment within its own territory or to account for the global impacts of the activities carried out on its territory.[38]

International forest law has the ability to directly, or at least indirectly, influence domestic forest law. International forest law needs to create global forest objectives that represent global interests in forest estates as compared to the current approach that preserves sovereignty and recognizes individual state interests. One international mechanism that has evolved to achieve wide-ranging implementation has been the National Forest Policies at state levels, which incorporate the criteria and indicator approach to measuring progress towards sustainable forest management.[39]

B. Legal Instruments as Tools for Sustainable Forest Management

Dovers and Connor's work identifies the main types of instruments used in environmental regulation as being: research and development; information and education; economic instruments; self-regulation and voluntarism; direct or 'command' regulation; legislation and quasi-legislation; institutional and organizational reform; and deliberate inaction.[40] Achieving sustainable forest management requires a combination of these instruments. Dovers and Connor suggest that while the choice of policy instrument is of crucial significance in ensuring that the objective of the regime is met, most regimes often base their decisions upon convenience, disciplinary bias or familiarity rather than informed judgement.[41] To make an informed judgement concerning instrument choice for sustainable forest management, it is necessary to identify all the values and interests covered by the concept. It is possible to create three main categories

- *Protective forest values*, which have generally been protected by restrictive legislative schemes (however, the emergence of forest markets is providing new avenues for financing forest conservation)

---

38 Ellis, Jay, and Stepan Wood (2006), 'International Environmental Law' in Richardson, Benjamin J. and Stepan Wood (eds), *Environmental Law for Sustainability*, Oxford; Portland, Oregon: Hart Publishing, pp. 343–344.

39 This is discussed in more detail in Chapter 5.

40 Dovers, Stephen and Robin Connor (2006), 'Institutional and Policy Change for Sustainability' in Richardson, Benjamin J. and Stepan Wood (ed.), *Environmental Law for Sustainability*, Oxford; Portland, Oregon: Hart Publishing, pp. 21–29.

41 Ibid. 28.

- *Productive forest values*, which are generally regulated by codes of conduct by both government and the forest industry
- *Social forest values*, which are probably best protected through legislative schemes.

Biodiversity forest values, ecological forest values and social forest values will generally be recognized through protective legislative instruments. These instruments may be generic or location-specific, depending upon the particular forest values being protected. New methods of legal regulation are being developed to protect biodiversity and ecological and social forest values. Payment for forest services means that traditional methods of protecting forest areas (for example, creating conservation parks) are gradually being expanded to include market-based approaches to achieve ecological outcomes. It is now recognized that national park status alone is not sufficient to ensure the protection of ecological forest values.[42] Forest reserves and parks will often require expenditure on maintenance and upkeep that the public sector is unwilling or unable to finance fully. Creating economic interests in forest areas may mean that the protection and associated maintenance of forest areas can be carried out at the expense of private investors whilst traditional legislative instruments would still be required to protect the area from future development.

Productive forest values, such as regeneration capacity and vitality have generally been regulated under legislative instruments and voluntary codes or standards. Forest regulation was initially introduced to ensure a continual supply of timber. This goal must now take into consideration sustainability and the adoption of silvicultural principles. Legislative instruments often define forest areas where timber processing is lawfully allowed to take place. Generic legislation generally exists for plantation areas, in addition to site-specific legislation dealing with areas where native forest harvesting occurs. Forest legislation often deals with issues such as definitions and application processes at the holistic or macro level dealing, allocating responsibility for implementing the instrument to a lower level, and prescribing certain requirements. Codes of conduct and industry standards are instruments that change frequently and are often not legally binding but which do deal with the micro details of how forest operations are to be conducted. The definition of regeneration capacity and vitality as elements of sustainable forest management should be incor-

---

[42] Hayes, Tanya, and Elinor Ostrom (2005), 'Conserving the World's Forests: Are Protected Areas the Only Way?', *Indiana Law Review*, **38** (3), 595–617.

porated into both broad, binding legislative instruments, and also into voluntary specific instruments.

The law may seek to protect local interests in forest areas by creating a site-specific piece of legislation, recognizing the local interest in the forest area. Local interests in forest areas are diverse, and range from rights to use the forest for daily livelihood requirements to rights to harvest timber and generate income from the forest area. Informal and formal agreements between governments and local communities may also be useful for recognizing local forest interests. National interests in forest areas will be identified from national surveys and research. These interests will usually be given legislative protection, and accompanying regulations may be created to detail the specifics of these interests. Governments may create policy documents that set goals for forest areas in the future, or which indicate a government's intention for future forest management. These policy documents will often be indicative of future forest legislative action. Formal and informal agreements may be reached between private forest landowners and governments, and agreements may also be made between individual states within a nation concerning forest use.

6. ASSESSING THE EFFECTIVENESS OF GLOBAL FOREST REGULATION

According to the *Macquarie Dictionary*, the ordinary meaning of the term 'effective' is to 'produce the intended or expected results'. To assess the effectiveness of a law, it is necessary to measure how successful the law has been in solving the problem that it was designed to address.[43] When assessing the effectiveness of global agreements, there should be a measure of tangible environmental improvement, and effectiveness should be judged on environmental results. McGrath suggests that reforms to improve the effectiveness of a law should focus on achieving real results by adopting a problem-solving approach that includes the formation of collaborative partnerships.[44]

To examine whether global forest governance is achieving its intended outcomes, it is necessary to clarify what the intended or expected global forest outcomes are. As discussed earlier in this chapter, the achievement

---

43 McGrath explores the concept of 'effectiveness in some detail in McGrath, Chris (ed.) (2010), *Does Environmental Law Work? How to Evaluate the Effectiveness of an Environmental Legal System*, Saarbrücken, Germany: Lambert Academic Publishing, Ch. 2.

44 Ibid.

of sustainable forest management is the intended outcome of all international forest policy. To measure progress towards achieving this target, a number of criteria will be used. The following criteria will be applied in the concluding chapter when assessing the effectiveness of global forest governance

- The existence of clear global forest objectives, with a preference for measured target goals.
- Evidence of good global forest governance arrangements (integrated, accountable, representative, equitable and adaptive).
- Evidence of implementation of the concept (definitional processes, reporting upon progress made and future progress needed).
- Improvement in forest conditions – a tangible improvement in the quality and quantity of forest ecosystem services.

In terms of examining the above criteria, it will be necessary to examine the theory of sustainable forest management, the law of sustainable forest management and, importantly, the practice of sustainable forest management activities.

7. CONCLUSION

This chapter has examined the concept of sustainable forest management. While some progress has been in creating tools for defining and measuring the concept, there is still a large gap when it comes to measuring the implementation of the concept. The international governance arrangements underpinning sustainable use and management of forest resources are exceptionally fragmented; they cause extreme inefficiency as a result of the duplication and overlap of international forest norms and standards. The arrangements concerning the global governance of forest resources are in dire need of a reform that is aimed at streamlining and strengthening obligations. At the national level, the governance arrangements necessary for implementing sustainable forest management are complex. This chapter explored a number of areas of law relevant to the concept. The range of economic, social and environmental values associated with forests means that a number of different areas of law are required to develop the concept. Similarly, a range of legal instruments should be used by those responsible for regulating forest areas. The governance of sustainable forest management is complex and an integrated approach to regulation is required.

Table 4.1 Criteria and indicator processes

Name of Process	Date Process Originated	States in Process
The Pan-European Forest Process on Criteria and Indicators for Sustainable Forest Management.	This process was initiated in Helsinki, Finland in June 1993 and finalized in Lisbon, Portugal in June 1998.	Albania, Austria, Belarus, Belgium, Bosnia-Herzegovina, Bulgaria, Croatia, Czech Republic, Denmark, Estonia, European Community, Finland, France, Georgia, Germany, Greece, Hungary, Iceland, Ireland, Italy, Latvia, Liechtenstein, Lithuania, Luxembourg, Monaco, Netherlands, Norway, Poland, Portugal, Romania, Russian Federation, Slovak Republic, San Marino, Slovenia, Spain, Sweden, Switzerland, Turkey, Ukraine, United Kingdom and Yugoslavia.
Non-European Process on Criteria and Indicators for Sustainable Forest Management (known as the Montreal Process).	The process was initiated in Geneva, Switzerland in June 1994 and endorsed in February 1995 by the Santiago Declaration.	Argentina, Australia, Canada, Chile, China, Japan, Republic of Korea, Mexico, New Zealand, Russian Federation1, United States of America and Uruguay.
Tarapoto Process on the Amazon Forest's Sustainability Criteria and Indicators.	This process took place in February 1995 at a workshop held in Tarapoto, Peru.	Bolivia, Brazil, Colombia, Ecuador, Guyana, Peru, Suriname and Venezuela.
The Dry Zone Africa Process on Criteria and Indicators for Sustainable Forest Management.	This took place in Nairobi, Kenya in November 1995.	Burkina Faso, Cape Verde, Chad, Gambia, Guinea Bissau, Mali, Mauritania, Niger, Senegal, Djibouti, Eritrea, Ethiopia, Kenya, Somalia, Sudan, Uganda, Angola, Botswana, Democratic Republic of Congo, Lesotho, Malawi, Mauritius, Mozambique, Namibian, Seychelles, South Africa, Swaziland, Tanzania, Zambia, Zimbabwe.

Table 4.1 (continued)

Name of Process	Date Process Originated	States in Process
The Near East Process on Criteria and Indicators for Sustainable Forest Management.	This took place in Cairo, Egypt in October 1996.	Afghanistan, Algeria, Azerbeijan, Bahrain, Cyprus, Djibouti, Egypt, Islamic Republic of Iran, Iraq, Jordan, Kuwait, Kyrgyz Republic, Lebanon, Libya, Malta, Mauritania, Morocco, Oman, Pakistan, Qatar, Kingdom of Saudi Arabia, Somalia, Sudan, Syria, Tadjikistan, Tunisia, Turkey, Turkmenistan, United Arab Emirates and Yemen.
Lepaterique Process of Central America on Criteria and Indicators for Sustainable Forest Management.	This took place in Tegucigalpa, Honduras in January 1997.	Belize, Costa Rica, El Salvador, Guatemala, Honduras, Nicaragua and Panama.
International Tropical Timber Process for Criteria and Indicators.	This took place in Yokohama, Japan in March 1992.	(55 countries: 29 producers, 25 consumers and the European Union): Australia, Austria, Belgium/Luxembourg, Bolivia, Brazil, Cambodia, Cameroon, Canada, Central African Republic, China, Colombia, Cote d'Ivoire, Democratic Republic of Congo, Denmark, Ecuador, Egypt, European Union, Fiji, Finland, France, Gabon, Germany, Ghana, Greece, Guyana, Honduras, India, Indonesia, Ireland, Italy, Japan, Liberia, Malaysia, Myanmar, Nepal, The Netherlands, New Zealand, Norway, Panama, Papua New Guinea, Peru, Philippines, Republic of Congo, Republic of Korea, Spain, Suriname, Sweden, Switzerland, Thailand, Togo, United

Name of Process	Date Process Originated	States in Process
		Kingdom, Unites States of America and Venezuela.
The African Timber Organization Process for Criteria and Indicators.	This took place in Libreville, Gabon in January 1993.	Angola, Cameroon, Central African Republic, Congo, Cote d'Ivoire, Democratic Republic of Congo, Equatorial Guinea, Gabon, Ghana, Liberia, Nigeria, Sao Tome et Principe and Tanzania.
Regional Initiative for the Development and Implementation of National Level Criteria and Indicators for the Sustainable Management of Dry Forests in Asia.	This took place in Bhopal, India December 1999.	Bangladesh, Bhutan, China, India, Mongolia, Myanmar, Nepal, Sri Lanka, and Thailand.

Note: The different processes have been created to reflect the different criteria and indicators required for different forest bio-regions. Classification by forest biome does increase the level of fragmentation from a governance perspective. In the event that a global body were created to steer international forest regulation, it should be encouraged, where possible, to streamline the different criteria and indicator processes.

Source: This table is based on information from a report of the Food and Agriculture Organization of the United Nations. An electronic copy of the report is available at http://www.fao.org/DOCREP/004/AC135E/AC135E00.HTM (accessed on 30 November 2006).

PART III

International regulation by public forest institutions

International forestry regulation involves a complicated matrix of public and non-state institutions. There is no one apparent body responsible for overseeing a global forest agenda. This results in international forest regulation being exceptionally fragmented and in a large number of duplicate and overlapping processes competing not only for regulatory legitimacy, but also for funding. Unlike other environment-specific international regimes, the international forest regime lacks a supreme institutional authority and accompanying legally binding convention. Such a situation has allowed for innovative regulatory approaches to emerge, most notably creating a space for non-state international forestry regulation.[1]

The number of public international, transnational, non-state, NGO, and international forest research and policy bodies is overwhelming. It would be an impossible task to assess all the bodies engaged in the creation of international sustainable forest management policy adequately in one book. This book examines five international forest governing regimes in detail. Three of them are institutions under public governance arrangements: the United Nations Forum on Forests; the regulation of forests under the International Climate Change regime; and the regulation of forest resources under World Bank policy. These three public bodies were selected for examination due their differing focuses and methods of regulation as well as their dominance in influencing international and national forest policy. The United Nations Forum on Forests focuses on all forest values, the International Climate Change regime focuses on the carbon

---

[1] See generally on the expansion of private international governance: Falkner, Robert (2003), 'Private Environmental Governance and International Relations: Exploring the Links', *Global Environmental Politics*, **3** (2), 72–87, 79.

values of forests, while the World Bank forestry policy explores the linkages between forests and poverty.

The table below sets out the major public international institutions and their instruments governing forest use and management.

P3.1 Regulation of forests by United Nations institutions

Institution	*Key Forest Instruments*
United Nations Forum on Forests	Non-Legally Binding Authoritative Statement of Principles for a Global Consensus on the Management, Conservation and Sustainable Development of all Types of Forests (1992) Intergovernmental Panel on Forests/ Intergovernmental Forum on Forests Proposals for Action (1995–2000). Non-Legally Binding Instrument on all Types of Forests (2007)
International Climate Change Regime	Land Use, Land-Use Change and Forestry Guidelines Decision 11/CP.7 2001 CDM Forest Credit Guidelines Decision 14/CP.10 2004 REDD policy (in development)
World Bank	The World Bank Forest Strategy 2002
Food and Agriculture Organization (Forestry)	FAO Ministerial Meeting on Forest Declaration: 1995, 1999, 2005 Voluntary Code on Fire Management 2007 Voluntary Code on Responsible Management of Planted Forests 2007 State of the World's Forest Annual Report
International Tropical Timber Organization	International Tropical Timber Agreement 1994 International Tropical Timber Agreement 2006
International Wetlands Regime	Convention on Wetlands of International Importance especially as Waterfowl Habitat 1971
International Biological Diversity Regime	Convention on Biological Diversity 1992 Forest Biological Diversity COP 6 Decision V1/22 2002
International Desertification Regime	United Nations Convention to Combat Desertification 1994

The two non-state regimes explored in this book are forestry markets schemes and forestry certification schemes. These regulatory approaches were selected as they are the two main forms of private forest regulation. For consistency, and for methodological purposes, the institutional structure – the principal instruments of the body and its progress in implementation – is examined for each of the five international forest regulatory approaches.

The distinction between public regulatory regimes and non-state regulatory regimes within this book has been drawn so as to demonstrate the different governing arrangements involved in these two forms of regulation. The term 'public regulatory regime' has been used to describe bodies within the United Nations system. The Charter of the United Nations created six principal organs: the General Assembly, the Security Council, the Economic and Social Council, the Trusteeship Council, the International Court of Justice and the Secretariat. Each of the three public bodies examined within this book are regulated under different organs of the United Nations.[2] The term 'non-state' has been used to describes regimes independent of the United Nations system.

The United Nations Charter defines the purpose and membership requirements of these processes. Article 1(3) provides that the purpose of the United Nations is to achieve international co-operation in solving international problems of an economic, social, cultural or humanitarian character. This is broadly enough stated to include environmental issues, though these were not separately listed when the Charter was drafted in 1945. In addition, Article 2(1) provides that the organization is based on the principle of the sovereign equality of all of its members. These two features of the United Nations system clearly influence the nature of obligations created by its bodies. Article 4(1) of the United Nations Charter provides that membership of the United Nations is open to all peace-loving states that accept the obligations contained in the present Charter and, in the judgement of the organization, are able and willing to carry out these obligations. These purpose and membership clauses mean that public international regulation is fundamentally very different in nature from domestic legal regulatory arrangements.

The United Nations Forum on Forests falls under the Economic and Social Council of the United Nations (ECOSOC). The international forest regime lacks a legally binding instrument and for this reason, no United Nations Secretariat has been created to implement the instrument. The United Nations Forum on Forestry therefore reports directly to the Economic and Social Council over implementation of its mandate. The creation of the Forum on Forests as a body under the Economic and Social Council was intended to enhance the international profile for for-

---

2 The United Nations website provides a useful visual diagram on the United Nations System: http://www.un.org/aboutun/chart_en.pdf.

ests.[3] However, as will be discussed further in Chapter 5, the creation of the body itself has not significantly altered the international governance of sustainable forest management.

The World Bank is a specialized agency also falling under the umbrella of the Economic and Social Council of the United Nations. Article 57 of the United Nations Charter allows for such agencies to be created by intergovernmental agreement. These agencies are then brought into a relationship with the United Nations system through Article 63 of the Charter. Under this article, the relationship between the United Nations and the specialized body is defined by an agreement between them.

The Secretariat to the United Nations Framework Convention on Climate Change sits within the UN Entities, Secretariat, division of the UN. The Secretariat of the Climate Change Convention provides organizational support and technical expertise to the regime.

---

[3] Humphreys, David (ed.) (2006), *Logjam: Deforestation and the Crisis of Global Governance*, London: Earthscan Publications, 91.

5. The role of the United Nations Forum on Forests

1. INTRODUCTION

The United Nations Forum on Forests is a unique member of the United Nations family. This is due not only to its status[1] within the United Nations framework, but also because of the political history leading to its current form and mandate. The body is probably better known for its shortcomings rather than its achievements. These shortcomings are worthy of examination, as the challenges causing deadlock within this setting reappear when other intergovernmental bodies attempt to create international forest policy.[2] The purpose of the forum is to provide an intergovernmental policy-negotiating platform that promotes the management, conservation and sustainable development of all types of forests, and which strengthens long-term political commitment to this end.[3]

The forum's sole focus on forest-related issues could explain the lack of progress made by this institution. Other intergovernmental institutions have appeared to be more successful in devising and implementing forest policy when they have dealt with forests in the context of another issue, for example, in relation to climate change or binding rules and methodologies concerning the use of forest sinks and sources.[4] The World Bank has tackled the forest policy challenge by linking forests and livelihood

---

[1] It is the only subsidiary body of the ECOSOC with universal membership and UNFF actions are reported to ECOSOC as opposed to an independent secretariat: Humphreys, David (ed.) (2006), *Logjam: Deforestation and the Crisis of Global Governance*, London: Earthscan Publications.

[2] Saunders, Jade, Hohannes Ebeling and Ruth Nussbaum, ProForest (2008), *Reduced Emissions from Deforestation and Degradation: Lessons from a Forest Governance Perspective*.

[3] General Assembly, 'Report of the United Nations Conference on Environment and Development', UN Doc A/CONF.151/26 (Vol. III) (14 August 1992) annex III, available at http://www.un.org/documents/ga/conf151/aconf15126-3annex3.htm (accessed 26 March 2012), principle 1.

[4] See Chapter 6.

concerns.[5] It appears that states are more willing to negotiate in the forest context when there will be dual benefits: a forest benefit and some other benefit, either financial or social. This type of dual focus seems to remove some of the political tensions that arise when forest issues are addressed in isolation. Traditionally, within forestry negotiations, there are stakeholders advocating for binding, target-driven regulatory approaches, who are counter-balanced by a strong opposition viewing the introduction of any such regulation as an infringement of state sovereignty.[6]

2. THE INSTITUTIONAL STRUCTURE OF THE UNITED NATIONS FORUM ON FORESTS

A. The Development of an International Forest Institution

The United Nations Forum on Forests (UNFF) is the only international institution that focuses solely on forest issues. As discussed above, this focus may explain the lack of progress made by the body. Following the Rio negotiations in 1992, an international forest body was created known as the Intergovernmental Panel on Forestry that operated from 1995 to 1997. After this, the body was renamed the International Forum on Forestry and operated from 1997 to 2000. Both of these international forest institutions carried out similar roles to the United Nations Forum on Forests, namely the provision of an international platform for forest negotiations.

In 2000, UNFF was created under ECOSOC Resolution/2000/35.[7] When the forum was established, it was envisaged that the body would continue in operation until 2005. In 2005, at the seventh session of UNFF, a new forest instrument was drafted. In addition to this, a multi-year program of work was drafted to implement this new forest instrument. This multi-year program of work has had the effect of informally continuing the forum until 2015.[8] No formal negotiations took place to extend

---

[5] See Chapter 7.

[6] Humphreys, above n 1, Ch. 5.

[7] An electronic copy is available at http://daccess-ods.un.org/TMP/9318769.html (accessed 22 May 2007).

[8] At the seventh session of the United Nations Forum on Forestry in 2007, a multi-year program was devised. This programme was adopted as the seventh session of the Forum, and has had the effect of continuing its existence until 2015. See United Nations Forum on Forestry (2007), 'Multi-year Programme of Work 2007–2015', available at http://www.un.org/esa/forests/multi-year-work.html

the duration of the forum but it seems to be accepted that the body will continue to operate for the foreseeable future, despite its lack of progress or clear leadership.

The forum is inclusive of all 'forest values' including productive and protective forest areas and all types of forest biome (such as tropical, boreal, native, and plantation forest types). The forum, therefore, focuses on the concept of a sustainable forest management that recognizes the ecological, social and economic values attributed to forest areas.[9] As discussed in Chapter 4, the concept of sustainable forest management is the mutually accepted goal of all international forest-related regimes. Because sustainable forest management involves the recognition of all forest values and services, the application of this concept often leads to conflict or confusion as the recognition of one forest value may be incompatible with the recognition of another. This difficulty has been identified and dealt with through the evolution of sustainable forest management methodologies and implementation tools.[10]

Examples of this are forest management practices and guidelines that have developed to incorporate the three thematic forest values into the management of forest areas. This means that productive forest areas are no longer only managed with economic forest values in mind. Instead, concerns about environmental sustainability – such as biodiversity, and soil and land health – are taken into consideration. In addition to this, any social values (such as the interests of special interest groups or a local community) are also included in the forest management plans. Taking an integrated management approach produces the dual benefits of increased site productivity, due to improved environmental health, and improving stakeholder and community relationships.

Another example of an integrated sustainable forest management practice appears in the management policy and plans of protected and conserved areas. Increasingly, it is becoming better understood that fencing off forest ecosystems will not improve their health.[11] Instead, active management of the conserved or protected area can lead to improved environmental outcomes. Payment schemes for ecosystem services and

---

(accessed 1 December 2009).

[9] Hickey, Gordon (2008), 'Evaluating Sustainable Forest Management', *Ecological Indicators*, **8** (2), 109–114, 111.

[10] Cubbage, Frederick, Patrice Harou and Erin Sills (2007), 'Policy Instruments to Enhance Multi-functional Forest Management', *Forest Policy and Economics*, **9** (7), 833–851.

[11] Hayes, Tanya, and Elinor Ostrom (2005), 'Conserving the World's Forests: Are Protected Areas the Only Way?', *Indiana Law Review*, **38** (3), 595–617.

environmental banking and trading schemes have developed to finance the management of these areas, which provide an economic incentive for managing the ecological values of the site.[12]

B. Role of the United Nations Forum on Forestry

The main objective of the Forum is to promote 'the management, conservation and sustainable development of all types of forests, and to strengthen long-term political commitment to this end'.[13] The functions of the Forum to implement this objective have been defined as[14]

- To facilitate the implementation of forest-related agreements and foster common understanding about sustainable forest management
- To provide continuing policy development and dialogue between governments to address forest issues and emerging areas of concern in a holistic, comprehensive, and integrated manner
- To enhance cooperation and coordination on forest-related issues
- To foster international cooperation
- To monitor, assess, and report on the progress of the above functions and objectives
- To strengthen political commitment to the management, conservation and sustainable development of all types of forests
- To enhance the contribution of forests to the achievement of internationally agreed development goals (e.g. the Millennium Development Goals)
- To encourage and assist countries to develop and implement forest conservation and rehabilitation strategies, increase areas of forest under sustainable forest management, and reduce forest degradation for the benefit of indigenous peoples and local communities whose livelihoods depends on forests

---

[12] Markets for environmental services are going through a growth period. Areas that were previously managed by being given conservation status are now being better managed under payment for ecosystem service arrangements. See Chapter 8 for further detail.

[13] Resolutions and Decisions of the Economic and Social Council (2000), '2000/35 Report on the Fourth Session of the Intergovernmental Forum on Forests', available at http://www.un.org/esa/forests/pdf/2000_35_E.pdf (accessed 26 March 2012).

[14] The list of functions is based on information obtained from the United Nations Forum on Forestry website. The functions have not been reproduced in full. Please refer to United Nations Forum on Forestry (2007), 'About UNFF', available at http://www.un.org/esa/forests/about.html (accessed 30 July 2012)

- To strengthen interaction between the United Nations Forum on Forestry and relevant regional and subregional forest-related mechanisms, institutions and instruments, organizations and processes.

The functions of the Forum are weakly worded. No accountable or goal-bound function is prescribed within the list of functions. For example, the function dealing with monitoring, assessment and reporting does not actually require or stipulate that reporting, monitoring or assessment take place – it merely suggests that, in the event of progress, reporting be carried out. It would, perhaps, be more realistic and useful for the Forum to reduce the number of functions to a more manageable workload. Many of the functions listed above discuss integration, interaction, and fostering cooperation among international forest-related processes.

Perhaps the key objective of the Forum should be the promotion of integrated action aimed at implementing sustainable forest management. This objective could then be made more accountable by introducing a number of specific functions that could be used to assess progress towards meeting this objective, for example: identify major international forest-related processes; identify inconsistencies or overlaps[15] among these major processes; direct stakeholders to the relevant processes; create regional outreach knowledge centres; and create an all-encompassing forest reporting framework that could be used to report to all major forest processes.[16] These types of functions are measurable and, therefore, allow for an increase in the accountability of the Forum.

C. Governing Bodies of the United Nations Forum on Forestry

The Forum operates through a bureau and secretariat. The bureau consists of one chairperson and four vice chairpersons in accordance with the principle of equitable geographical distribution. This principle aims to ensure that power is jointly held and exercised. Bureau members are

---

[15] It is recognized that the forest-related policy of a number of international institutions overlaps with that of other institutions. See Rosendale, Kristin (2001), 'Overlapping International Regimes: The Case of the Intergovernmental Forum on Forests (IFF) between Climate Change and Biodiversity', *International Environmental Agreements: Politics, Law and Economics*, **1** (4), 447–468.

[16] A number of these functions are currently performed by the Collaborative Partnership on Forestry – an international forestry research body. As a research body, the primary focus is the output of policy and information. The UNFF should act as an administrator of this research and try to ensure the implementation of the partnership's findings.

elected to office at the end of Forum sessions and are elected from the membership pool. The bureau has several responsibilities, which include: following up decisions made at Forum sessions; preparing for subsequent sessions; and managing and organizing during sessions. No further information is available which defines the parameters of what the 'follow-up' requirements are, and it seems that bureau members are not required to track implementation or progress towards implementation. The bureau members for the eighth session of the Forum are Australia, Cuba, Indonesia, Latvia and Morocco.[17]

The secretariat is compact in size, consisting of 10 full-time staff members. The same secretariat also serves the Collaborative Partnership on Forests (this body is discussed in greater detail below). The secretariat is responsible for organizing the logistical preparation for meetings, the timely preparation of documents and information, and servicing the meetings of the forum and bureau. The secretariat also supports and facilitates inter-session activities (such as meetings of experts or government-led initiatives).[18] The role of the secretariat is thus mainly administration and support.

D. Membership of the United Nations Forum on Forestry

Membership of the Forum is open to all states and all nations are encouraged to send representatives to Forum negotiations.[19] As with most international negotiating processes, a state representative is elected to attend the negotiations and represent their nation's global forest interests. The Forum is meant to provide a central place to discuss all issues associated with forest use and management. It has, however been suggested that the Forum has an inflexible policy agenda and that certain issues are sometimes left unaddressed.[20] Typically, contentious issues such as state sovereignty, finance, capacity-building, and technology transfer dominate forest negotiations.

Decisions at UNFF negotiations are reached by consensus. Individual states often contribute innovative and feasible policy options but consensus within the UNFF is difficult to achieve and progress remains slow. This type of decision-making process has led commentators to suggest

---

[17] United Nations Forum on Forestry (2007), 'Member States and Bureau', available at http://www.un.org/esa/forests/about.html (accessed 2 December 2008).

[18] United Nations Forum on Forestry (2007), 'UNFF Secretariat', available at http://www.un.org/esa/forests/secretariat.html (accessed 2 December 2008).

[19] Ibid.

[20] For example, the issue of genetically modified tree species was raised by NGOs at the 4th and 5th sessions, but this issue still remains unaddressed. See Humphreys, above n 3, 91.

that the UNFF instruments are based on the lowest common denominator and are, therefore, ineffective in changing the use and management of global forest resources.[21]

E. Multi-stakeholder Engagement

The Forum allows for a degree of multi-stakeholder engagement, although previous Forum sessions have had inadequate stakeholder engagement. Stakeholders have traditionally contributed to sessions by making opening statements. After this, there is a period for questions and interaction.[22] Any accredited delegate or observer may make a verbal intervention. It is worth noting that no resolution has ever been agreed as a result of multi-stakeholder dialogue.[23] The format of this engagement means that stakeholders are engaged through a process of statement submission and presentation, after which the interests of these groups is forgotten during the state versus state process of positional bargaining.

The combative nature of the various state forest interests results in little time being allocated for the inclusion of other non-state stakeholder interests. There have also been cases where stakeholders' interests have not been accurately portrayed to the plenary, leading to disillusionment among the parties seeking the right for their interests to be heard and recognized.[24] This had led to NGOs arguing that 'unless there are radical changes to ensure the effective consideration of the proposals and views of the major groups, the organization of these events should be discouraged'.[25] This is a direct call for more effective stakeholder engagement within the processes of the regime.

Nine major groups have participated in multi-stakeholder dialogues within the Forum:

- business and industry;
- children and youth;
- farmers and small landowners;
- indigenous peoples;
- non-government organizations;

[21] Humphreys above n 1, 103 and Humphreys, David (2003), 'The United Nations Forum on Forests: Anatomy of a Stalled International Process', *Global Environmental Change*, **13** (4) 319–323.

[22] Humphreys, above n 1, 96–7.

[23] Ibid.

[24] Ibid.

[25] Ibid. 97.

- scientific and technological communities;
- women;
- workers;
- trade unions.

At the sixth session of the Forum, in 2007, it was decided that stakeholder consultation would occur as side events to the main negotiations. Although this move effectively acknowledges that the current model of stakeholder engagement is flawed, it is not particularly helpful as it sends a message that stakeholder engagement is not important enough for inclusion in the main sessions. The inclusion of stakeholders must be for a clear purpose, linked back to achieving a specified outcome. Groups merely presenting individual interests that are in direct contradiction to another group's interests will only to add to the animosity that already exists between these groups. The Forum must decide for what purpose these stakeholders are engaged, and how their input will be incorporated.[26]

The inclusion of multi-stakeholders interests is important and must occur in ways that are meaningful – as opposed to implementing stakeholder engagement in order to 'tick the box'. Regular consultation with stakeholder groups should take place between official Forum negotiations, with positions summarized for inclusion in Forum negotiations. A process that responds to these summarized positions must then be created.[27] This could be yet another target-driven task assigned to the bureau or secretariat of the Forum. This would ensure that future policy outputs by the Forum incorporate and respond to feedback provided by forest stakeholders.

F. Collaboration with External International Forest Institutions

The Forum recognizes that forest-related issues are interconnected with many other environmental issues. This means that many intergovernmental institutions have overlapping mandates concerning international resource management. For this reason, the Forum is a member of an informal voluntary international forest research network. The Collaborative

---

[26] For further information on stakeholder engagement and dialogue, see Unerman, Jeffery (2007), 'Stakeholder Engagement and Dialogue', in Unerman, Jeffrey, Jan Bebbington and Brendan O'Dwyer (eds), *Sustainability Accounting and Accountability*, London; New York: Routledge, pp. 86–89.

[27] A multi-stakeholder process is described in Hemmati, Minu (ed.) (2002), *Multi-Stakeholder Processes for Governance and Sustainability: Beyond Deadlock and Conflict*, London; Sterling, VA: Earthscan Publications, 211. This process involves five phases: i) context; ii) framing; iii) inputs; iv) dialogue/meetings; and v) outputs.

Partnership on Forests (CPF) consists of 14 intergovernmental institutions that all work on forest-related issues. This network encourages dialogue and interchange about forest research findings and opens lines of communication between members in relation to policy developments and lessons. The CPF has an open membership, and the 14 major groups involved in the partnership are:

- Centre for International Forestry Research (CIFOR);
- Food and Agricultural Organization of the United Nations (FAO);
- International Tropical Timber Organization (ITTO);
- International Union of Forest Research Organizations (IUFRO);
- Secretariat of the Convention on Biological Diversity (CBD);
- Secretariat of the Global Environment Facility (GEF);
- Secretariat of the United Nations Convention to Combat Desertification (UNCCD);
- United Nations Forum on Forestry Secretariat (UNFF);
- Secretariat of the United Nations Framework Convention on Climate Change (UNFCC);
- United Nations Development Program (UNDP);
- United Nations Environment Program (UNEP);
- World Agroforestry Centre (ICRAF);
- World Bank;
- World Conservation Union (IUCN).

The CPF does not provide a forum for inter-governmental institutions to meet and discuss conflicting or duplicating policy. Instead, it examines all relevant policy and legal instruments concerning forest areas and attempts to find common approaches between relevant instruments. The CPF has produced three documents, which have received approval and praise from the Secretariat of UNFF:

- 'International Forestry Directory';[28]
- 'Sourcebook on Funding for Sustainable Forest Management';[29]
- 'Streamlining Reporting'.[30]

---

[28] Collaborative Partnership on Forestry (1998), 'International Forestry Directory', available at http://www.fao.org/forestry/site/1662/en (accessed 16 June 2007).

[29] Collaborative Partnership on Forestry (2007), 'Sourcebook on Funding for Sustainable Forest Management', available at www.fao.org/forestry/site/7148/en (accessed 6 June 2007).

[30] Collaborative Partnership on Forestry (2006), 'Streamlining Reporting', available at www.fao.org/forestry/site/6308/en (accessed 6 June 2007).

From a governance perspective, the report on *Streamlining Reporting* is exceptionally valuable. A major challenge for many states is meeting the ever-growing demand for the production of international environmental reports. This is particularly an issue in the forest context, due to fragmented governance arrangements. The streamlining project was commenced 'to propose ways to reduce the forest-related reporting burden, for example, through reducing and streamlining reporting requests, synchronizing reporting cycles, harmonizing data collection methods and increasing data comparability and compatibility, and facilitating the accessibility and flows of existing information'.[31]

The report on streamlining distinguishes between the reporting requirements of actions (implementation of policy), situations, and trends (reporting forest statistics).[32] It defines sustainable forest management as consisting of seven thematic elements (these are based on previous work carried out by criteria and indicator processes):

- extent of forest resources;
- biological diversity;
- forest health and vitality;
- productive functions of forest resources;
- protective function of forest resources;
- socio-economic functions;
- legal, policy and institutional frameworks.[33]

The CPF recommends that all international forest reporting guidelines be based on these seven thematic elements so that data prepared for one inter-governmental institution can be transferred for other reports. In the future, the CPF hopes to establish a 'Common Information Framework', which will be a database consisting of states' reports that are easily accessible and searchable (that is, updated regularly). Once this common information framework is established, and states provide the relevant information to it, reporting requirements will decrease for individual states. Information on this database can then be used to meet the reporting requirements of many different international forest processes. The creation of the CPF undoubtedly increases the interaction and communication between inter-governmental institutions. Future work carried out by the

---

31 Ibid.

32 Collaborative Partnership on Forestry (2003), *Towards a Common Information Framework for Forest-Related National Reporting to International Processes*

33 Ibid.

CPF could lead to greater integration of forest regulatory programs and reduce the overlap and inconsistency that currently exists.

3. INTERNATIONAL INSTRUMENTS CREATED BY THE UNITED NATIONS FORUM ON FORESTRY

A. The Forest Principles 1992

The UNFF and its predecessor institutions (IPF and IFF) have been unable to create any legally binding instruments concerning the use and management of the world's forests. At the UNCED negotiations held in Rio de Janeiro, The Forest Principles 1992 were developed. These principles are known as the *Non-Legally Binding Authoritative Statement of Principles for a Global Consensus on the Management, Conservation and Sustainable Development of all Types of Forests*.[34] The first three words of the title demonstrate the lack of mutuality between the states concerning the use and management of the world's forests.

The Forest Principles 1992 were drafted as goals or guidelines to guide states when creating domestic forest policy. They were never intended to be a 'command and control' style of instrument. In the preamble to the principles in paragraph (d) it is provided that: 'These principles reflect a first global consensus on forests. In committing themselves to the prompt implementation of these principles, countries also decide to keep them under assessment for their adequacy with regard to further international cooperation on forest issues'.[35] This statement explains the objective of the principles, which was to identify some common ground in relation to forest use and management and to agree to work together in the future on forest related issues at an international level. Seven general themes can be identified from Forest Principles 1992

1. The concept of sovereignty and the rights associated with sovereignty[36]
2. Consultation and involvement with forest stakeholders such

---

[34] United Nations Conference on Environment and Development, *Non-legally Binding Authoritative Statement of Principles for a Global Consensus on the Management, Conservation and Sustainable Development of all types of Forests*, UN Doc A/Conf 151/126 (14 August 1992).

[35] Ibid.

[36] Ibid. principles 1 (a), 2 (a) and 14.

as indigenous groups,[37] local community groups,[38] and women's groups[39]

3. The principle of integration and calls for it to be incorporated into ecological considerations,[40] institutional considerations,[41] and policy considerations.[42] The principles of sustainable development are also included[43]
4. The recognition and valuation of all forest values and services (such as ecological values,[44] conservation values,[45] social values,[46] indigenous values[47] and trade values)[48]
5. Exploring the relationship between native forest and plantation forest[49]
6. Increased development assistance [50] and funding for developing countries to assist in implementing forest projects[51]
7. The development of global[52] and national[53] regulatory frameworks to support and encourage the implementation of the Forest Principles.

The Forest Principles do not contain any enforcement mechanisms. Forest negotiations at Rio were tense; and any suggestion of including an enforcement mechanism would have stalled the negotiations and prevented any instrument from being created. Weiss suggests that 'nonbinding legal instruments set forth norms that states and other actors may observe even though they are not strictly required to do so. They create expectations that may shape behaviour and avoid disputes'.[54] These types of instrument

---

[37] Ibid. principles 5 (a) and 12 (d).
[38] Ibid. principles 2 (d) and 7 (a).
[39] Ibid. principle 5 (b).
[40] Ibid. principle 8 (e).
[41] Ibid. principles 13 (d) and (e).
[42] Ibid. principle 3 (b).
[43] Ibid. principle 3 (c).
[44] Ibid. principle 4.
[45] Ibid. principle 8 (f).
[46] Ibid. principle 9 (b).
[47] Ibid. principle 12 (d).
[48] Ibid. principles 13 (a) and 14.
[49] Ibid. principles 6 (b) and (d).
[50] Ibid. principle 11.
[51] Ibid. principles 7 (b), 9 (a) and 10.
[52] Ibid. principle 3 (b).
[53] Ibid. principles 3 (a), 5 (a), 6 (b), 8 (f) (h) and 9 (c).
[54] Weiss, Edith Brown (1999), 'Understanding Compliance with International Environmental Agreements: The Bakers Dozen Myth', *University of Richmond Law Review*, **32** (5), 1555–1589.

give states a high degree of flexibility and do not require strict compliance. The Forest Principles 1992 do not compel states to report on forest conditions, nor to achieve a specific target (for example reducing logging of native forest); they merely require states to bear the principles in mind when drafting domestic forest policy. Perhaps it was thought that setting minimal obligations on states was more realistic as there was at least some chance that states might comply if the obligations were not viewed as onerous (as compared with setting highly ambitious goals that could not be attained). The limited obligations are also attributable to a number of states' concerns about interference with state sovereignty rights.

B. International Panel on Forestry / International Forum on Forestry Proposals for Action

The United Nations Forum on Forestry was created to implement the IPF/IFF proposals for action. In total, there are 270 proposals for action designed to achieve sustainable forest management. Australia has prepared a report which summarizes the IPF/IFF proposals for action.[55] The mandate of the IPF was to pursue consensus and formulate options for further actions to combat deforestation and forest degradation, as well as to promote the management, conservation and sustainable development of all types of forests.[56] The IPF proposals for action were drafted by the IPF secretariat, ITFF[57] member organizations, NGO groups, and as a result of inter-session initiatives.[58]

The mandate of the IFF was to facilitate implementation of the IPF proposals for action and to further the policy dialogue on a number of issues that were left pending from the IPF process.[59] The IFF working

---

55 Australian Department of Agriculture, Fisheries and Forestry, *Summary of Proposals for Action: A Tool to Assist Countries to Measure Progress and Establish Priorities for Sustainable Forest Management* (2000).

56 Ibid. 8.

57 The Interagency Task Force on Forests (IFFF) comprised eight international organizations and was responsible for creating the IPF work agenda.

58 Humphreys, above n 1, 29.

59 Department of Agriculture, Fisheries and Forestry – Australia (AFFA); Program on Forests (PROFOR) at the World Bank (2003), 'Implementing the Proposals for Action of the Intergovernmental Panel on Forests and the Intergovernmental Forum on Forests: A Tool to Assist National-level Assessment of Progress and Priorities for Action Toward Sustainable Forest Management, Developed in Support of the United Nations Forum on Forests', available at http://www.profor.info/profor/Documents/pdf/IPF_IFFtext.pdf (accessed 26 March 2012).

agenda was divided into three categories: the first was concerned with the implementation and review of the IPF proposals for action; the second investigated matters left pending from the IPF (such as the relationship between trade and the environment); and the third addressed governance-related issues (such as international institutions and organizations).[60] At the end of the IPF/IFF lifespan, the IPF had created 150 proposals for action, and the IFF had created an additional 120.

The summary report prepared by Australia categorizes the proposals into four thematic areas.[61]

1. Implementation of international forest-related decisions within countries:
   - national forest programs;
   - implementation of IPF and IFF proposals for action;
   - forest information and awareness;
   - underlying causes of deforestation and forest degradation;
   - future supply and demand of forest products;
   - protected areas and forest conservation;
   - forests in environmentally critical areas;
   - impact of airborne pollution on forests;
   - forest research and development;
   - traditional forest-related knowledge;
   - requirements for countries with low forest cover.
2. International cooperation in financial assistances and technology transfer:
   - provision of financial resources to developing countries and countries with economies in transition;
   - technology transfer and capacity building.
3. Trade and environment in relation to forest products and services:
   - market access and transparency.
4. Forest-related work of international organizations and multilateral institutions and instruments:
   - involvement of countries in international programs; work of international organizations.

The political climate of forest negotiations during the creation of the IPF/IFF proposals for action meant that any new forest instrument created

---

60 Humphreys, above n 1, 66.

61 Department of Agriculture, Fisheries and Forestry – Australia (AFFA) Program on Forests (PROFOR) at the World Bank, above n 59.

would be non-binding in nature and a voluntary instrument. The IPF/IFF proposals for action were designed to provide policy guidance for the development of international forest instruments.[62] As the name of the IPF/IFF 'proposals for action' suggests, they are meant to stimulate action in the implementation of forest initiatives. Therefore, these proposals are not in themselves enforceable. Any reporting requirements suggested by the proposals were voluntary in nature.[63] Compliance with the IPF/IFF proposals has been, unsurprisingly, weak.

The most significant outcome from the proposals for action has been the concept of 'National Forest Programmes'. This concept has been widely embraced by the international forest community and the adoption and implementation of national forest policies at the domestic level has been significant:[64] 138 countries have national forest policies at the planning or implementation phase.[65] Overall, the IPF/IFF proposals for action are recognized as useful in that they contributed a comprehensive list of forest issues and suggestions for addressing these issues. The IPF/IFF proposals for action are, however, criticized for being insufficiently focused and for being repetitive. It has also been suggested that the proposals for implementation has been hampered by their convoluted and complex nature.[66]

C. Forest Principles 2007

The Non-Legally Binding Instrument on All Types of Forests is the official name of the Forest Principles 2007. Due to the ongoing political issues present in forest negotiations, they add little to existing international policy on global forest use and management. The first principle provides, 'the instrument is voluntary and non-legally binding'.[67] Statements such as this are normally followed by a qualifying statement, which encourages state implementation and political support for the principles. The lack of a qualifying statement could be interpreted as effectively lessening the authority of the Forest Principles 2007.

---

62 Humphreys, above n 1, 29.

63 Ibid. 69.

64 See section 4.4 which discusses the Australian National Forest Policy.

65 United Nations Forum on Forestry (2004), *Ad Hoc Expert Group on Consideration With a View to Recommending the Parameters of a Mandate for Developing a Legal Framework on All Types of Forests* 7.

66 Ibid. 6.

67 United Nations Forum on Forestry, *Non-legally Binding Instrument on All Types of Forests*, 2nd Comm, 62nd sess, Agenda Item 54, UN Doc A/C.2/62/L.5 (22 October 2007) annex, principle 2(a).

The purpose of the instrument is stated to be threefold[68]

- To strengthen political commitment and action at all levels to effectively implement sustainable management of all types of forests and to achieve the shared global objective on forests
- To enhance the contribution of forests in achieving internationally agreed development goals, including the Millennium Development Goals – in particular, with respect to poverty eradication and environmental sustainability
- To provide a framework for national action and international cooperation.

Three main themes can be identified in the Forest Principles 2007.[69] These are: the recognition of the multiple benefits that forests provide; the forest/poverty nexus; and the concept of sustainable forest management. The call for recognition of multiple forest benefits is present throughout the instrument.[70] This theme draws attention to the understanding that forest policy can no longer be viewed in isolation from broader development and economic policy. Forests are increasingly recognized as part of the solution to a number of problems: climate change mitigation policy; poverty reduction policy; and the health of other ecosystems (such as biodiversity, water, and soil health).

Principle 6(j) discusses the importance of recognizing all forest values and extending this recognition to the marketplace. Understanding forest values and assigning a market value to all forest services will continue to feature heavily in future forest negotiations at the international level, both public and private. The Forest Principles 2007 do not provide any guidance about how this valuation is to be done. Presumably, these methods of valuation will be similar to the credit schemes established under the climate change and biological diversity conventions. Forests are set to play an ever-increasing role in the offsetting of emissions. In the future, the international forest regime and the international climate change regime will need to align their priorities and agree upon a valuation method for forest areas.

---

68 Ibid. principles 1(a), (b) and (c).

69 There are other themes such as the promotion and use of science and technology, stakeholder engagement, engagement with public and private sector, but the three themes identified above are the three developments presently taking precedence in forest policy dialogues.

70 United Nations Forum on Forestry above n65, principles 1 (b), 4, 5 (2), 6 (d), 6 (j), 6 (t), 6 (w).

The second theme is the recognition that forestry and poverty issues are interconnected. This is aligned with current research and policy by other international bodies such as the World Bank, the Food and Agricultural Organization, and the United Nations Development Program.[71] Addressing forest issues in connection with livelihood issues provides a pragmatic pathway for change. Often, the factors leading to forest degradation and poverty will overlap and, to make significant headway on either issue, will need to be addressed collectively.

The Forest Principles 1992 allude to the concept of 'sustainable forest management', but do not directly identify the concept.[72] Following the creation of the Forest Principles 1992, a number of international forest processes, and academics,[73] embraced the concept of sustainable forest management as the overarching goal for forest use and management. The first statement of the Forest Principles provides '. . . that sustainable forest management contributes significantly to sustainable development and poverty reduction'. Earlier drafts of the Forest Principles 2007 contained a definition of sustainable forest management. These earlier drafted adopted the criteria and indicator definition of sustainable forest management. According to this definition, sustainable forest management means:

> The stewardship and use of forests and forest land in a way, and at a rate, that maintains their biodiversity, productivity, regeneration capacity, vitality and their potential to fulfil, now and in the future, relevant ecological, economic and social functions, at local, national, and global levels, and that does not cause damage to other ecosystems.[74]

As discussed in Part One, many different international institutions have now accepted this as being the definition of sustainable forest management. The Forest Principles 2007 failed to specify a meaning for 'sustainable forest management' instead providing that 'sustainable forest management is a dynamic and evolving concept, that aims to maintain and enhance the economic, social and environmental values of all types of forests, for the benefit of present and future generations'.[75]

---

[71] See Chapter 7 for a discussion on the World Bank Forest Policy, which has a strong focus on the poverty/forest link.

[72] See Part One for further analysis on this point.

[73] Kant, Shashi (2004), 'Economics of Sustainable Forest Management', *Forest Policy and Economics*, **6** (3–4), 197–203.

[74] Ad Hoc Group to Consider the Content of the Non-legally Binding Instrument (2006), *Composite Draft Text for a Non-legally Binding Instrument on All Types of Forests*.

[75] United Nations Forum on Forestry, above n 65, principle 4.

This flexible definition can be interpreted in two ways. First, that the Forest Principles statement is essentially the same as the criteria and indicator definition of sustainable forest management, but has been left purposely flexible so as to include relevant new developments. Or, alternatively, the lack of an agreed definition for sustainable forest management suggests that forest negotiations at the United Nations level produce limited positive outcomes. Failing to specify the meaning of 'sustainable forest management' could be interpreted as undermining the authority of the Forest Principles 2007.

Earlier drafts of the Forest Principles 2007 also contained the seven thematic elements of sustainable forest management adopted from the criteria and indicator processes. In the final instrument, these seven thematic elements have been incorporated as a footnote to principle 6(b). The failure to include widely accepted criteria and indicators that detail the requirements for sustainable forest management again demonstrates the political tensions in forest negotiations at the United Nations level.

Reporting requirements are discussed in principle 9. The wording of this principle is weak, and highlights the voluntary nature of the principles:

> Member States should submit, on a voluntary basis, taking into account availability of resources and the requirements and conditions for the preparation of reports for other bodies or instruments, national progress reports as part of their regular reporting to the United Nations Forum on Forestry.[76]

This principle gives other international reporting requirements precedence over its own. It does not elaborate on what information would be required if a voluntary report were made, nor suggest any recommended periods for making these voluntary reports. This lack of clarity could result in states submitting a report to the UNFF that has already been submitted to a different international institution whose priorities are unaligned with that of the Forum. The Forum's governing bodies are then placed in the position of trying to use varying information as reported by each country. In 2011 and 2015, the Forum's secretariat is tasked with collating the voluntary reports submitted by states for inclusion in the Forum's sessions.[77]

The seventh session report[78] made reference to Resolution 2006/49[79]

---

[76] Ibid. principle 9.

[77] Economic and Social Council of the United Nations, *United Nations Forum on Forestry: Report of the Seventh Session*, UN ESCOR, 7th sess, Supp No 22 (24 February 2006 and 16–27 April 2007) 16.

[78] Ibid.

[79] Ibid.

with respect to reporting requirements. Resolution 2006/49, at paragraph 19, provides that 'countries should, on a voluntary basis, submit national reports to the Forum, in accordance with a timetable established by the Forum'. The UNFF 9 negotiations provided clarity on the purpose and requirements of submissions made to UNFF and directed that the:

- The Secretariat and Collaborative Partnership on Forests develop a simple template for states to communicate their progress towards the introduction of policies and measures to implement the four global objectives of forests;[80] and
- Extended a formal invitation to the FAO to include in its State of the World Forests Reports data regarding the implementation of the Forest Principles 2007 and the four global objectives on forests in future UNFF reports.[81]

The most positive outcome from the Forest Principles 2007 was the creation of the four global objectives on forests contained in principle 5

1. Reverse the loss of forest cover worldwide through sustainable forest management, including protection, restoration, afforestation and reforestation, and increase efforts to prevent forest degradation.
2. Enhance forest-based economic, social and environmental benefits, including by improving the livelihoods of forest dependent people.
3. Increase significantly the area of protected forests worldwide and other areas of sustainably managed forests, as well as the proportion of forest products from sustainably managed forests.
4. Reverse the decline in official development assistance for sustainable forest management and mobilize significantly increased new and additional financial resources from all sources for the implementation of sustainable forest management.

If these objectives are considered to be common global aims, future negotiations at UNFF should explore their implementation. Implementation of the first principle could use mechanisms that already exist within the

---

[80] Economic and Social Council of the United Nations, *United Nations Forum on Forestry: Report of the ninth session (24 January – 4 February 2011)*, UN ESCOR, 9th sess, Supp No 22, UN Doc E/2011/42, e/cn.18/2011/20 (1 May 2009 and 24 January to 4 February 2011) at para 3 of 'The Assessment of Progress Made on the Implementation of the Non-legally Binding Instrument on All Types of Forests and Towards the Achievement of the Four Global Objectives on Forests.

[81] Ibid. para 6.

criteria and indicator process to achieve sustainable forest management. Further implementation of the first global objective could be assisted through the implementation of Kyoto Protocol afforestation and reforestation guidelines.

Implementation of the second and fourth objectives requires development assistance and technology transfers to help developing countries. This issue has proved controversial in all forest negotiations. The REDD+ mechanism may prove to have an integral part in ensuring the financing of policies and measures to achieve these four goals. The Collaborative Partnerships on Forests handbook on global funding for forest projects should be used in this process. The Environment Director of the World Bank commented at the Forum's seventh session that the four global objectives of the Forum were similar to the Bank's forests strategy. It was emphasized by the World Bank that traditional official development assistance (ODA) was not adequate to address the economic cost associated with the implementation of the four goals. Furthermore, alternative and new forms of financial support would need to be identified to compensate for the decrease in traditional ODA finance.[82] It should be noted that due to a significant contribution by Japan, the traditional forest ODA finance actually increased from 2000 to 2007.[83]

The Global Environment Facility (GEF) assists developing countries to protect the global environment. This facility is a provider of traditional ODA finance. The GEF does not have an individual forest program but funds forest projects through biodiversity, climate change and land degradation programs. Within the land degradation program, a project theme of sustainable land management has been created that is broad enough to encompass projects related to sustainable forest management.

To achieve the third global objective, that is to increase the area of protected forest worldwide, voluntary targets should be set prescribing individual targets for states to work towards. To ensure that more timber is sourced from sustainably managed areas, the Forum should publicly support and endorse genuine sustainable forest management certification programs, such as the Forest Stewardship Council certification process. Recognizing certification, a valuable tool at the United Nations level, has the potential to influence the intended target of certification programs: the consumer.

---

[82] Ad Hoc Group to Consider the Content of the Non-legally Binding Instrument, above n 44, 36.

[83] Smula, Markku, Advisory Group on Finance of the Collaborative Partnership on Forests, and Program on Forests of the World Bank (2008), *Financing Flows and Needs to Implement the Non-Legally Binding Instrument on All Types of Forests* 30.

At UNFF 7, a timetable of work was created entitled the Multi-Year Program of Work for 2007–2015.[84] The UNFF negotiations are to take place every second year with: UNFF 8 in 2009; UNFF 9 in 2011 (International Year of Forests); UNFF 10 in 2013; and UNFF 11 in 2015. The main task between 2009 and 2015 is to achieve the four global objectives on forests and implement the Forest Principles 2007. Each Forum session has been allocated a theme

- UNFF 8 (2009) – Forests in a changing environment (climate change and forest loss) and means for implementation for sustainable forest management
- UNFF 9 (2011) – Forests for people, livelihoods and poverty eradication
- UNFF 10 (2013) – Forests and economic development
- UNFF 11 (2015) – Forests: progress challenges and the way forward for the international arrangement on forests.

2011 was declared the International Year of the Forest in a bid to raise public awareness of the significant contributions forests make around the world and to highlight the many challenges facing the world's forests and forest-dependent people. Year of the Forest activities included: global launch and closing at UN Headquarters; creation of Year of the Forest logo and website; hosting a forest film festival; appointment of forest spokespersons; public service announcements on TV and radio; photo, film and art forest competitions; and the creation of a UN forest stamp collection. These activities were supplemented by government activities such as tree planting and celebrations for forest day, along with initiatives by international environmental institutions.[85] The precise impact of such events is difficult to measure but their educational value, and the awareness raising role of these events, is important. Raising the profile of global forest issues is a crucial first step in increasing public, government and international institutional support for progressing the policies and measures of the international forest regime.

---

[84] United Nations Forum on Forestry (2007), 'Multi-year Programme of Work 2007–2015', available at http://www.un.org/esa/forests/multi-year-work.html (accessed 1 December 2009) 18–19.

[85] Secretary General, Economic and Social Council, *Preparations for International Year of the Forest 2011*, 9th sess, Agenda Item 6, UN Doc E/CN.18/2011/7 (8 November 2010).

4. ARGUMENTS FOR AND AGAINST A FOREST CONVENTION

Since the Rio meeting in 1992, there has been an ongoing dialogue within international negotiations, and within academic literature, about the lack of binding instruments in the international forestry arena. Many critics of the current international arrangement on forests argue that a lack of a binding legal instrument is responsible for a lack of development and implementation of international forest standards and rules. Other proponents suggest that a binding legal instrument will not necessarily increase the enforceability or implementation of forest standards and norms. They use other binding international legal instruments, such as the Convention on Biological Diversity and the Framework Convention on Climate Change, to support this argument.

A. Support for a Legally Binding International Forest Instrument

The main arguments supporting the creation of legally binding international instruments concerning forests (a forest convention) include:

- Existing global governance has been largely ineffectual and extremely limited in scope[86]
- The current regulation provides an *ad hoc* and fragmented approach to regulation. A forest convention could address overlap and inconsistency if all under one approach[87]
- A forest convention could strengthen existing multilateral environmental agreements (such as the Convention on Biological Diversity and the Framework Convention on Climate Change)[88]
- The lack of a binding instrument has increased the fragmentation of forest governance with resultant duplications and inefficiencies[89]
- There is need to harmonize the multiple and potentially conflicting international instruments impacting on forests.[90]

---

86 Vanderzwagg, D. and D. Mackinlay (1996), 'Towards a Global Forest Convention: Getting out of the Woods and Barking up the Right Tree' in Johnson, S. (ed.), *Global Forests and International Environmental Law*, Boston: Kluwer Law International, pp. 1–4.

87 Humphreys, above n 1, 42.

88 Ibid.

89 Vanderzwagg and Mackinlay above n 86, 1–4.

90 Ibid.

- The absence of harmonization means that states are in a weak position to impose domestic standards that would achieve sustainable forest management[91]
- The need for a global forest convention that could play a valuable role in facilitating international trade in forest products to the benefit of both developing and developed countries[92]
- The need for a forest convention that would demonstrate high level political support and commitment to solving forest-related concerns and issues at the international level[93]
- A forest convention could increase funding for forest projects in developing countries[94]
- Individual state forest governments departments may be interested in obtaining a forest convention as it would assist in elevating the status of their department.[95]

B. Arguments Against a Legally Binding International Forest Instrument

At the international level, major players in international forestry, including the USA, Brazil and Russia, are strongly against the creation of legally enforceable duties in relation to forests. These 'super powers', along with many smaller states, consistently argue against a legally binding instrument, and advocate the preservation of state sovereignty over natural resources to support their position. Commentators in the area have suggested the following arguments against the creation of binding commitments concerning forestry

- The creation of a forest convention could lead to political complications and turf wars with other legally binding instruments – for example, the Convention on Biological Diversity and the Framework Convention on Climate Change[96]
- The creation of a forest convention has the potential to undermine other legal instruments and add further uncertainty in international

---

[91] Lipschutz, Ronnie D. (2001), 'Why is There No International Forestry Law? An Examination of International Forestry Regulation, Both Public and Private', *Journal of Environmental Law and Policy*, **19** (1), 153–179, 159.

[92] Vanderzwagg and Mackinlay above n 86, 1–4.

[93] Humphreys, above n 1, 42.

[94] Saunders, Phillip (1996), *Global Forests & International Environmental Law*, Boston: Kluwer Law International, Ch. 7, 305.

[95] Humphreys, above n 1, 42.

[96] Ibid.

forest regulation. This uncertainty would result in an increase in regulation, not in action[97]

- The harmonization of forestry practices creates a form of cultural imperialism, negatively affecting cultural and social benefits connected with forests[98]
- A forest convention would impact upon state sovereignty and limit a state's sovereign use of its natural resources[99]
- A legally binding instrument does not necessarily increase compliance with the legal obligations and duties created by it[100]
- Certain states are only interested in creating a forest convention to create legal rules on tariffs and trade barriers against timber exports[101]
- The negotiations to create a convention would be lengthy and could consume significant financial and technical resources, thereby slowing much needed action on the ground. Even if a new instrument were to be adopted, it may not enter into force quickly, while effective implementation would take even longer[102]
- The divisions within the international community may lead to a weak set of norms (reflecting the lowest common denominator). This risk is particularly significant given the current lack of substantive global consensus on key issues.[103]

C. Forest Convention Drafts

Two forest convention drafts have been created during this international/ academic debate over the requirement for a legally binding instrument. The Food and Agriculture Organization of the United Nations (FAO) created the first blueprint for a legally binding instrument concerning

---

97 Ibid.

98 Lipschutz, above n91, 159.

99 Ibid.

100 For example, compliance levels with Convention on Biological Diversity and Framework Convention on Climate Change.

101 The United States of America argues that Canada only supports the notion of a forest convention so as to secure itself larger access to timber markets. For more information, see Humphreys, above n 1, 42.

102 Tarasofsky, Richard (1999), 'Assessing the International Forest Regime: Gaps, Overlaps, Uncertainties and Opportunities' in Tarasofsky, Richard (ed.), *Assessing the International Forest Regime*, Gland, Switzerland: IUCN – The World Conservation Union, pp. 1–3, 4.

103 Ibid.

forests.[104] This draft was compiled around 1990 and was based on three basic principles:[105] the guarantee of state sovereignty; the operation of the stewardship ethic; and the acceptance of the concept of burden-sharing. The operation of these three principles allowed tropical forest countries to act as global stewards of forests representing the international community – which, in turn, agreed to share the burden of conservation (the burden here, being the loss of economic development through logging of forest areas).[106] The three-prong approach advocated by the document is logical and does address forest concerns, however this design has had little impact on the design of the present regime.

Another blueprint for a forest convention emerged from GLOBE (the Global Legislators Organization for a Balanced Environment).[107] The group made a number of useful suggestions concerning the operation of the convention including the creation of

- A conference of the parties to the convention
- An authority tasked with preparing international standards and guidelines
- The creation of an international compliance and monitoring body
- The creation of a international forest information network, and
- The creation of a forest products council with a mandate to facilitate development of international trade standards for forest products.[108]

The UNFF should adopt the institutional reforms suggested by the draft GLOBE forest convention and adopt this design to implement the Forest Principles of 2007. However, to fully incorporate the suggestions made by the GLOBE model, further integration of forest bodies would be needed to carry out certain activities (for example, monitoring compliance with international obligations). If a forest convention is ever agreed upon and

---

104 Humphreys, David (ed.) (1996), *Forest Politics: The Evolution of International Cooperation*, London: Earthscan Publications, 87.

105 See the following report that discusses approaches undertaken by FAO: Peck, T.J. and E.G. Richards (2005), 'Fifty Years of International Co-operation in Forestry', available at http://www.unece.org/timber/docs/publications-other/JCHistory.pdf (accessed 11 February 2010).

106 Humphreys, above n 1, 87.

107 A copy of the GLOBE Forest Convention is no longer available online. GLOBE is still in operation and its objective is to support ambitious and political leadership on issues of climate and energy security, land use change and ecosystems and economic and population growth. See, generally, http://www.globeinternational.org/ (accessed 11 February 2010).

108 Vanderzwagg and Mackinlay, above n 86, 32.

created, the institutional operations proposed by the GLOBE draft should be considered.

The forest convention debate moves international forest dialogue in an unsuitable direction. Instead of discussing the issues that undermine the current system of regulation (state sovereignty, financial and development assistance, technology transfer), participants are distracted by strong arguments for and against a convention. A suitably negotiated forest convention would be of great benefit to the world's forests, however, the current climate of forest dialogue and negotiations is not going to support this seemingly radical ideal for a forest convention. Before a forest convention is created, the underlying issues identified above must be recognized and attempts made to solve the issues they present.

5. IMPLEMENTATION OF UNITED NATIONS FORUM ON FOREST PROGRAMS

A. Implementation to Date

It is generally accepted that there has been weak implementation and compliance with the major forest instruments created by UNFF and its predecessor bodies.[109] Generally, the lack of compliance and implementation is attributed to the politics surrounding the international forestry negotiations (including issues connected to state sovereignty), the lack of authority conferred upon the Forum, the lack of finance to implement the objectives of the Forum, and the lack of binding mandates created by the Forum.[110]

UNFF and predecessor bodies created the following legal instruments

- Non-legally Binding Authoritative Statement of Principles for a Global Consensus on the Management, Conservation and Sustainable Development of all Types of Forests 1992 (Forest Principles 1992)
- Intergovernmental Panel on Forests/ Intergovernmental Forum on Forests Proposals for action (1995–2000)
- Non-legally Binding Instrument on all Types of Forests 2007 (Forest Principles 2007).

---

109 Humphreys, above n 1, 91.
110 Ibid. 106.

There is a common belief within international forestry networks, domestic policy counterparts, and academic circles, that the Forest Principles 1992 were largely ignored by domestic regimes. For international law to be effective, there must be domestic implementation of international commitments. The subsequent proposals for action created between 1995 and 2000 were an attempt to address this. The proposals for action created the concept of 'National Forest Policies'. These policies are considered to be the most successful outputs of the international forest regime; as they define forest policy and examine implementation and resources requirements. UNFF was formed to implement the proposals for action generated by the Intergovernmental Panel on Forests and the Intergovernmental Forum on Forests.

During the UNFF 9 negotiations the following reporting statics were provided: 'Since 2002, 86 countries have submitted national reports, ranging from 14 in 2002 to 54 in 2005. Three countries submitted national reports for all six sessions, with 24 countries submitting reports for between three and five sessions. However, 41 countries submitted national reports for only one of the six sessions for which they had been requested to submit reports.'[111] Such reporting rates are relatively low and are reflective of the authority of the regime and the lack of clarity provided by the regime in specifying its reporting requirements and purpose.

At the outset, it was decided that the progress of the Forum would be reviewed after five years.[112] In 2005, a report was prepared reviewing the effectiveness of the international arrangement on forests[113] using information from voluntarily submitted reports from countries and other organizations on the implementation of the proposals for action, and from voluntarily submitted questionnaires. Thus, those countries that did not respond were not included in the report's findings.

The report was arranged in three main sections: thematic areas of proposals for action; effectiveness of international arrangement on forests; and conclusions. The first section of the study identified 16 thematic areas

---

[111] Secretary General, Economic and Social Council, *Assessment of Progress Made on the Implementation of the Non-legally Binding Instruments on all Types of Forests and Towards the Achievement of the Four Global Objectives on Forests*, 9th sess, Agenda Item 3, UN Doc E/CN/18/2011/2 (15 November 2010) para 7.

[112] Resolutions and Decisions of the Economic and Social Council, above n 13.

[113] Department of the Economic Secretariat of the United Nations Forum on Forestry, *Review of the Effectiveness of the International Arrangement on Forests*, 5th sess, Agenda Item 5, UN Doc CN.18/2005/6 (11 March 2005).

covered by the proposals for action. These thematic areas are listed below, with a brief synopsis of the findings[114]

- *National Forest Policies* – the overall finding was that National Forest Policies are useful tools for devising domestic forestry regulation. In general, it was found that stakeholder consultation needs to be improved, and that forest priorities need to be integrated with broader development priorities[115]
- *Public Participation* – certain countries were identified as having advanced frameworks for public participation in forest policy and planning (Switzerland, Norway, and Finland). In general, it was found public participation opportunities often exist for areas of forest land that are under public ownership but, frequently, the public is given no right to participate in negotiations concerning the use of private forest land
- *Deforestation and forest degradation* – the overall global trend shows a decrease in forest cover. All countries that responded (except for Europe) reported a decrease in forest cover and a number of countries identified the underlying cause of deforestation within their jurisdictions[116]
- *Traditional forest related knowledge* – the traditional forest knowledge registries or databases of Australia, Canada, India, New Zealand, Peru, the Philippines and the South Pacific region were identified . Overall, it is recognized that beneficial traditional knowledge is still being amassed
- *Forests and related scientific knowledge* – it was found that science has a made significant contribution to forest management and forest ecosystem health. The traditional avenues of knowledge dissemination were identified as meetings, training sessions, and publication of findings in various places. New avenues of dissemination were electronic-based, websites, interactive online forums, and mailing lists
- *Forest health and productivity* – air pollution was identified as a

[114] Ibid.

[115] It was noted that, in Sweden, quantifiable targets were set in the National Forest Policy; this, therefore, creates legally binding forest commitments in Sweden.

[116] For example: Cambodia identified illegal logging; El Salvador identified a high level of population in small geographical area; Sudan identified agricultural expansion; the USA identified a shift in population from rural to metropolitan areas; and Korea stated that deforestation was unavoidable, as 64% of the country is covered by forests, and essential infrastructure and development necessitates a degree of deforestation.

major threat to forest health, now and in the future. Storms, fires and climate change were also identified as threats to forest health and productivity

- *Criteria and Indicators* – the value of criteria and indicators as tools for assessing progress towards sustainable forest management was acknowledged. It has been suggested that national forest policies could be better aligned with existing criteria and indicator frameworks
- *Economic, social and cultural values of forests* – it was found that, in general, there is a lack of data about the economic value of all forest services. It is recognized that timber prices do not accurately reflect the true value of forests, and that policy reform is needed to address this[117]
- *Conservation* – it was found that many countries have significant tracks of forest areas under some form of legal protection. In developing countries, improving capacity and revenue for forest conservation should be addressed as a priority
- *Monitoring, assessment, reporting, concepts, terms, and definitions* – it was found that many countries have well-established forestry inventory systems; again, developing countries require assistance and capacity building to support them in implementing forest monitoring systems
- *Rehabilitation in low forest cover* – a number of afforestation initiatives were identified by the report
- *Rehabilitation in degraded lands* – it was found that the establishment of plantations and agroforestry practices using native species was a common approach when attempting to rehabilitate degraded land
- *Future forest interests* – some responses showed a consideration of future forest trends, including long-term predications concerning timber supply and demand, and adaptation of the forest economy. It was also stated that national forest policies should make reference to future forest interests
- *Financial resources* – financing sustainable forest management and payment for the non-market outputs of forests remain critical factors for achieving successful implementation of the proposals
- *International trade and sustainable forest management* – it was found

[117] Some examples of policies that are attempting to address the current undervalue include increasing rent on public forest land, tightening up the collection of revenues, and taking account of environmental consideration in both the setting of payments and the application of contractual conditions to concessions.

that processes outside the forum (such as the FLEG[118] and forest certification processes) have made more progress in creating standards for international trade in timber products
- *International cooperation capacity building and technology transfer* – numerous examples of bilateral and multilateral partnerships were reported. The role of NGOs in assisting with capacity-building and technology transfer was recognized.

The second section of the report collated responses on the overall effectiveness of international arrangements for forests. Switzerland suggested that resolutions negotiated during Forum sessions have had limited impact on the implementation of proposals for action and, furthermore, that Forum sessions did not prompt action for implementation.[119] The European Union suggested that formal forest policy development at the global level has been hindered by a lack of participation and the absence of an implementation framework.[120] Norway suggested that the proposals for action have enhanced policy development and dialogue at the international level and contributed to European collaboration on forest policy. Norway noted that the effects of the proposals for action on policy development and dialogue at the domestic level were less obvious.[121]

The third part of the report contained overall conclusions based on the submissions received. Generally, it was considered that the advances were made in national forest policy formulation, extending stakeholder engagement and the development of criteria and indicators. It should be noted, however, that stakeholder engagement no longer takes place at the Forum's main sessions, and there is no clear process or purpose for stakeholder engagement (as discussed earlier in this chapter). So, arguably, no real advances have been made in the context of stakeholder engagement at the international level. As to the development of criteria and indicators, the creation of these tools actually emerged from multi-lateral intergovernmental negotiations that were hosted separately from the Forum negotiations. As such, the Forum cannot claim credit for the development of this sustainable forest management tool.

Part three of the report identifies the following areas in which the international arrangement on forests has failed to deliver: addressing

118 FLEGT stands for Forest Law Enforcement Governance and Trade Scheme.

119 Department of Economic Secretariat of the United Nations Forum on Forestry, *Review of the Effectiveness of the International Arrangement on Forests*, 5th sess, Agenda Item 5, UN Doc CN.18/2005/6 (11 March 2005).

120 Ibid. 73.

121 Ibid. 74.

deforestation and illegal forestry practices; assisting in the development of domestic institutional forestry frameworks; creating sectoral links with other national policy processes; and creating a mechanism to value forests' non-market values.[122] The report also revealed that many of responses noted the lack of a legally binding instrument concerning the use of the world's forests.

The Forest Principles 2007 are due to be implemented from 2009 to 2015 (as stated in the Multi-year Programme of Work).[123] According to the Multi-Year Program of Work, the Forum will meet every second year and a theme has been assigned to each session. In 2009, at the Forum's eighth meeting, implementation of sustainable forest management was discussed. In preparation for this session, a number of reports were prepared concerning mobilizing finance for sustainable forest management.[124] These reports will be discussed in more detail below.

While the program of work provides some direction and structure for future Forum sessions, the effectiveness of the program will depend upon the creation of agreed objectives arising from session negotiations, and on activities that take place in between sessions to implement these objectives. If no clear activities or goals are set for action in between each session meeting, there is high risk that the Forum's sessions will continue to operate purely as a 'talkfest'. Small, identifiable targets around the relevant assigned theme must be set for each Forum.

B. Improving Implementation

To address the lack of implementation of the international forest regimes instruments, three barriers must be overcome. Firstly, financial support must be provided to developing countries to allow implementation of sustainable forest management principles and practices. The forum recognizes the necessity of increasing international financial forest revenue and has commissioned reports exploring options on how to generate increased revenue for sustainable forest management. Secondly, the capacity of developing countries to implement and enforce sustainable

---

122 Ibid. 97.

123 See http://www.un.org/esa/forests/multi-year-work.html.

124 Simula, Markku, Advisory Group on Finance of the Collaborative Partnership on Forests and Program on Forests of the World Bank, *Financing Flows and Needs to Implement the Non-Legally Binding Instrument on All Types of Forests* (2008), and El Lakany, Hosny, Michael Jenkins, and Michael Richards, PROFOR, *Background Paper on Means of Implementation (Contribution by PROFOR to discussions at UNFF-7)* (2007).

forest management principles and practices must be addressed. For forest programs to be successful, there must be an underlying political structure that is secure, transparent and accountable. In the absence of such a system, implementation of sustainable forest management will continue to be weak. Thirdly, there needs to be policy integration of forest issues into broader sustainable development and development strategies.

Implementation lessons

In 2008, Ghana became the first country to systematically implement the Forest Principles 2007 as a result of assistance provided by Germany. The Ghana Project Pilot indicates that the Forest Principles 2007 provide a framework and performance assessment tool for national and international cooperation related to the forests of Ghana.[125] Three countries, in Africa, Latin America and Asia, will also participate in this pilot scheme in the future, once funding for extending this initiative is secured.[126] The Ghana project was entitled 'Moving Forward in the Implementation of the Non-legally Binding Instrument on All Types of Forests' and was aimed at identifying Ghana's gaps, strengths and weaknesses with respect to the national policies and measures set forth in the forest instrument, as well as determining priorities and initiating measures to address these gaps and weaknesses.[127]

The following four topics and accompanying activities were selected as priority measures in Ghana

1. Promote cross-sectoral coordination
    - Activity: enhance the visibility of the forestry sector's contribution to national development planning
2. Strengthen forest law enforcement
    - Activity: enhance the watchdog role of communities in combating corruption and other illegal activities
3. Develop financing strategies for sustainable forest management
    - Activity: enhance the access of local stakeholders to funding sources that support socio-economic activities in the forestry sector
4. Integrate national forest programs into relevant national development plans and poverty reduction strategies

---

[125] Secretary General, Economic and Social Council, *Assessment of Progress Made on the Implementation of the Non-legally Binding Instruments on All types of Forests and Towards the Achievement of the Four Global Objectives on Forests*, 9th sess, Agenda Item 3, UN Doc E/CN/18/2011/2 (15 November 2010) para 20.

[126] Ibid. para 21.

[127] Ibid. para 15.

- Activity: enhance the capacity of district assemblies in the development and implementation of district forestry programs.[128]

A number of challenges were identified during the implementation of the Forest Principles 2007 in Ghana including: insufficient capacity and funding to carry out collaborative forest management and to strengthen forest law enforcement; unsustainable agriculture practices occurring in forest areas due to lack of policy and evidence of a lack of inter-ministerial collaboration to harmonize conflicting land use policies; widespread poverty among rural forest dependent communities who lack viable income or lifestyle alternatives; and lack of data in the national budget concerning the forest sectors contribution to national development.[129] The Ghana project is a unique international forest program due its focus on all forest values, which is implemented through the concept of sustainable forest management. UNFF should continue to implement such programs as a way of translating international forest goals into domestic action in regions where implementation would otherwise remain elusive without international support.

Finance

Two reports have been prepared that focus on the issue of international finance for the implementation of the Forest Principles 2007.[130] Both reports suggest that increased investment in sustainable forest management is dependent upon two prerequisites being addressed: land tenure and governance concerns.[131] First, secure land tenure and forest use rights must be established. Secure land and property rights increase private investor confidence and provide a degree of security for the investment (being the value of the land). Official development assistance for forest programs also value secure land tenure rights because this provides assurance that the work carried out as a result of the funding will continue even after the project is finished. It is noted that many externally funded projects aimed at improving land tenure often underestimate the time that the process of land tenure reform take.[132]

---

[128] Ibid. para 17.

[129] Ibid. para 19.

[130] Simula, Markku, Advisory Group on Finance of the Collaborative Partnership on Forests and Program on Forests of the World Bank, *Financing Flows and Needs to Implement the Non-Legally Binding Instrument on All Types of Forests* (2008), and El Lakany, Hosny, Michael Jenkins, and Michael Richards, PROFOR, *Background Paper on Means of Implementation (Contribution by PROFOR to discussions at UNFF-7)* (2007).

[131] Simula, above n 130, 72.

[132] Ibid.

Governance

Secondly, improving institutional capacity and weak forest governance frameworks should be a priority for sustainable forest management funding. Without adequate governance models, the provision of funding to implement forestry-related works will be undermined since the integrity of the project cannot be assured. Governance can be defined as the structures, functions, processes and organizational traditions that have been put in place within the context of a program's authorizing environment to ensure that it is run in such a way that it achieves its objectives in an effective and transparent manner.[133] Weak forest governance acts as a barrier to public and private forest investment. In particular, in the forest context, there is a pressing need to improve monitoring capacity and enforcement mechanisms to prevent illegal forest activities. Furthermore, institutional capacity needs to be improved to ensure that forest areas are not further encroached on due to agricultural expansion and land development.

The reports on sustainable forest management finance discuss existing and emerging forest revenue streams. In terms of existing forest revenue, El Lakany distinguishes between public and private finance.[134] Public forest revenue comes from domestic government budgets and official development assistance initiatives. Domestic funding can come in the form of direct investment and intervention, market interventions, or governance law and enforcement. Private forest investment comes from investment activities, portfolio investments, loans, and philanthropic funds. El Lakany identifies the following emerging forest financial mechanisms: payment for ecosystem services schemes; carbon markets; community/company partnerships; and conservation funding provided by NGOs. It is also suggested that the private sector can be engaged though the use of eco-security,[135] forest insurance,[136] the 'Equator Principle', and due diligence.[137]

Simula contends that examining sustainable forest management as a financing objective involves recognition of the dual nature of sustain-

---

133 Simula, above n 131, 69.

134 El Lakany, above n 131, 8.

135 The aim of eco-securitization is to enable access to international finance through a forest funding instrument or bond in which the returns match the biological and sustainable growth profile of natural growth forestry. For further information, El Lakany, above n 131, 25.

136 Commercial investment is dependent upon insurance being provided. Therefore, obtaining insurance for forest investment would be a way to motivate private sector investment. For further information see ibid.

137 The 'Equator Principles' require that all loans over $50 million adhere to the environmental and social safeguard policies of the IFC. For further information, see ibid. 26.

Table 5.1 External financial flows to forests[138]

Source	2000–02	2005–07	% Change
Public Sector			
Bilateral	959.3	1,103.4	+15.0
Multilateral	335.0	806.7	+140.8
Total	1,294.3	1,910.1	+47.6
Private Sector			
Foreign investment	400.0	516.0	+29.0
Other private funding	n.a.	n.a.	Increase
NGO, philanthropic	n.a.	n.a.	Probable Increase

able forest management.[139] Forests provide public goods such as climate regulation and biodiversity stocks, while forests also serve private interests by generating financial capital. Traditionally, public forest services have been funded by the public sector while private sector forest services have been funded by the private sector. In the future, the cost of public forest sector investment needs to be shared between the two sectors. Payment for ecosystem services provides a model for public/private investment in forest areas.

Simula's report finds that, in general, there has been a modest increase in all forms of forest finance (see the table below). Despite this, the gap between the funding available funding and funding needs is still wide.[140] It is suggested that to increase official development assistance funding for forests a priority should be the inclusion of forest issues in poverty reduction policies and other development priorities policies. This way, finance mobilized to address cross-sectoral issues (such as capacity-building and poverty) can be directed into forest projects which seek to improve these goals alongside forest goals. Additionally, country aid policies could be lobbied to include forests as a priority – because often forests are only mentioned indirectly.[141]

138 Ibid. 28.
139 Simula, above n 130, 22.
140 Ibid. 71.
141 Ibid. 25.

There are a number of emerging market mechanisms which have the potential to mobilize finance for forestry projects. These include the existing Kyoto flexibility mechanism, the REDD+ initiative and the payment for ecosystem services markets that are emerging. All of these market mechanisms have institutional and other barriers that must be overcome to increase take up and participation in their schemes.

6. CONCLUSION: RECOMMENDATIONS FOR REFORM

The future mandate of the Forum needs to be established. Reform of the institution is desired by many sectors of society, and is most likely required to attract future United Nations operational funding. The Forum should be reformed to carry out one of the roles described below

1. The institution recognizes its inability to create binding legal commitments and forest targets. Instead, the Forum should focus on acting as a coordinator for global, international and intergovernmental programs on forests. Within this new role, the institution would focus on aligning the research priorities of major global forest initiatives and implementing a streamlined singular reporting framework that could be used to meet most international reporting requirements. In this coordination role, the body would identify thematic areas for regulation and focus on connecting relevant stakeholders with the forest governing body most useful to them.
2. The institution should attempt to win back the respect of its members and stakeholders by taking a hard-line approach and create a specific program of work, binding commitments, and binding targets. This type of action would be controversial, and it is likely that a number of parties would withdraw from it. The withdrawal of these members would allow the program to move forward and develop. These parties may then be induced to come back into the regime at a later date by market and non-market incentives. The forum could follow the lead of the climate change regime by creating binding commitments for developed countries in its first commitment period, with a view to increasing capacity within developing countries so that future commitment periods create obligations for these parties.

Currently, the Forum serves as an international 'talkfest' on forest-related issues and concerns. Long-formed alliances continue to operate within the negotiations, resulting in political power plays and stand-offs. Any major policy advances are soon turned into weakly worded aspirational statements to obtain consensus from the majority of parties.[142] The creation of non-legally binding aspirational statements serves no practical role in the global regulation of forests. Parties attend the negotiations, make strong self-serving forest statements, and then return home to continue to operate a 'business as usual' forestry sector. Major reform must take place because the effort put into negotiations in the past has proved to be time-consuming, fruitless and expensive.[143]

142 Davenport, Deborah, and Peter Wood (2006), 'Finding the Way Forward for the International Arrangement on Forests: UNFF-5,-6 AND-7', *Review of the European Community and International Environmental Law*, **15** (3), 316-326, 319.

143 Ibid.

6. Regulation of forests by the international climate change regime

Forests are vitally important in addressing climate change because forest ecosystems affect the climatic conditions experienced on the earth's climate, through the absorption of carbon in wood, leaves and soil. This carbon is released to the atmosphere when forests are burned or during forest clearance and harvesting. Quantifying the role of forests as sources of carbon emissions and in their role as carbon sinks has become essential to understanding the global carbon cycle.[1]

1. INTRODUCTION

A. Background on the Climate Change Regime

The United Nations Framework Convention on Climate Change was negotiated at the Earth Summit (Rio 1992) and entered into force on 21 March 1994. The 1992 Convention attributes the high presence of greenhouse gases in our atmosphere to industrialized nations.[2] It also directs industrialized nations to take responsibility for this[3] and creates a framework for addressing climate change at the international level.[4] The convention boasts almost universal membership with 192 parties having ratified the instrument.

1 Dresner, Simon, Paul Etkins, Kate McGeevor and Julia Tomei (2007), 'Forests and Climate Change: Global Understanding and Possible Responses', in Freer-Smith, P., M. Broadmeadow and J. Lynch (eds), *Forestry and Climate Change*, Wallingford, Oxon: CABI Publishing, pp. 33–48.

2 The United Nations Framework Convention on Climate Change, opened for signature on 4 June 1992, 1771 UNTS 107. See the opening statement, which provides '*noting* that the largest share of historical and current global emission of greenhouse gases has originated in developed countries, and that per capita emissions in developing countries are still relatively low'.

3 It does this in Article 3 by adopting the principle of 'common but differentiated responsibilities' and stating in 3(1) that developed country parties should take the lead in combating climate change and the adverse effects thereof.

4 Convention on Climate Change, above n 2, Arts 7–26.

The ultimate objectives of climate change are found in Article 2 of the Convention which requires the

> stabilization of greenhouse gas concentrations in the atmosphere at a level that would prevent dangerous anthropogenic interference with the climate system. Such a level should be achieved within a time frame sufficient to allow ecosystems to adapt naturally to climate change, to ensure that food production is not threatened and to enable economic development to proceed in a sustainable manner.

Article 2 provides that this objective applies not only to the Convention, but also to any related legal instrument that the Conference of the Parties (COP) adopts. This creates one focused objective for the regime, which is addressed though different legal policies, instruments and measures. The objective is framed as an environmental quality standard.[5] It does not prohibit the emission of greenhouse gases; rather it seeks to restrict these activities when they exceed a certain threshold (that of dangerous anthropogenic interference). The objective also sets a timeline for when the environmental standard must be met, requiring that the changes take place so as to not affect ecosystem adaptation, food security, or economic development occurring in a sustainable manner.

The international climate change regime is ambitious in its scope and design. During the first commitment period, the effectiveness of the protocol in reducing global emissions will be revealed. The first commitment period is focused on mitigating climate change but future commitment periods are likely to focus on both mitigation and adaptation to climate change. Sceptics of the regime anticipate that the global target of reducing emissions by 5 per cent from 1990[6] levels will not be met. Current predictive modelling shows that if no mitigation action is taken, an increase in greenhouse gas emissions is certain.[7] International climate change policy

---

[5] Farhana, Yamin and Joanna Depledge (2006) *The International Climate Change Regime: A Guide to Rules, Institutions and Procedures,* Cambridge University Press, 61.

[6] Convention on Climate Change, above n 2, Art. 3.

[7] Metz, B., O. Davidson, P. Bosh, R. Dave, and L. Meyer (eds) (2007), *Summary for Policymakers in Climate Change 2007: Mitigation, Contribution of Working Group III to the Fourth Assessment Report of the Intergovernmental Panel on Climate Change*, Cambridge, United Kingdom; New York, NY, USA: Cambridge University Press. This report finds that a range of policies, including those on climate change, energy security and sustainable development, have been effective in reducing greenhouse gas emissions in different sectors and many countries. The scale of such measures, however, has not yet been large enough to counteract the global growth in emissions.

will have to evolve to address unexpected and challenging scenarios. The advent and seriousness of these challenges will be dependent upon the regime's ability to control or limit emissions in the early commitment periods. For the sake of present and future generations, one must hope that the regime achieves some level of success in its early commitment periods.

B. The Role of Forests in the Climate Change Regime

About two-thirds of the globe's terrestrial carbon, excluding what is held in rocks and sediments, is sequestered in standing forests, forest understorey, plants, leaves, forest debris and in forest soils.[8] Forest ecosystems use five methods to store carbon. This occurs through above-ground biomass, below-ground biomass, litter, dead wood and the soil's organic carbon. As the forest biomass experiences growth, the amount of carbon stored within the forest stock increases. As forest ecosystems slowly decompose, the stored carbon is released back into the atmosphere. This means that the amount of carbon stored continually changes during the lifetime of the forest ecosystem.[9]

When land that contains forest ecosystems is converted and is used for another purpose (for example, agricultural or rural development), there is a net loss of carbon stored on the land. This is because most other land use sequesters less carbon than forest ecosystems. If the previous forest site is cleared and burned, most of the stored carbon will be released back into the atmosphere.[10] Carbon is stored in forest ecosystems in a number of ways and the extent of land clearance and the method used to clear the land will vary the amount of carbon released back into the atmosphere. Future climate change policies will also consider the ability of long-life wood products to store carbon and will likely provide some recognition for the carbon stored in such goods. Wood products continue to store carbon until the point of decomposition of the material. This means that, if forest stock is converted to wood products, the carbon stored in the wood will not be immediately released back into the atmosphere.

The variance in the amount of carbon stored in a forest ecosystem – and in the subsequent use of previously forested land, presents uncertainties for scientists who attempt to measure the amount of carbon stored. This scientific uncertainty has created legal uncertainty, with international

---

8 Sedjo, Roger and Masahiro Amano (2006), Resources for the Future Washington, *Forest and Biological Carbon Sinks after Kyoto*.

9 Ibid. 3.

10 Ibid. 4.

negotiations attempting to grapple with the incorporation of forest sink activities within the international climate regime.

Forests are regulated under the Convention due to their ability to absorb and store carbon. There is no particular reference made to the individual biome of forests (for example, boreal, temperate or tropical), or to the classification of the forest (for example, primary forest or plantation forest). The term 'forest' is used in Article 4 of the Convention. After studying the Convention, it is clear that it is not intended to limit the application of the instrument to a certain type of forest. Negotiations following the creation of the Convention have attributed a definition to 'forest', and this definition is inclusive of all forest types.

2. THE INSTITUTIONAL STRUCTURE OF THE INTERNATIONAL CLIMATE CHANGE REGIME

A. Implementing Bodies of the Convention

The governance structure underlying the implementation of the international climate change regime is comprehensive and involves three bodies: the Secretariat, the Conference of the Parties (COP), and two permanent subsidiary bodies. The Secretariat is charged with providing organizational support and technical expertise during negotiations. It is also responsible for facilitating the flow of authoritative information on the implementation of the Convention.[11] The Secretariat's precise duties are outlined in Article 8 of the Convention: notably, subsection (e) requires the Secretariat to ensure the necessary coordination with the secretariats of other relevant international bodies. Climate change will impact upon all sectors of society, thus integration, input and cooperation from key stakeholder representatives groups is crucial for the regime's success.

The COP is the principle body created by the convention. It meets annually unless otherwise specified.[12] Article 7 of the Convention outlines the purpose and function of the COP. Its mandate is wide-ranging and requires it, among other listed duties: to examine the obligations of the parties in light of the experience gained in the implementation and

---

[11] United Nations Framework Convention on Climate Change (2007), 'The Secretariat', available at http://unfccc.int/secretariat/items/1629.php (accessed 7 January 2008).

[12] United Nations Framework Convention on Climate Change (2007), 'Meetings', available at http://unfccc.int/meetings/items/2654.php (accessed 7 January 2008).

evolution of scientific and technological knowledge;[13] to seek to mobilize financial resources;[14] and to seek the services and cooperation of competent international organizations.[15] Practically, the COP is responsible for adopting decisions and requesting further research or information, thus guiding the direction and form of the regime.

The research body of the international climate change organization is the Intergovernmental Panel on Climate Change (IPCC). This body was created by the United Nations Environment Program and World Meteorological Organization to provide an objective source of information about climate change. The IPCC provides scientific, technical and socio-economic information to policy makers from an unbiased and (some suggest) conservative standpoint. The IPCC has produced four assessment reports. These have significantly influenced development of the international climate change regime and its ongoing operation and implementation.[16]

The Convention also established two permanent subsidiary bodies known as the Subsidiary Body for Scientific and Technological Advice (SBSTA) and the Subsidiary Body for Implementation (SBI). Article 9 of the Convention creates the SBSTA, whose main purpose is to provide timely information and advice on scientific and technological matters relating to the Convention. The SBSTA is made up of government representatives selected as being competent in their relevant field of expertise. Article 10 of the Convention creates the SBI, which seeks to ensure compliance with Article 12 of the Convention. Article 12 requires parties to provide a national inventory of emissions and removals[17] and a general description of steps taken by a party to implement the Convention.[18] The SBI is open to the participation of all parties and is comprised of government representatives who are experts on matters related to climate change.

B. Membership to the Convention

The Convention recognizes three different categories of membership, each attracting different levels of commitment. This categorization of membership is consistent with the international concept of common but

---

13 Convention on Climate Change, above n 2, Art. 7 (a).
14 Ibid. Art. 7 (h).
15 Ibid. Art. 7 (l).
16 Intergovernmental Panel on Climate Change (2004), 'About the IPCC', available at http://www.ipcc.ch/about/index.htm (accessed 7 January 2008).
17 Convention on Climate Change, above n 2, Art. 12 (1) (a).
18 Ibid. Art. 12 (1) (b).

differentiated responsibility. The concept is based upon a shared notion of fairness.[19] Within the climate change regime, it is recognized that developed countries are disproportionately responsible for the high level of greenhouse gas emissions present in the atmosphere and, therefore, have an historical responsibility to make amends. In addition to this historical responsibility, developed countries also have the greatest capacity to implement activities aimed at reducing greenhouse gas emissions.

The concept of common but differentiated responsibility manifests clearly within the Convention.[20] This principle acknowledges that every state has a common responsibility in relation to the global environment. Differentiated responsibilities arise due to the unique circumstances of each state.[21] A number of justifications exist for differentiated treaty obligations including:[22]

- the recognition of differing capacities and capabilities among states;
- the recognition of historic entitlements and legitimate expectations;
- responsibility for mitigation, therefore, being attributed on the basis of past actions;
- the notion that the international community is under an obligation to take into account the special needs and circumstances of developing countries;
- the notion that the international community has entered a new stage of international cooperation that obliges the more developed states to take on more burdensome obligations; and
- the notion that differentiated responsibilities provides an inducement to hesitant states (particularly developing ones) to participate in multilateral environmental agreements.

Creating a mutual responsibility with differentiated standards also imparts an obligation on capable states to assist less capable states. This

---

19 Harris, Paul G. (1999), 'Common but Differentiated Responsibility: The Kyoto Protocol and United States Policy', *New York University Environmental Law Journal*, **7** (1), 27–48, 28.

20 Article 3(1) provides that all states should protect the climate system for the benefit of present and future generations of humankind, on the basis of equity and in accordance with their 'common but differentiated responsibilities and respective capabilities – accordingly the developed country parties should take the lead in combating climate change the adverse effects thereof'.

21 Melkas, Erika (2002), 'Sovereignty and Equity within the Framework of the Climate Regime', *Review of European Community and International Environmental Law*, **11** (2), 115–128, 123.

22 Ibid.

has the potential to improve environmental conditions within less capable states, hence providing a general benefit to the global community. It is thought that fewer environmental obligations for less capable and less responsible states will undermine the operation of global initiatives. Therefore, it is better for all states to have similar responsibilities, while acknowledging that some states will require more assistance with their responsibilities than others.[23]

In accordance with the principle of common but differentiated responsibility, Article 3 of the Convention obliges developed countries to take the lead in combating climate change and the adverse effects thereof.[24] Developing countries that bear a disproportionate or abnormal burden are given special recognition within the convention.[25] Formally, the differentiated responsibilities are recognized by creating a three-tier membership approach. The three types of membership are:[26]

1. *Annex I Parties*. These countries are responsible for creating a regulatory domestic framework aimed at controlling their emissions output. Countries that were members of the Organization for Economic Co-operation and Development (OECD) in 1992, plus countries with economies in transition (EIT) including the Russian Federation, the Baltic states, and several Central and Eastern European states are Annex I parties.[27]
2. *Annex II Parties*. These countries are required to provide financial assistance to enable developing countries to undertake emission reduction activities and help developing countries to adapt to the adverse effects of climate change. They are also obliged to take all practical steps to promote the development and transfer of environmentally friendly technologies to economy in transition parties and developing

[23] Melkas, Erika (2002), 'Sovereignty and Equity within the Framework of the Climate Regime', *Review of European Community and International Environmental Law*, **11** (2), 115–128, 123.

[24] Convention on Climate Change, above n 2, Art. 3(1).

[25] Ibid. Art. 3(2).

[26] Ibid.

[27] *Parties with Annex 1 obligations are*: Australia; Austria; Belarus; Belgium; Bulgaria; Canada; Croatia; Czech Republic; Denmark; European Economic Community; Estonia; Finland; France; Germany; Greece; Hungary; Iceland; Ireland; Italy; Japan; Latvia; Liechtenstein; Lithuania; Luxembourg; Monaco; Netherlands; New Zealand; Norway; Poland; Portugal Romania; Russian Federation; Slovakia; Slovenia; Spain; Sweden; Switzerland; Turkey; Ukraine; United Kingdom of Britain and Northern Ireland; and United States of America.

parties. The Convention creates a financial mechanism within Article 11, whose funding is channelled through to these purposes. A number of Annex I parties (the OECD members), therefore, attract Annex II obligations in addition to their Annex I obligations.[28]

3. *Non-Annex I Parties.* These are mostly developing countries, and countries that are recognized as being especially vulnerable to the adverse impacts of climate change (low-lying coastal areas, and areas prone to desertification and drought). In addition, countries that rely heavily on income from fossil fuel production and commerce are given non-Annex 1 status. These countries do not have any formal obligations under the Convention.[29]

---

28 *Parties with Annex II commitments are:* Australia; Belgium; Canada; Denmark; European Economic Community; Finland; France; Germany; Greece; Iceland; Ireland; Italy; Japan; Luxembourg; Netherlands; New Zealand; Norway; Portugal; Spain; Sweden; Switzerland; United Kingdom of Great Britain and Northern Ireland; and the United States of America.

29 *Parties with non-Annex I commitments are:* Afghanistan; Albania; Algeria; Angola; Antigua and Barbuda; Argentina; Armenia; Azerbaijan; Bahamas; Bahrain; Bangladesh; Barbados; Belize; Benin; Bhutan; Bolivia; Bosnia and Herzegovina; Botswana; Brazil; Burkina Faso; Burundi; Cambodia; Cameroon; Cape Verde; Central African Republic; Chad; Chile; China; Colombia; Comoros; Congo; Cook Islands; Costa Rica; Cuba; Cyprus; Côte d'Ivoire; Democratic People's Republic of Korea; Democratic Republic of Congo; Djibouti; Dominica; Dominican Republic; Ecuador; Egypt; El Salvador; Equatorial Guinea; Eritrea; Ethiopia; Fiji; the Former Yugoslavia Republic of Macedonia; Gabon; Gambia; Georgia; Ghana; Grenada; Guatemala; Guinea; Guinea-Bissau; Guyana; Haiti; Honduras; India; Indonesia; Iran; Israel; Jamaica; Jordan; Kazakhstan; Kenya; Kiribati; Kuwait; Kyrgyzstan; Lao People's Democratic Republic; Lebanon; Lesotho; Liberia; Libyan Arab Jamahiriya; Madagascar; Malawi; Maldives; Mali; Malta; Marshal Islands; Mauritania; Mauritius; Mexico; Micronesia; Mongolia; Montenegro; Morocco; Mozambique; Myanmar; Namibia; Nauru; Nepal; Nicaragua; Niger; Nigeria; Niue; Oman; Pakistan; Palau; Panama; Papua New Guinea; Paraguay; Peru; Philippines; Qatar; Republic of Korea; Republic of Moldova; Rwanda; Saint Kitts and Nevis; Saint Lucia; Saint Vincent and the Grenadines; Samoa; San Marion; Sao Tome and Principe; Saudi Arabia; Senegal; Serbia; Seychelles; Sierra Leone; Singapore; Solomon Islands; South Africa; Sri Lanka; Sudan; Suriname; Swaziland; Syrian Arab Republic; Tajikistan; Thailand; Timor-Leste; Togo; Tonga; Trinidad and Tobago; Tunisia; Turkmenistan; Tuvalu; Uganda; United Arab Emirates; United Republic of Tanzania; Uruguay; Uzbekistan; Vanuatu; Venezuela (Bolivian Republic of); Viet Nam; Yemen; Zambia; Zimbabwe.

3. INTERNATIONAL INSTRUMENTS CREATED BY THE INTERNATIONAL CLIMATE CHANGE REGIME CONCERNING FORESTS

A. Rules Concerning Forest Practices

Forests are regulated under the international climate change regime in a number of ways, with different forest activities attracting the operation of different forest policies. The United Nations Convention on Climate Change sets out a number of broad obligations related to mitigating the adverse risks of climate change. A number of these obligations refer specifically to forest activities. The Kyoto Protocol specifies the types of forest activities that are suitable for inclusion in accounting frameworks created by the Protocol. These forest activities are then further explained in the Land Use, Land Use Change, and Forestry Guidelines. These are the modalities used by parties when accounting for forest sink or source activities that occur within their own jurisdictions and for preparing national emission inventories.

The Clean Development Mechanism (CDM) forestry rules come into operation when an Annex I party invests in a forest project in a developing country and wants to include this activity in its accounting practices. The Reduced Emissions from Deforestation and Degradation (REDD) initiative is broader in scope and covers a range of donor-funded forest practices that occur in developing countries. It is not yet clear if funded REDD activities will be accredited for inclusion within emission reporting schemes.

In summary, analysis of forest regulation by the climate change regime can be undertaken by reference to the:

- The United Nations Framework Convention on Climate Change (UNFCCC)
- The Kyoto Protocol (deals with afforestation and reforestation)
  - Land Use, Land-Use Change and Forestry Guidelines
  - Clean Development Mechanism Forest Credit Guidelines
- REDD Instrument (instrument in development to deal with deforestation)

During the first commitment period, these separate accounting practices in relation to forestry activities will continue. Later commitment periods may see an amalgamation of rules concerning forest activity eligibility and accounting practices.

B. United Nations Framework Convention on Climate Change

The Framework Convention sets out the definitions, objectives, principles, implementing bodies, and procedural rules necessary for the climate change regime to operate. Other instruments, most notably the Kyoto Protocol and the Cancun Adaptation Framework specify how mitigation and adaptation measures and policies are to be implemented. The Convention states in the opening plenary of the instrument that the parties to this instrument are:

> Aware of the role and importance of terrestrial and marine ecosystems as sinks and reservoirs of greenhouse gases.[30]

The Convention does not directly spell out the role of sinks in addressing climate change but there is reference to them within Articles 1, 3, and 4 of the instrument.[31] Article 1 provides definitions for the Convention including that a 'sink' is defined as any process, activity or mechanism that removes greenhouse gases. This definition is wide enough to encompass a range of earth cycle processes that absorb carbon (such as services performed by trees, grasslands and soils).

Article 3 contains the principles of the Convention. These include common but differentiated responsibilities found in Article 3(1) and (2) and the Precautionary Principle found in Article 3(3). Sinks are also recognized in Article 3(3), where it is stated that all 'policies and measures should take into account different socio-economic contexts, be comprehensive, cover all relevant sources, sinks and reservoirs of greenhouse gases'.[32]

Article 4(d) of the Convention creates an obligation in relation to forest areas:

> [To] promote sustainable management, and promote and cooperate in the conservation and enhancement, as appropriate, of sinks and reservoirs of all greenhouse gases not controlled by the Montreal Protocol, including biomass, forests and oceans, as well as other terrestrial coastal and marine ecosystems.[33]

---

30 Convention on Climate Change, above n 2.

31 Rosenbaum, Kenneth, Dieter Schone, Ali Mekouar, Food and Agriculture Organization of the United Nations (2004), – 'Climate Change and the Forest Sector: Possible National and Subnational Legislation' *FAO Forestry Paper No 144*, 5.

32 Convention on Climate Change, above n 2.

33 Ibid.

Under Article 4(1), all parties must:[34]

- Develop and update inventories of emissions and removals of greenhouse gases (this includes emissions that occur as result of deforestation and removals that occur as a result of forest management activities)[35]
- Develop programs to mitigate climate change, including efforts to address emissions and sinks (which would include forests and their soils)[36]
- Promote technologies that lead to lower greenhouse gas emissions in a number of sectors including the forestry sector[37]
- Promote sustainable management of sinks and reservoirs (forests are again specifically mentioned)[38]
- Prepare to adapt to the impacts of climate and develop appropriate plans for areas that might be subject to flooding, drought or desertification.[39]

Forest-dependent developing countries are also given special recognition under the Convention. Article 4(8)(c) requires parties to consider the impact of the Convention's obligations on developing countries with forested areas. It encourages funding, insurance and technology transfer for these countries. The reasoning underlying this obligation comes from what would be the otherwise unfair burden that forest-dependent developing countries face in addressing concerns related to climate change.

C. Kyoto Protocol

The Kyoto Protocol is the vehicle in which the mitigation obligations created within the Framework Convention are operationalized. The objective of the Kyoto Protocol can be found in Article 3 which requires at its core 'greenhouse gas stabilisation and reduction commitments for industrialised (Annex I) countries meant to add up to a 5% reduction in aggregate greenhouse gas emission compared to 1990 levels'.[40] The Kyoto

---

34 Rosenbaum, above n 31, 5.

35 Convention on Climate Change, above n 2, 4 (1) (a).

36 Ibid. 4 (1) (b).

37 Ibid. 4 (1) (c).

38 Ibid. 4 (1) (d).

39 Ibid. 4 (1) (e).

40 Verheyen, Roda (2005), *Climate Change Damage and International Law: Prevention, Duty and State Responsibility*, Leiden; Boston: Martinus Nijhoff Publishers, 110.

Protocol sets individual legally binding emission reduction targets for 37 industrialized nations and the European Community in Annex B to the agreement. The range of targets varies, with the European community adopting the most stringent reduction of 8 per cent while other countries such as Australia and Iceland were able to increase their emissions from 1990 levels by 8 per cent and 10 per cent respectively.

Article 3 of the Kyoto Protocol provides that the first commitment period of the agreement will operate from 2008–2012. During the first commitment period, parties are required to demonstrate compliance with their individual mitigation pledges as contained in Annex B of the Protocol. The second commitment period of the Kyoto Protocol is determined by a COP decision from the Durban COP Negotiations.[41] This decision determined that the second commitment period of the Kyoto Protocol will commence on I January 2013 and expire either on 31 December 2017 or 31 December 2020 (the expiry date to be determined at the 2012 COP negotiations in Qatar). The objective of the second commitment period is to 'ensure that aggregate emissions of greenhouse gases by parties included in Annex I are reduced by at least 25–40% below 1990 levels by 2020'. Annex I to the agreement sets out the new individual pledges of the Parties to the agreement.

The Protocol creates a framework designed to ensure that a 5 per cent reduction in global emissions is achieved by allocating to each Annex I party a certain number of 'assigned amount units' for the 2008–2012 period. Each assigned amount unit represents a right to emit one tonne of carbon dioxide or equivalent.[42] If parties exceed their allocation, they can create additional assigned amount units by implementing projects under the Clean Development Mechanism or the Joint Implementation Mechanism. This creates a system whereby parties to the Kyoto Protocol are able to meet their targets by implementing domestic policies to reduce emissions, or by investing in projects in other countries that lead to reduced emissions. This flexible approach allows parties to minimize the transaction costs associated with the implementation of activities to reduce emission and create new low-emissions technologies.

Under Article 3 of the Protocol, parties are able to take into account the

---

[41] Conference of the Parties (2011), 'Outcome of the Work of the Ad Hoc Working Group on Further Commitments for Annex I Parties under the Kyoto Protocol at its sixteenth session', available at http://unfccc.int/files/meetings/durban_nov_2011/decisions/application/pdf/awgkp_outcome.pdf (accessed 21 March 2012).

[42] All gas emissions are converted into carbon dioxide measurements.

effect of domestic forest activities within their emission calculations. This allows parties to meet their specified target of greenhouse gas reduction by including reductions that occur as a result of forestry activities. This, therefore, rewards parties that carry out practices to enhance the sustainable management of their forest areas. This section also requires that parties account for: emission changes resulting from deforestation; forest activities that act as sinks of greenhouse gas emissions; and forest activities that act as sources of greenhouse gas emissions.

Article 3(3) of the Protocol specifies the particular forest activities that can be included for accounting purposes. These are defined as afforestation and reforestation. Notably this definition excludes 'deforestation', however more recent COP decisions suggest that deforestation will be an included activity within the second commitment period of the regime.[43] Under Article 3(3), these three activities must be included in parties' reports, and must be reported in a manner that is transparent and verifiable.

Article 3(4) of the Protocol deals with the sink and source emissions arising from land use activities. Actions undertaken in forest estates (for example deforestation) have the potential to cause emissions and as such are considered a source emission activity. Other actions such as sustainable forest management, afforestation, and reforestation are considered to be activities providing a sink for emissions. Parties are not obliged to report upon land use change practices during the first commitment period but they will be required to report upon such activities in the second commitment period. Those parties that voluntarily report on sink and source practices must report upon both sink and source actions, and such parties cannot elect to report upon only the positive actions that have been undertaken. Due to the perceived complexity associated with measuring and reporting emissions from land use practice, most parties decided to opt out of reporting upon these activities. It has been suggested that that the 'opt in or out' process allows parties to select only those activities from which they expect to gain credits, and to ignore other activities through which they would incur debits.[44]

---

[43] United Nations Framework Convention on Climate Change, Decision CMP.7, Land Use, Land Use Change and Forestry (2011), available at http://unfccc.int/files/meetings/durban_nov_2011/decisions/application/pdf/awgkp_lulucf.pdf (accessed 14 March 2012).

[44] Schlamadinger, Bernhard, N. Bird, T. Johns, S. Brown, J. Canadell, L. Ciccarese, M. Dutschke, J. Fiedler, A. Fischlin, P. Fearnside, C. Forner, A. Freibauer, P. Frumhoff, N. Hoehne, M.U.F. Kirschbaum, A. Labat, G. Marland, A. Michaelowa, L. Montanarella, P. Moutinho, D. Murdiyarso, N. Pena, K.

Article 3(7) of the Protocol allows Annex I parties with net emissions in 1990 (from land use change and forestry) to add these emissions to their base year emissions for the determination of their assigned amount.[45] The operation of this clause means that parties that in 1990 had net emissions from land use change and forestry practices receive a benefit (because their actual assigned unit quota will increase). 'Net' means that the sustainable land practices are outweighed by unsustainable practices. Annex I parties that, in 1990, had net emissions from other sectors do not receive the benefit offered by the operation of Article 3(7). This article operates to reward those parties that previously generated net emissions from unsustainable land use.[46] Australia was responsible for inserting this article and it has been suggested that it was self-serving for Australia which, as a result, was able to increase its emission outputs by 8 per cent during the first commitment period.

D. Land Use, Land Use Change, and Forestry Guidelines

Articles 3(3), 3(4) and 3(7) of the Kyoto Protocol are implemented through the application of the LULUCFG.[47] The LULUCFG are consistent with the 1996 Revised IPCC Guidelines for National Greenhouse Gas Inventories,[48] being the authoritative guidelines that the parties have agreed to use for estimation of greenhouse gas emissions and removals. In 2006, the IPCC updated its Guidelines for National Greenhouse Gas Inventories to include a chapter on Agriculture, Forestry and other Land Use (material in this chapter is reflective of the LULUCFG).[49]

Pingoud, Z. Rakonczay, E. Rametsteiner, J. Rock, M.J. Sanz, U.A. Schneider, A. Shvidenko, M. Skutsch, P. Smith, Z. Somogyi, E. Trines, M. Ward, Y. Yamagata (2007), 'A Synopsis of Land Use, Land-Use Change and Forestry (LULUCF) under the Kyoto Protocol and Marrakech Accords', *Environmental Science and Policy*, **10** (4), 271–282, 280.

[45] Fry, Ian (2002), 'Twists and Turns in the Jungle: Exploring the Evolution of Land-Use, Land-Use Change and Forestry Decisions within the Kyoto Protocol', *Review of European Community and International Environmental Law*, **11** (2), 159–168, 161.

[46] Ibid.

[47] Intergovernmental Panel on Climate Change (2003), 'Good Practice Guidelines for Land Use, Land-Use Change and Forestry', available at http://www.ipcc-nggip.iges.or.jp/public/gpglulucf/gpglulucf_files/0_Task1_Cover/Cover_TOC.pdf (accessed 7 February 2012).

[48] Intergovernmental Panel on Climate Change (1996), 'Revised IPCC guidelines for National Greenhouse Gas Inventories', available at http://www.ipcc-nggip.iges.or.jp/public/gl/invs1.html (accessed 7 February 2012).

[49] Intergovernmental Panel on Climate Change (2006), 'IPCC Guidelines of

The LULUCFG clarified the role that land use practices play in carbon mitigation and the manner in which reductions from these activities are to be reported. It has been suggested that the modalities themselves are overly complex and that the name of guidelines is unnecessarily vague.[50] The guidelines cover forest and land management practices that lower greenhouse gas emissions. Each part of the title refers to an activity that achieves this. 'Land use' refers to land practices that effect emission levels; 'land use change' refers to practices where the purpose of land use is changed (and which consequently affects the level of emissions); and 'forestry' refers to activities carried out within forests that act as sinks or emitters of greenhouse gases.

The Marrakesh COP 7 negotiations in 2001 were responsible for moving the dialogue concerning land-based emission reduction activities forward. The decisions made at Marrakesh in 2001 are reflected in the LULUCFG, which were released in 2003. Parties were initially hesitant about including land use, land use change, and forestry practices within emission reduction reporting frameworks. In particular, parties were concerned about the potential for leakage-related issues to arise in forest carbon projects. Leakage occurs when a reduction of emissions occurs within one area while, simultaneously in another area, the activity generating emissions is instead carried out. In the forest context, leakage is a valid concern and can easily occur when deforestation is avoided in one location only to be displaced to a second location. This risk associated with leakage has the ability to undermine all the rules concerning land and forestry management, especially when one considers leakage on an international scale. For example, developed countries may comply with Kyoto reporting requirements making them more reliant on importing products from developing countries not regulated by Kyoto requirements.

Other concerns about land-based emission practices included: disagreement over the definitions of key concepts; concerns about scale; concerns about non-permanence; concerns about uncertainty; concerns

---

National Greenhouse Gas Inventories Online', available at http://www.ipcc-nggip.iges.or.jp/public/2006gl/index.html (accessed 14 December 2011).

50 Benndorf, R., S. Federicib, C. Fornerc, N. Penad, E. Rametsteinere, M.J. Sanzf, Z. Somogyi (2007), 'Including Land Use, Land-Use Change and Forestry in Future Climate Change Agreements: Thinking Outside the Box', *Environmental Science and Policy,* **10** (4), 283–284; Schlamadinger, B., N. Bird, T. Johns, S. Brown, J. Canadell, L. Ciccarese, M. Dutschke, J. Fiedler, A. Fischlin, P. Fearnside, C. Forner, A. Freibauer, P. Frumhoff, N. Hoehne, M.U.F. Kirschbaum, A. Labat, G. Marland, A. Michaelowa, L. Montanarella, P. Moutinho (2007), 'Options for Including Land Use in a Climate Agreement Post 2012: Improving the Kyoto Protocol approach', *Environmental Science and Policy*, **10** (4), 295–305, 297.

about creditability; reporting requirements; harvested wood products; and biodiversity. Each of these issues is analysed below. In addition to this, some commentators have suggested improvements to the land use, land use change, and forestry rules for the second commitment period (also examined below).

Definitions

The first challenge that arose was the attribution of legal definitions to key concepts within Article 3(3) (for example, forests, afforestation, reforestation and deforestation). The development of these definitions proved to be contentious, resulting in a drawn out process. Definitions were crucial because they would in turn determine the amount of land that Annex I countries could bring to account under Article 3(3).[51] Different stakeholders suggested their own definitions based upon formulas that suited their own individual needs.[52] Some of the issues taken into consideration when drafting the definition included: ensuring that the definition did not provide incentives for native forest to be replaced with plantation forest; whether a minimum tree canopy method took into account the wide variety of forests; and whether a minimum area space of forest was required.[53] A special report prepared by the Intergovernmental Panel on Climate Change also explored the issues surrounding the adoption of key definitions.[54]

At the seventh session of the COP in Marrakesh in 2001, the following key definitions were agreed upon. The relevant definitions are located in Decision 11 of the COP 7 Report.

- 'Forest' is a minimum area of land 0.05–1 hectares with tree crown cover (or equivalent stocking level) of more than 10–30 per cent and trees with the potential to reach a minimum height of 2–5 metres in maturity *in situ*. A forest may consist either of closed forest formations where trees of various storeys and undergrowth cover a high

---

[51] Ibid. 5.

[52] For further information on the political motivations see: Fry, Ian (2002), 'Twists and Turns in the Jungle: Exploring the Evolution of Land-Use, Land-Use Change and Forestry Decisions within the Kyoto Protocol', *Review of European Community and International Environmental Law*, **11** (2), 159–68.

[53] Ibid.

[54] Watson, Robert, Ian R. Noble, Bert Bolin, N. H. Ravindranath, David J. Verardo and David J. Dokken, Intergovernmental Panel on Climate Change, *IPCC Special Report Land Use, Land-Use Change and Forestry: Summary for Policymakers* (2000).

proportion of the ground, or open forest. Young natural stands, and all plantations that have yet to reach a crown density of 10–30 per cent or a tree height of 2–5 metres, are included under 'forest', as are areas normally forming part of the forest area that are temporarily unstocked as a result of human intervention (such as harvesting or natural causes) but which are expected to revert to forest.[55]

- 'Afforestation' is the direct human-induced conversion of land that has not been forested for a period of at least 50 years and forested through planting, seeding or the human-induced promotion of natural seed sources.[56]
- 'Reforestation' is the direct human-induced conversion of non-forested land to forested land through planting, seeding or the human-induced promotion of natural seed sources on land that was forested but has been converted to non-forested land. For the first commitment period, reforestation activities will be limited to reforestation occurring on those lands that did not contain forest on 31 December 1989.[57]
- 'Deforestation' is the direct human-induced conversion of forested land to non-forested land.[58]

Additionally at the Marrakesh COP negotiations, the optional Article 3.4 activities were clarified. Article 3(4) refers to land use change, forestry activities and agricultural practices. It was decided that the following activities fitted within Article 3.4:[59]

- Re-vegetation[60]
- Forest Management[61]

---

[55] UNFCCC, *Report of the Conference of the Parties on its Seventh Session, held at Marrakesh from 29 October to 10 November 2001*, 7th sess, UN Doc FCCC/CP/2001/13/Add.1 (21 January 2002), Decision 11, 58.

[56] Ibid.

[57] Ibid.

[58] Ibid.

[59] UNFCCC, *Report of the Conference of the Parties on its Seventh Session, held at Marrakesh from 29 October to 10 November 2001*, 7th sess, UN Doc FCCC/CP/2001/13/Add.1 (21 January 2002), Decision 11, 58.

[60] 'Re-vegetation' is a direct human-induced activity to increase carbon stock on sites through establishment of vegetation that covers a minimum area of 0.05 hectares, and does not meet the definition of afforestation or reforestation.

[61] 'Forest Management' is a system of practices for stewardship and use of forest land aimed at fulfilling the relevant ecological (including biological diversity), economic and social functions of the forest in a sustainable manner.

- Cropland Management[62]
- Grazing Land Management[63]

The inclusion of these four activities allows for more land management activities to be incorporated into accounting practices. It should be noted that forest management activities are capped, so that only 9 mega tonnes of carbon can be accounted for each year from this activity.[64] These four categories could be used to support the implementation of policies aimed at sustainable forest management and sustainable planning and development rules. Article 3(4), therefore, provides an incentive for government to introduce regulatory frameworks compatible with the Kyoto definitions, so that sustainable land management and forest practices can be included in national accounting practices. Forest certification frameworks assign certification status to producers who are following sustainable forest management practices. Therefore, links should be established between forest certification regimes and Kyoto reporting regimes so that these sustainable forest management practices can be included in Kyoto accounting procedures.

The Marrakesh COP negotiations also addressed a number of other concerns associated with land based emission reduction activities. There was a great concern that land and forest management practices were cheaper and quicker to implement than other sectoral initiatives aimed at addressing climate change. Additionally, there was scientific uncertainty as to the amount of carbon sequestered and the duration for which carbon was sequestered in land management and forest practices. Graichen[65] classifies these concerns into four categories: scale; non-permanence; uncertainty; and creditability.

---

62 'Cropland Management' is the system of practices on land on which agricultural crops are grown and on land that is set aside or temporarily not being used for crop production.

63 'Grazing land management' is the system of practices on land used for livestock production aimed at manipulating the amount and type of vegetation and livestock produced.

64 This cap was negotiated at the sixth and seventh conference of the parties' accords. For further detail see: Rosenbaum, Kenneth, Dieter Schone, Ali Mekouar, Food and Agriculture Organization of the United Nations (2004), 'Climate Change and the Forest Sector: Possible National and Subnational Legislation', *FAO Forestry Paper No 144*, 16.

65 Graichen, Patrick (2005), 'Can Forestry Gain from Emissions Trading? Rules Governing Sinks Projects Under the UNFCCC and the EU Emissions Trading System', *Review of the European Community and International Environmental Law*, **14** (1), 11–18, 11–12.

Scale concerns

There was concern about the considerable amount of carbon that is stored in soils and plants. The total value of carbon stored in these sources is thought to be substantial[66] and, consequently, it was feared that forest management and land enhancement activities would dominate emission reduction activities and that other sectoral mitigation approaches (for example, the energy sector) would suffer as a result. It was rightly argued that the development of low-carbon energy sources and industry practices must remain the central focus of the regime. Planting forests is seen as a cheaper mitigation alternative compared with transforming energy sectors and, as such, some proponents did not want to create any incentives that would detract from the development of cleaner technologies.

To overcome these concerns, it was decided to cap or limit the amount of credits that can be generated from this type of activity. Each party would be allocated a forest and land management credit limit. These limits are specified in Decision 11 of the seventh session of the COP.[67] To prevent parties from generating forest sink credits in developing countries, where it is cheaper to establish forest sinks, another cap was introduced. This cap was 1 per cent of the parties' greenhouse gas emissions, measured at the 1990 level. Forestry activities can only be used to offset 1 per cent of the 1990 level.[68] No limit was set for land management activities because it was realized that only a small amount of carbon is stored by these activities and that a cap was, thus, not necessary.

Non-permanence concerns

'Non-permanence' covers the variance of carbon stored in forests, soils, and other vegetation. This variance depends on the amount of carbon sequestered and the duration for which it is stored. Because carbon is stored in these products as a result of natural cycles, the amount and duration of carbon storage is never stable. Rather, the amount of carbon stored continually changes. This pattern of carbon sequestration presents difficulties when seeking to include these quantities within carbon accounting regimes.

The COP 7 decision in response to non-permanence concerns was that, if parties choose to include forest sinks in their accounting practices, they would be liable to account on two fronts: first, they must account for the

---

66 Ibid.

67 UNFCCC, *Land Use, Land-Use Change and Forestry*, Decision 11/CP.7, UN Doc FCCC/CP/2001/13/Add.1 (10 November 2001), available at http://maindb.unfccc.int/library (accessed 16 January 2008).

68 Graichen, above n 65, 11–12.

removal of the carbon as a credit; secondly, they must account for emissions that occur as a result of the natural life cycles of forests, soil, and vegetation processes. These accounting practices take place at the appropriate point in time (that is, when the carbon is stored and when the carbon is released).[69]

The relatively limited scope of the land use, land use change, and forestry activities means that the non-permanence risk associated with these activities is generally small for the first commitment period.[70] If, however, the global community were to adopt a full carbon accounting framework (that is, one that includes all emissions and removals on all managed land in all pools and of all relevant greenhouse gases, without temporal interruption), then the risk associated with non-permanence could significantly impact a country's ability to meet its commitments. On the other hand, the integrity of the atmosphere would be better preserved.[71]

Uncertainty concerns

There is a large degree of scientific uncertainty surrounding the capacity of forests, soils and other vegetation to sequester carbon.[72] There is though, more certainty about forests' ability to store carbon compared with that for soil and vegetation, due to published research findings on the topic.[73] More advanced carbon sequestration methodologies exist for different tree species, and the uncertainty concerns are lessened by their availability. Methodologies that measure soils and other vegetation carbon storage capacity are still in their infancy and, as such, it is difficult to include them

---

69 See at 1(g), where it states that reversal of any removal due to land use, land-use change and forestry activities must be accounted for at the appropriate point in time. UNFCCC, *Land Use, Land-Use Change and Forestry*, Decision 11/CP.7, UN Doc FCCC/CP/2001/13/Add.1 (10 November 2001), available at http://unfccc.int/resource/docs/cop7/13a01.pdf (accessed 6 August 2012)

70 Schlamadinger, Bernhard, N. Bird, T. Johns, S. Brown b, J. Canadell, L. Ciccarese, M. Dutschke, J.Fiedler, A. Fischlin, P. Fearnside, C. Forner, A. Freibauer, P. Frumhoff, N. Hoehne, M.U.F. Kirschbaum, A. Labat, G. Marland, A. Michaelowa, L. Montanarella, P. Moutinho, D. Murdiyarso, N. Pena, K. Pingoud, Z. Rakonczay, E. Rametsteiner, J. Rock, M.J. Sanz, U.A. Schneider, A. Shvidenko, M. Skutsch, P. Smith, Z. Somogyi, E. Trines, M. Ward, Y. Yamagata (2007), 'A Synopsis of Land Use, Land-Use Change and Forestry (LULUCF) under the Kyoto Protocol and Marrakech Accords', *Environmental Science and Policy*, **10** (4), 271–282, 274.

71 Schlamadinger et al, above n 70, 274.

72 Graichen, above n 65, 13.

73 Robert Watson et al. (2000), Intergovernmental Panel on Climate Change, *IPCC Special Report Land Use, Land-Use Change and Forestry: Summary for Policymakers.*

within accounting practices. Without clear scientific consensus on the carbon storage capacity of soils and vegetation, inclusion of these activities within accounting frameworks is problematic.

It was decided that treatment of these issues that were linked to lack of scientific consensus and uncertainty should be based on sound science.[74] This requires the introduction of a thorough monitoring system, which can monitor all greenhouse gas emissions and removals in identifiable areas of land. An independent review team must then review this system.[75] It has been suggested that the best available knowledge at the time be used and that the development of technologies to monitor these practices be encouraged.[76]

Creditability concerns

Creditability concerns relate to the concept of 'additionality'. This concept requires that an activity must be additional for it to be included in accounting frameworks. In relation to forest and land management activities, there was concern that these land and forest management activities were already taking place (the 'business as usual' scenario). Rewarding these practices means that no additional new activity has taken place, which effectively undermines the operation of a carbon accounting regime (because there has been no real change). It has been suggested that allowing land and forest management activities to be included within accounting frameworks opens up the possibility for corruption of the system through 'mere storytelling'. For example, claims might arise stating that a certain forest area was planned to be cleared, but that, in light of the accounting benefit, this clearing was no longer to take place. Determining the truthfulness of these types of claims would introduce a new layer of complexity to the regime.

In relation to the creditibility concerns, it was decided that the 'mere presence of carbon stock should be excluded from accounting practices'.[77]

---

[74] See at 1(a) which provides that the treatment of land use, land use change and forestry activities be based on sound science. UNFCCC, *Land Use, Land-Use Change and Forestry*, Decision 11/CP.7, UN Doc FCCC/CP/2001/13/Add.1 (10 November 2001), available at http://maindb.unfccc.int/library (accessed 16 January 2008) 54–63.

[75] Graichen, above n 65, 12.

[76] Graichen, Patrick (2005), 'Can Forestry Gain from Emissions Trading? Rules Governing Sinks Projects Under the UNFCCC and the EU Emissions Trading System', *Review of the European Community and International Environmental Law*, **14** (1), 11–18, 13.

[77] UNFCCC, *Land Use, Land-Use Change and Forestry*, Decision 11/CP.7, UN Doc FCCC/CP/2001/13/Add.1 (10 November 2001), available at http://maindb.unfccc.int/library (accessed 16 January 2008), 1 (d).

This has the effect of limiting the type of land and forest management activities that can be taken into account. This means that only specified activities – such as afforestation, reforestation, re-vegetation, forest management, cropland management, and grazing land management – can be used for accounting. It also means that not all existing carbon sinks are automatically incorporated into accounting practices. This requires that some 'additional activity' be carried out to account for these practices. It was also decided at this juncture that avoided deforestation should be excluded from accounting practice, due to additionality and evidentiary concerns.[78]

Reporting requirements

The operation of these modalities requires parties to identify 'Kyoto land'. This is land that is being used in accordance with land use, land use change, and forestry guidelines. Kyoto land must be identified in national inventories, which are then used to meet international reporting requirements. This has increased the reporting burden on parties because virtually every piece of organic carbon on specified Kyoto lands has to be accounted for.[79] This is a large undertaking for parties with vast organic carbon storage opportunities. This reporting requirement also assumes that parties have the capacity and technology to monitor organic carbon storage processes.

Harvested wood products

A further issue yet to be fully resolved relates to the storage of carbon that takes place in long-life wood products. Current accounting practices do not take into account the fact that carbon remains stored in harvested wood products. Under current accounting practice, it is assumed that once a tree is cut down the carbon stored in the wood is lost and, as such, must be duly reported as an emission.[80] However, these accounting practices do not equate with scientific understanding in the area which demonstrates that carbon is stored for varying periods in harvested wood products. Within the first commitment period, the carbon stored in harvested wood products is not eligible for inclusion within accounting inventories. Parties in a position to do so are able to voluntarily report on harvested

---

78 Graichen, above n 65, 13.

79 Fry, Ian (2007), 'More Twists, Turns and Stumbles in the Jungle: A Further Exploration of Land Use, Land-Use Change and Forestry Decision within the Kyoto Protocol', *Review of the European Community and International Environmental Law*, **16** (3), 341–355, 342.

80 Ibid. 163.

wood products in their national inventories.[81] Negotiations on the post-2012 period provide a means for possible inclusion of wood products in accounting practices.[82]

Biodiversity concerns

The inclusion of forest activities within the international climate change framework means that forest policy challenges are now being faced by the climate change regime. One of these issues is the compatibility of Kyoto forest practices and biodiversity. Forests contribute to climate change mitigation as terrestrial sinks and assist conservation efforts by acting as pools of biological diversity.[83] Therefore, a potential conflict could arise in the implementation of Kyoto forestry rules. By way of example in relation to afforestation and reforestation rules, a landowner will receive an economic incentive to plant species of trees that are high absorbers of carbon dioxide. These high carbon absorption tree species may, however, not lead to the highest level of biological diversity possible on the relevant parcel of land.

These concerns were addressed at the Marrakesh meeting of the Conference of the Parties, where it was decided that activities under Article 3(3) and 3(4) must contribute to conservation and biodiversity and to the sustainable use of natural resources. Parties must report on the administrative and legislative procedures that ensure this. It has been suggested that the reporting requirements do not go far enough, because they only require that parties provide information on national laws, and do not require information on tangible results for the conservation of biodiversity.[84]

---

81 Subsidiary Body for Scientific and Technological Advice, UNFCCC, *Intergovernmental Panel on Climate Change Guidelines for National Greenhouse Gas Inventories*, 26th Session, Agenda Item 7(a); Methodical issues under the Convention, UN Doc FCCC/SBSTA/2007/L.5, GE.07-70214 (15 May 2007).

82 Grêt-Ragamey, Ardrienne, Eugene Hendrick, Sebastian Hetsch, Kim Pingoud, Sebastian Rüter (2008), *Challenges and Opportunities of Accounting for Harvested Wood Products*, Swiss Federal Office for the Environment, UNECE/FAO and MCPFE.

83 Jacquemont, Frederic, Alejandro Caparros (2002), 'The Convention on Biological Diversity and Climate Change Convention 10 Years after Rio: Towards a Synergy of the Two Regimes?', *Review of the European Community and International Environmental Law*, **11** (2), 169–180.

84 Jacquemont, above n 83, 172. Also see Sagemuller, Imke (2006), 'Forest Sinks under the United Nations Framework Convention on Climate Change and the Kyoto Protocol: Opportunity or Risk for Biodiversity?', *Columbia Journal of Environmental Law*, **31** (2), 189–242.

Future land use, land use change, and forestry guidelines

The current regulatory framework accounting for land use, land use change and forestry has proved to be complex to understand and arduous to implement. This is largely due to the difficulty associated with creating a framework based upon incomplete knowledge and uncertainty (about terrestrial carbon storage and release). The complexity of the regime is also related to parties promoting their own self-interests, as opposed to incorporating best practice guidelines (such as meeting additionality requirements).[85] The following weaknesses can be identified in the current system[86] [87]

- The biased application of Article 3(7). This article allows parties – for whom land use change and forestry practices before 1990 led to a net source of emissions – to include emissions from deforestation in their actual assigned unit, thus raising their quota in the first commitment period. Not all parties meet this requirement, and so deforestation is not managed consistently in relation to determining actual assigned unit allowances. It is suggested that Article 3(7) be amended so that all parties' deforestation rates are inclusive within their actual assigned unit quotas.
- Creating a cap limit for the amount of forest management activities that can be included within accounting inventories acts as a disincentive to improved forest management. Improved forest management should be encouraged and caps should be reassessed or removed.
- Considerations related to harvested wood products should be included within the second commitment framework. The 2006 IPCC Guidelines for National Green House Gas (GHG) Inventories offer

---

[85] Fry, above n 79.

[86] A number of reasons have been given to support amendment of the current regime, such as the complexity of the rules and the time limits applied to the rules (that is, they are only valid for first commitment period). It has also been suggested that sectoral approaches may be an option for the second commitment period. Sectoral approaches focus on addressing a sector that it is a high emitter and creating a framework to address this particular sector (for example, the forestry sector or energy sector). For further information on these options, see Benndorf, R., S. Federicib, C. Fornerc, N. Penad, E. Rametsteinere, M.J. Sanzf, Z. Somogyi (2007), 'Including Land Use, Land Use Change and Forestry in Future Climate Change Agreements: Thinking Outside the Box', *Environmental Science and Policy*, **10** (4), 283–284.

[87] Schlamadinger et al., above n 70, 297.

several potential reporting approaches to including harvested wood products within reporting guidelines.[88]

The 2011 UNFCCC negotiations held in Durban clarified the role of forest sink and source activities for the second commitment period. Decision CMP.7, 'Land Use, Land Use Change and Forestry'[89] suggests that parties will have mandatory obligations to report upon forest sink and source GHG emissions in the second commitment period. This is a significant change from the first commitment, which allowed parties to opt in or opt out of reporting on sink and source activities related to land use.

Decision CMP.7 clarifes the types of forest activities that can be included within existing implementation mechanisms. Noteworthy changes include

- Deforestation activities to be included within the ambit of Article 3 and 4 of the Kyoto Protocol. This means that countries will be able to account for avoided deforestation activities undertaken domestically.[90]
- Carbon stored in harvested wood products to be recognized in the second commitment period.[91]

E. Clean Development Mechanism and Forestry

Flexibility mechanisms

The Kyoto Protocol creates three flexibility mechanisms that can be used by the parties to meet their own international commitments. These mechanisms are designed to assist Annex I parties to meet their emissions targets at least cost. These three mechanisms are: the clean development mechanism; joint implementation; and emissions trading. The Clean Development Mechanism (CDM) has developed special rules for forestry projects.[92] CDM allows developed countries (Annex I parties) to invest

---

88 Ibid.

89 CMP is the acronym used to designate decisions emerging from the Conference of the Parties serving as a meeting of the Parties to the Kyoto Protocol.

90 United Nations Framework Convention on Climate Change, Decision CMP.7 (2011), 'Land Use, Land Use Change and Forestry', available at http://unfccc.int/files/meetings/durban_nov_2011/decisions/application/pdf/awgkp_lulucf.pdf (accessed 14 March 2012), Art. 2.

91 Ibid. Art. 16.

92 United Nations Framework Convention on Climate Change, *Decision*

in emission reduction programs in developing countries and record this activity in their national emissions inventory. This is beneficial, as it reduces transaction costs for developed countries and lowers emissions in developing countries.

CDM sink politics

The inclusion of forest sinks within the operational modalities of the CDM proved to be the most controversial[93] of all forest negotiations within the regime. Opponents of the inclusion of forest sinks within the CDM argued that it created a loophole which developed countries could use to postpone the implementation of domestic policies and measures aimed at reducing greenhouse gas emissions. This would, therefore, undermine the development of competitive climate-friendly technologies.[94] Alliances that sought to include forest activities in the mechanism argued that investment in the forestry sector was a crucial step in developing measures to address global deforestation. They also argued that investment in forest projects provided social benefits to local communities by providing diversified income streams and other socio-economic benefits.[95] Ultimately, forest practices were incorporated within the CDM although the contentious nature of the inclusion led to the creation of cumbersome operational modalities.[96]

There is a range of legal issues related to implementation of forest projects within the CDM

13/CP.10, Incorporation of the Modalities and Procedures for Afforestation and Reforestation Project Activities under the Clean Development Mechanism Guidelines under Article 7 and 8 of the Kyoto Protocol, UN Doc FCCC/CP/2004/10/Add.2 (19 April 2005).

93 The USA and developing countries with high tropical forest cover formed an alliance to argue for the inclusion of sinks. The European Union, India, Indonesia and the Association of Small Island States formed an alliance seeking a more responsible approach to forest sink inclusion. For further information on these alliances, see: Boyd, Emily, Esteve Corbera and Manuel Estrada (2008), 'UNFCCC Negotiations (pre Kyoto to COP 9): What the Process Says about the Politics of CDM-Sinks', *International Environmental Agreements*, **8** (2), 95–108, 106.

94 Ibid.

95 Ibid.

96 A study examined scientific knowledge and international policy development in the CDM sink arena. It found that scientific knowledge is not used to inform policy development: political motivations guide policy development and science plays catch up trying to fill in the scientific uncertainties created by the policy development. See: Palm, Matilda, Madelene Ostwald and John Reilly (2008), 'Land Use and Forestry Based CDM in Scientific Peer-reviewed Literature Pre and Post COP 9 in Milan', *International Environmental Agreements*, **8** (3), 249–275.

- Defining eligible forest practices
- Creating a 'cap' that regulates the amount of sink credits generated
- Creating unique certified emission reduction credits for forest practices
- Creating operational modalities for CDM forest practices
- Defining land and property rights in carbon
- Providing market demand and transaction costs associated with sinks in the CDM
- Developing capacity to implement forest practices in developing countries.

Eligible forest activities

To generate certified emission reduction credits from forest activities, the forest activity must fall into the afforestation or reforestation category. This means that the conservation of existing forests, such as through REDD practices, are not eligible activities under the CDM. Activities that fall within the category of afforestation or reforestation include: agro-forestry; monoculture or mixed industrial plantations; forest landscape restoration projects; community forest projects; and other afforestation or reforestation projects that focus on timber production, biomass energy or watershed management.[97] These activities have been described as[98]

- Agro-forestry – systems of mixing agricultural or horticultural crops and livestock with woody perennials
- Monoculture or mixed industrial plantations – forest areas planted with selected tree species
- Forest landscape restoration – this helps to restore many of the goods and services that enhance ecological integrity and provide tangible benefits to local people living in degraded or deforested landscapes. Activities include natural regeneration, tree planting and agro-forestry
- Community forestry – this emphasizes the social dimension of forestry and its contribution to sustainable livelihoods for rural people
- Biomass energy – projects which generate energy using organic matter that is available on a renewable basis. Biomass can be sourced from agricultural crops, timber and organic waste.

---

[97] Manguiat, Maria Socorro, Z. Verheyen, Roda Mackensen and Gerald Jens Scholz (eds) (2005), *Legal Aspects in the Implementation of CDM Forestry Projects*, Gland, Switzerland: IUCN in collaboration with GTZ and UNEP, 7–8.

[98] Ibid.

Cap limit on forest sinks

During the first commitment period, parties can claim 1 per cent of their 1990 base year emission rate using credits generated from sinks activities under the mechanism. Parties can claim 1 per cent for each year of the first commitment period, meaning that a total of 5 per cent clean development mechanism credits for sinks can be included in accounting inventories.[99] This cap was introduced to placate parties concerned that sink credits would flood the global emission market.

Certified emission reduction sink credits

To recognize the non-permanent nature of carbon storage in terrestrial carbon stocks, new categories of certified emission reduction credits were devised. Two categories of credits can be generated from afforestation and reforestation activities:[100]

- temporary certified emission reduction credits; and
- long-term certified emission reduction credits.

Temporary certified emission reduction credits have to be replaced every five years. Long-term certified emission reduction credits have a lifespan of 60 years but must still be verified every five years. The limited life span of afforestation and reforestation credits increases their transaction costs. Temporary credits have to be replaced at the end of their five-year lifespan. Forest credits are seen as a risky investment as replacing the credits at the end of their lifespan may cost the investor significantly more. This, therefore, acts as a disincentive to investing in sink credits under the clean development mechanism.

While the long-term credits have a longer lifespan, they still need to be replaced in time and this may mean that, when the purchase of the replacement credits takes place, their price may be higher. Additionally, the verification process every five years will also involve transaction costs. Comparatively, credits generated from energy activities do not suffer from these issues about impermanence and, as such, may require higher up-front investment. However, overall, they may be viewed as a preferable way to meet emission reduction targets.

---

99 This cap came from negotiations at the 6th and 7th Conference of the Parties ; see Rosenbaum, above n 31, 17.

100 Ibid. 17–8.

Operational modalities

Certified emission reduction credits are obtained through a complex and lengthy process. Critics of the process suggest that the process inhibits the number of projects in the pipeline and, importantly, reduces the number of projects that are reducing greenhouse gas emissions.[101] To obtain credits under the clean development mechanism, a number of steps must be taken[102]

- Background research to ensure suitable selection of the project site – a number of issues related to land tenure should be considered at this stage: what is the legal status of the land? Who uses or occupies the land? Who will act on behalf of the land? Does the law require the present use of the land to be part of the project design? If land use change is to occur, is approval and re-zoning required? What other permits or clearances need to be obtained to carry out the project?
- Submission or product design documentation
- Selection of baseline methodology – the methodology is used to determine the amount of emissions removed by the project. Parties may create their own baseline methodology (which must be approved) or use a methodology that already exists. Methodologies already exist for grassland to forest, cropland to forest, wetland to forest, and settlement to forest
- Validation of project by the Conference of the Parties to the Kyoto Protocol
- Registration of a validation report submitted by the Designated Operational Entity – this is an independent accredited body that assesses the amount of emissions removed by the project
- Monitoring of the project area to ensure compliance with project design documentation
- Verification of the emission reduction – parties nominate the first verification date and re-verification then takes place every five years
- Certification – the Designated Operational Entity certifies that a project activity achieved the net anthropogenic greenhouse gas removals
- Issue of certified emission reduction credits – where the amount of credits issued is based on the calculation reached at certification.

---

[101] Rosenbaum, above n 31.
[102] Manguiat, above n 97, 9–11.

Property rights

The rights attached to certified emission reduction credits are generally protected by contract law.[103] The country that sponsors the project does not usually obtain ownership of the land where the project is carried out. Generally, the contract specifies the obligations associated with the land during the contract term. The contracts normally contain specific clauses to ensure compliance with the objective of carbon sequestration, and usually also address: plantation maintenance; replanting of seedlings that die off; harvest modalities; issues of risk (fire and pests); and the financial arrangements agreed between the parties.

In host countries where a land registry is in operation, the certified emission reduction credits may be registrable with the land registrar.[104] This provides extra security for the sponsoring country should the project land be sold and transferred to another party. Alternatively, where the legal system does not recognize the certified emission reduction as a registrable interest, the contract may contain a clause that makes the certified emission reduction credit a separate alienable right. This means that the owner of the land can transfer the project land without transferring the rights associated with the certified emission reduction credits.[105]

Capacity development

Because the clean development mechanism is implemented in developing countries, the success of the project will be dependent upon the host country's regulatory capacity. A strong regulatory capacity is essential for ensuring that the integrity of the project is maintained. The table below identifies six pre-conditions that are required for the implementation of the clean development mechanism.[106]

---

[103] Manguiat, above n 97, 20.

[104] For example, in Ghana, the right to certified emission reduction credits is potentially recognizable as a right or interest in or over land capable of being registered in the register of the Land Title Registry. See Manguiat, above n 97, 50.

[105] In the Philippines, it may be possible to classify a certified emission reduction as a forest resource, and the right to certified emission reductions as immovable property. As a forest resource, certified emission reductions could be transferred not only to Philippine nationals, but to foreigners as well and, as an immovable property, the interest in the right could be more securely protected than if the certified emission reductions were treated as movable property. See Manguiat, above n 97, 50.

[106] Minany, Peter, Hans Bressers, Margaret Skutsh and Michael McCall (2007), 'National Forest Policy as a Platform for Biosphere Carbon Management: The Case of Community Forestry in Cameroon', *Environmental Science and Policy*, **10** (3), 204–218, 207.

Table 6.1 Pre-conditions for implementing the clean development mechanism

CDM Requirements	Compatibility question or determinant
Institutional Development	What is the nature of the designated national authority? How do current institutions support CDM projects?
Eligibility	What is the country definition of a forest? What is the country definition of a low-income community or individual?
Additionality	How are the land and resource tenure rights implemented? How does forest policy cater for forest resource management risk?
Acceptability	What are the sustainable development criteria and indicators for the host country? What laws regulate the use of potentially invasive species and genetically modified organisms in the forestry sector?
Externalities	How are environmental and social impact assessment regulation applied in resource management?
Certification	What procedural regulation exists for CDM project approval?

F. Reduced Emissions from Deforestation and Degradation

As described above, forest activities under the Kyoto Protocol are restricted to afforestation and reforestation. Throughout the Kyoto forest negotiations, issues related to the inclusion of 'avoided deforestation' remained unresolved.[107] Initially, some proponents wanted avoided deforestation included within the clean development mechanism forest activities. This notion was rejected due to the existence of a number of scientific uncertainties and methodological gaps. The purpose of an avoided deforestation mechanism is to provide incentives to developing countries to reduce the level of deforestation occurring at unsustainable levels. This mechanism would be prohibitive because it restricts the sovereign use of parties' land. In some instances, parties may have to improve the remaining forest estate and this will attract positive obligations of management.

Different terminology is used to describe REDD activities. This chapter will use the term REDD+ as this is the term most commonly used when

[107] Fry, above n45.

discussing the initiative. The scope of a legally binding instrument for REDD could take any of the following forms

- RED: Reducing emissions from (gross) deforestation. this option only includes changes from 'forest' to 'non-forest' land cover types[108]
- REDD: as above, plus forest degradation, or shifts to lower carbon-stock densities within the forest[109]
- REDD+: as above, plus the term includes afforestation, poverty alleviation, biodiversity conservation, and improved forest governance[110]
- REDD++: as above, but it also includes emissions from other land conversion (for example agriculture).[111]

Development of policy

REDD+ negotiations are progressing within the climate change regime to develop a legally binding instrument that defines REDD+ targets, policies and methodologies.[112] Outside of the climate change regime, REDD+ activities are being implemented by REDD+ demonstrating bodies such as the UN-REDD Program and the Forest Carbon Partnership Facility. REDD+ policy has evolved over a number of COP negotiations. The first time that REDD+ reached the COP negotiation platform was in 2005 at the 11th Meeting of the COP in Montreal when Costa Rica and Papua New Guinea[113] submitted a proposal entitled 'Reducing Emissions from Deforestation in Developing Countries: Approaches to Stimulate Action'. The submission made a compelling case why emissions from deforestation

---

108 Minany, Peter Akong, Stefan Jungcurt, Vanessa Meadu, Deborah Murphy (2009), 'THE REDD negotiations: Moving into Copenhagen', International Institute for Sustainable Development, available at http://www.iisd.org/pdf/2009/redd_negotiations.pdf (accessed 14 December 2011).

109 Ibid.

110 Blom, Benjamin, Terry Sunderland and Daniel Murdiyarso (2010), 'Getting REDD to Work Locally: Lessons Learned from Integrated Conservation and Development Projects', *Environmental Science and Policy*, **13** (2), 164 –172, 165.

111 Ibid.

112 For a concise summary of REDD+ negotiations, see: Parker, C., A. Mitchell, M. Trivedi, N. Mardas and K. Sosis (2009), 'The Little REDD+ Book: An Updated Guide to Governmental and Non-governmental Proposals for Reducing Emission from Deforestation and Degradation', available at http://unfccc.int/files/methods_science/redd/application/pdf/the_little_redd_book_dec_08.pdf (accessed 14 December 2011).

113 Supported by eight other countries: Bolivia, the Central African Republic, Chile, the Republic of Congo, the Democratic Republic of Congo, the Dominican Republic and Nicaragua.

should be included within climate regulation and demonstrated that the climate change regime had authority to create policy on deforestation under the convention. The proposal suggested the creation of an optional protocol to the UNFCCC dealing with deforestation or, in the alternative, expanding the definition of forest activities under the LULUCFG. It was suggested that the methodological issues preventing the development of REDD policy be referred to the Subsidiary Body of Scientific and Technical Advice (SBSTA)[114] for further investigation. This proposal received broad support and created the impetus for the development of an international policy within the climate change regime to reduce emissions from deforestation – referred to at this point in time as RED.

COP 13 Bali 2007

The SBSTA provided reports on the methodological issues associated with recognizing deforestation activities at a number of SPSTA sessions: the SBSTA 25th session (Nairobi, 2006); 26th session (Bonn, 2007); and at the 27th session (Bali, 2007). These SBSTA reports led to a decision at COP 13 in Bali in 2007 to formally include RED on the policy agenda. The COP decision encouraged parties to undertake RED demonstration activities, leading to formal approval for the pilot RED projects that started to take place across the globe[115] and multiple requests for the mobilization of finance to support efforts to address deforestation. The COP decision also provided some guidance on how such activities were to occur

- Demonstration activities to occur with the approval of the host party
- Estimates of reductions or increases in emissions should be founded on results-based, demonstrable, transparent and verifiable data, and estimated consistently over time
- Parties are encouraged to use the LULUCFG as the basis for estimating and monitoring emissions
- Emission reduction from national demonstration activities should be assessed on the basis of national emissions from deforestation and forest degradation

---

114 This body is established under Article 9 of the UNFCCC. The purpose of the SBSTA is to provide the COP with timely information and advice on scientific and technological matters relating to the convention. It is comprised of government representatives who are competent in their relevant field of expertise.

115 United Nations Framework Convention on Climate Change, *Decision 2/CP.13 Reducing Emissions from Deforestation in Developing Countries: Approaches to Stimulate Action*, UN Doc FCCC/CP/2007/6/Add.1 (14 March 2008).

- Sub-national demonstration activities should be assessed within the boundary used for demonstration and for associated displacement of emissions
- Reductions in emissions or increases resulting from the demonstration activity should be based on historical emissions, taking into account national circumstances
- Sub-national approaches, where applied, should constitute a step towards the development of national approaches, reference levels and estimates
- Demonstration activities should be consistent with sustainable forest management, noting *inter alia*, the relevant provisions of the UNFF, the UNCCD and the CBD
- Experiences in implementing activities should be reported and made available through the Web platform
- Reporting on demonstration activities should include a description of the activities and their effectiveness, and may include other information
- Independent expert review is encouraged.

There has been much contention over the development of REDD+ policy and the consideration of indigenous groups and forest dwellers' community interests in forest estates where such activities are taking place.[116] The host country should not provide approval for such activities unless there is free prior, and informed consent of the group whose forest area is to be part of REDD+ activities.[117] There is a moral obligation on the part of the global community to design REDD+ in a way that is equitable for indigenous, poor and forest-dwelling communities. The objective of a funded forest program should be to provide local communities and governments with the support, technology and financial assistance necessary to sustainably manage their natural environment and to improve their own livelihoods thorough the sale of carbon and other ecosystem products. In order for such projects to be considered equitable it is essential to ensure that carbon payments do in fact flow through to local communities and farmers and that this process is not impeded by corrupt governance practices.[118]

---

[116] Griffiths, Tom, Forest People Programme (2008), *Seeing 'REDD'? Forests, Climate Change Mitigation and the Rights of Indigenous Peoples and Local Communities*, 27.

[117] Hiraldo, Rocío and Thomas Tanner (2011), 'Forest Voices: Competing Narratives over REDD+', *IDS Bulletin*, **42** (3), 42–51.

[118] Crittenden, J. and Wilder, M. (2008), 'Bringing the Forest to Market:

Principles two to seven provide some guidance on the methodological issues associated with forest activities. It is interesting that a distinction is drawn between national and sub-national forest activities and the means by which displacement of emissions is to be dealt with in either of these instances. Most REDD+ demonstration activities would currently fall within the sub-national category, which means that those in charge of implementing the pilot studies must be careful to ensure that sustainable forest management in the project area is not leading to forest degradation or deforestation in another region. Principle eight could be seen as implying that REDD+ activities are not to be undertaken on the basis of managing the forest carbon value alone and that all other values and services associated with the forest area must also be recognized. Such an interpretation is made on the basis of the inclusion of the term 'sustainable forest management' and reference to other international instruments concerned with the regulation and protection of ecological and social forest values.

COP 15 Copenhagen 2009

The Copenhagen negotiations were considered to be a failure by many but one of the positive stories to emerge from them was a decision by the COP concerning RED. This decision was responsible for changing the acronym of policy in this area from RED to REDD, which has the effect of including activities that prevent forest degradation within the scope of the program. Decision 4/CP.15, entitled 'Methodological Guidance for Activities Relating to Reducing Emissions from Deforestation and Forest Degradation and the role of Conservation, Sustainable Management of Forests and Enhancement of Forest Carbon Stocks in Developing Countries', requested developing country parties to report upon the following issues:

- to identify the drivers of deforestation resulting in emissions and also the means to address these drivers; and
- to identify activities within the country that result in reduced emissions and increased removals and stabilization of forest carbon stocks.

Developing countries were also requested to develop a transparent national forest monitoring system that used a combination of remote sensing and ground-based forest carbon inventory approaches. It is not directly recog-

---

Structuring Avoided Deforestation Projects', *The Carbon & Climate Law Review*, **2** (3), 273–280, 280.

nized within the text that developing countries parties may lack technical and other capacity to carry out this assessment, but despite it not being expressly called for, a number of international bodies with an interest in REDD+ have stepped forward to assist countries in meeting this requirement (referred to as demonstrating institutions).[119] The decision also calls for increased coordination in policy development from all international organizations, non-governmental organizations and other stakeholders working in these areas to prevent duplication and enhance the synergy of their efforts. In particular, the decision encourages the effective engagement of indigenous people and local communities in monitoring and reporting, but this clearly falls short of ensuring that indigenous people and local communities have consented to and benefit from the project.

COP 16 Cancun 2010

The Cancun negotiations provided clarification on the nature and processes required for REDD+ activities. The Cancun COP report provides in paragraph 70 for the types of REDD+ activities that are encouraged. While paragraph 71 requests developing countries to advance systems and capacities required to implement these activities. Annex I of the agreement sets out a number of REDD+ safeguards.[120] The REDD+ activities encouraged under paragraph 70 are: reducing emissions from deforestation; reducing emissions from forest degradation; conservation of carbon stocks; sustainable management of forests; and the enhancement of forest carbon stocks. The types of REDD+ activities are therefore much broader then the forest activities provided for under the LULUCFG and the CDM forest guidelines.

Paragraph 71 requests developing country parties to establish (in accordance with national circumstances) the following systems and capacities: a national strategy plan; a national forest emission level or forest reference level; a robust and transparent national forest monitoring system; and a system of safeguards for respecting sovereignty. The activities described under paragraph 71 are the types of activities being funded by REDD+ demonstration bodies (discussed in more detail below) in order to prepare developing countries for participation in the REDD+ regime. The Durban COP negotiations stated that forest reference emission levels and forest reference levels should be expressed in tonnes of carbon dioxide

---

119 See Potvin, Catherine and A. Bovarnick (2008), 'Reducing Emissions from Deforestation and Forest Degradation in Developing Countries: Key Actors, Negotiations and Actions', *Carbon and Climate Law Review*, (3), 264–273, 268.

120 Report of the Conference of the Parties on its sixteenth session, FCCC/CP/2010/7/Add.1.

equivalent per year, as a benchmark for assessing each country's performance in implementing REDD+ activities from paragraph 70.[121]

The Annex I safeguards build on the guidance provided at Bali and require the following conditions to be addressed in REDD+ projects

1. The activities referred to in paragraph 70 of this decision should
   a. Contribute to the achievement of the objectives set out in Article 2 of the Convention
   b. Contribute to the fulfilment of the commitments set out in Article 4, paragraph 3, of the Convention
   c. Be country-driven and consider the options available to Parties
   d. Be consistent with the objective of environmental integrity and take into account the multiple functions of forests and other ecosystems
   e. Be undertaken in accordance with national development priorities, objectives and circumstances and capabilities, and respects sovereignty
   f. Be consistent with Parties' national sustainable development needs and goals
   g. Be implemented in the context of sustainable development and reducing poverty, while responding to climate change
   h. Be consistent with the adaptation needs of the country
   i. Be supported by adequate and predictable financial and technological support, including support for capacity building
   j. Be results based
   k. Promote sustainable management of forests.
2. When undertaking the activities referred to in paragraph 70 of this decision, the following safeguards should be promoted and supported
   a. That actions complement or are consistent with the objectives of national forest programs and relevant international conventions and agreements
   b. Transparent and effective national forest governance structures, taking into account national legislation and sovereignty
   c. Respect for the knowledge and rights of indigenous peoples and members of local communities, by taking into account relevant

---

[121] Conference of the Parties (2011), 'Draft Decision -/CP.17, Guidance on Systems for Providing Information on how Safeguards are Addressed and Respected Modalities Relating to Forest Reference Emission Levels and Forest Reference Levels as Referred to in Decision 1/CP.16', available at http://unfccc.int/files/meetings/durban_nov_2011/decisions/application/pdf/cop17_safeguards.pdf (accessed 21 March 2012), para 7.

international obligations, national circumstances and laws, and noting that the United Nations General Assembly has adopted the United Nations Declaration on the Rights of Indigenous Peoples

d. The full and effective participation of relevant stakeholders, in particular indigenous peoples and local communities, in the actions referred to in paragraphs 70 and 72 of this decision
e. That actions are consistent with the conservation of natural forests and biological diversity, ensuring that the actions referred to in paragraph 70 of this decision are not used for the conversion of natural forests, but are instead used to incentivize the protection and conservation of natural forests and their ecosystem services, and to enhance other social and environmental benefits
f. Actions to address the risks of reversals
g. Actions to reduce displacement of emissions.

Regulatory framework challenges

The inclusion of 'avoided deforestation' has remained unresolved due to the existence of party interests. Advocates for avoided deforestation base their argument on the significant contribution that deforestation plays in global emission levels (estimated to be around 10–25 per cent).[122] Parties in favour of avoided deforestation use these estimates to demonstrate the need for incentives to reward parties that reduce their level of deforestation. Opponents of the 'avoided deforestation' idea acknowledge the significant contribution that deforestation plays in global emissions levels, but remain unconvinced that a sound scientific and regulatory framework can be created to deal with the complex nature of issues raised by 'avoided deforestation' negotiations. These issues include that of permanence, measurement, additionality, undermining the carbon market, sovereignty, and leakage.[123] To include avoided deforestation within existing Kyoto markets, the six issues above must be resolved.

As with all Kyoto forest negotiations, the issue of permanence arises. In relation to avoided deforestation and degradation, permanence concerns

---

[122] The Stern Review estimates 18 per cent. See Stern, N. (2006), 'Stern Review on the Economics of Climate Change', available at http://www.occ.gov.uk/activities/stern.htm (accessed 26 November 2008).

[123] Fry, above n45. Also see Skutsch, M., N. Bird, E. Trines, M. Dutschke, P. Frumhoff, B.H.J. de Jong, P. van Laake, O. Masera and D. Murdiyarso (2007), 'Clearing the Way for Reducing Emissions from Tropical Deforestation', *Environmental Science and Policy*, **10** (4), 322–334, 329–30.

are warranted.[124] This is because ensuring the existence of forest areas in perpetuity has proven to be challenging in developing countries due to a wide range of factors related to the underlying causes of deforestation and degradation. Countries may over-estimate their ability to control deforestation, and history has indicated that deforestation is a complex problem to address. There are some means that can be used to account for non-permanence; these include various insurance options, renewal or temporary crediting, and banking carbon credits as a risk buffer.[125]

To measure changes in forest stocks, a baseline must be identified. Establishing a forest baseline in developing countries requires the capacity to measure and monitor forest stocks. Countries such as Brazil and India have strong global information systems and remote sensing capacity but in most developing countries this type of capacity is unusual.[126] Appropriate technology transfer and development assistance will be required for remote sensing and ground-based accounting practices to take place in accordance with sound science and related guidelines.[127]

Concerns about project additionality for avoided deforestation and degradation are also warranted. A framework will need to be devised that avoids giving credits to projects that would have occurred in any event.[128] 'Avoided deforestation' projects must be able to demonstrate that they are contributing to long-term emission reduction schemes.

As with the inclusion of other forest activities under Kyoto, it is once again suggested that 'avoided deforestation' may act to undermine the carbon market.[129] This will occur in circumstances where forest credits are cheaper than other sectoral credits and, because of this cost advantage, more investment is made in forest projects. However this would only occur if REDD+ credits were recognized by the Kyoto Protocol.

There is a possibility that any Kyoto project implemented outside national boundaries could involve state sovereignty concerns.[130] REDD+

---

124 Skutsch, above n 123, 329.

125 Fry, above n 45, 172.

126 Skutsch, above n 123, 330.

127 The Australian Government is supportive of market-based approaches aimed at addressing deforestation in developing countries. Support has been demonstrated through the provision of funds to carry out pilot projects and funds for research into policy and science development. The name of the project is the International Forest Carbon Initiative. For further information see http://www.climatechange.gov.au/international/publications/fs-ifci.html (accessed on 27 November 2008.

128 Fry, above n 45, 177.

129 Fry, above n 45, 172.

130 Ibid., 177.

will require parties to adopt new forest management practices; and this will (to some extent) erode state sovereignty. In relation to forest reserves, the rights and responsibilities associated with this concept have proved to be contentious during all forest negotiations (but especially within the United Nations Forum on Forestry platform). However, with the introduction of markets for environmental services, state sovereignty concerns may not be as dominant as in previous international regimes. This is because parties will be economically rewarded for infringements of their sovereignty, and involvement within these markets will be optional.

The concept of 'leakage', perhaps, presents the most formidable challenge for avoided deforestation regimes. Leakage occurs when mitigation actions in one area result, directly or indirectly, in emission increases in another area.[131] In the avoided deforestation context, this is a real threat as deforestation may be displaced, and rewarding the initial avoided deforestation is pointless if the deforestation occurs in another location. Certain developing nations may prove to have superior capacity conditions for implementing Kyoto forest projects and, as such, deforestation activities may then be transferred to countries with weaker capacity – resulting in no net global gain in emission reductions.[132] Without adequate safeguards within the international timber market requiring the provision of sustainably grown timber, global demand for wood products will contribute to leakage concerns.

Two measures have been suggested to address leakage concerns.[133] First, there are demand-side measures that could be introduced which create obligations for importers of timber products. This obligation would require that timber products be sustainably harvested. Existing schemes could be used to enforce this obligation (such as the certification programs discussed in Chapter 8, and the FLEGT initiative discussed in Chapter 7). Secondly, there could be development of sub-national leakage frameworks that measure forest leakage within the national context. This would require the introduction of a baseline and monitoring system. However, the introduction of a sub-national leakage framework will not prevent leakage occurring outside national boundaries. To address inter-state leakage, a global baseline and monitoring system would be needed.[134]

---

131 Ibid., 173, and Jung, above n 145, 206.

132 Combes Motel, P., R. Pirard, and J.L. Combes (2008), 'A Methodology to Estimate Impacts of Domestic Policies on Deforestation: Compensated Successful Efforts for "Avoided Deforestation" (REDD)', *Ecological Economics*, **68** (3), 680–691.

133 Fry, above n45, 175.

134 Saunders, Jade, Hohannes Ebeling and Ruth Nussbaum, ProForest (2008),

Possible REDD+ governance models

In order for REDD+ credits to be included within emission reduction accounting frameworks, accounting rules will need to be established. The Centre for International Sustainable Development Law has suggested the creation of a carbon reservoir mechanism.[135] The objective of this mechanism would be to financially reward any activity that permanently protects parts of the carbon stock held in forests. The model they suggest involves a 'cap and trade' type approach. Under this system, parties are allocated credits based on the level of tropical forest and carbon storage capacity. Parties can then use these credits to meet their own international obligations, or sell these credits to other parties that will then use them to meet their international obligations. This type of system creates an incentive to permanently protect forest area by creating income through the sale of credits.[136]

The proposals by Papua New Guinea and Brazil seek to include REDD initiatives through a baseline and credit scheme. The proposal by Papua New Guinea[137] requires the establishment of a baseline rate of deforestation (converted into carbon emissions) and the subsequent negotiation of a carbon credit for the amount it reduces deforestation rates, which can then be sold on the international carbon market.[138] The proposal by Brazil is similar, except that it requires compensation to be paid to countries for 'avoided deforestation'. Under the Brazilian proposal, tropical forest countries that reduce their emissions, compared to a baseline, would receive compensation from developed country parties that is based on the average value of carbon on the carbon market. This type of modelling is currently referred to as 'compensated reduction' within international negotiations.[139]

Reduced Emissions from Deforestation and Degradation: Lessons from a Forest Governance Perspective.

[135] Prior, S. and C. Streck, Centre for International Sustainable Development Law (2006), *Submission to the COP UNFCCC in Response to the Call for Views on the Issue of Avoided Deforestation.*

[136] This proposal, along with many others, was submitted for consideration at the 11th Conference of the Parties to the United Nations Framework Convention on Climate Change, held in Bali in November 2007.

[137] Made on behalf of Costa Rica, Nicaragua and Papua New Guinea, supported by the Central African Republic, the Dominican Republic and the Solomon Islands.

[138] Dresner, Simon, Paul Etkins, Kate McGeevor and Julia Tomei (2007), 'Forests and Climate Change: Global Understanding and Possible Responses', in Freer-Smith, P., M. Broadmeadow. and J. Lynch (eds), *Forestry and Climate Change*, Wallingford, Oxon: CABI Publishing, pp. 33–48, 38.

[139] Skutsch, above n 123.

The baseline and credit approach for REDD is not recommended by the Centre for International Sustainable Development Law.[140] This is due to the difficulty associated with calculating the baseline level where a standard level of deforestation must be established. There is no such thing as a standard level of deforestation that can be calculated upon a year-to-year basis. A number of other issues are also identified as being problematic with the baseline approach[141]

- The occurrence of unplanned and illegal logging may be excluded from baseline calculations;
- In periods where actual deforestation is less then the calculated baseline level, a reward is given where no additionality factor has been met (that is, there has been no change from business as usual);
- In a period of increased deforestation, the baseline would need to be modified to recognize this. This may encourage increased deforestation activities before the start of the scheme;
- If an average baseline were calculable, this would not actually represent future events.

There are difficulties associated with both a cap and trade system and a baseline credit system. Both proposals require improved capacity to monitor forest conditions to ensure compliance with the schemes.

4. IMPLEMENTATION OF FOREST INITIATIVES WITHIN THE INTERNATIONAL CLIMATE CHANGE REGIME

A. Implementation of LULUCFG

Because reporting on land-based sinks and source activities was optional during the first commitment period, there has been limited use of the LULUCFG. These guidelines will be more heavily used in the second commitment period, which will oblige parties to report upon land-based emissions and removals of emissions from land-based activities. At the COP meeting in Cancun in 2010 the LULUCFG were revisited and Annex 1 parties were requested to submit information on their forest management systems to be reviewed by an independent panel for

---

140 Prior, above n 135.

141 Ibid.

verification.[142] Parties were requested to provide information on: areas under forest management; emissions and removal from forest management; forest characteristics (age, structure, and rotation length), historical and assumed harvesting rates, harvested wood products and disturbances in the context of force majeure. Parties were also required to provide a description of domestic policies adopted and implemented in regard to forest management.

B. Implementation of Clean Development Mechanism Project for Forests

As at May 2011, there were 22 CDM forest projects, out of a total 3000 CDM registered projects.[143] Late finalization of CDM modalities in 2003 impacted upon the implementation of forest sink projects in the first commitment period. The average tree rotation cycle is around 5–15 years, which requires parties to establish such projects early in any given commitment period. Other impediments to wider implementation include the temporary nature of the forest carbon credits and the challenges associated with meeting the methodological requirements for CDM implementation.[144] Furthermore, the European Union decided to exclude certified emission reduction credits for sinks from the EU trading scheme; thus reducing some of the potential global demand for forest sink credits.[145]

C. REDD+ Implementation

REDD+ preparation and pilot activities are being implemented at a rapid pace. The institutions doing this are referred to as demonstration bodies and are responsible for implementing REDD+ strategies at the local level and reporting back on the challenges and lessons learnt from these experiences. The UN-REDD platform identifies two institutions as playing a large role in the implementation of demonstration REDD activities. These

---

[142] United Nations Framework Convention on Climate Change, Decision 2/CMP.6 The Cancun Agreements: Land Use, Land Use Change and Forestry. FCCC/KP/CMP2010/12/Add.1.

[143] Chokkalingam, Unna and Anvradha, Vanniarachchy, (2011), 'CDM AR did not Fail and is not Dead', available at http://www.forestcarbonasia.org/articles/cdm-ar-did-not-fail-and-is-not-dead/ (accessed 8 February 2012).

[144] Jung, Martin (2005), 'The Role of Forestry Projects in the Clean Development Mechanism', *Environmental Science and Policy*, **8** (2), 87–104.

[145] Boyd, Emily, Esteve Corbera and Manuel Estrada (2008), 'UNFCCC Negotiations (pre Kyoto to COP 9): What the Process says about the Politics of CDM-sinks', *International Environmental Agreements*, **8** (2), 95–108.

two bodies are the Forest Carbon Partnership Facility (FCPF) and the UN-REDD Program.

The FCPF is housed within the World Bank's Carbon Finance Unit and became operational in 2008.[146] The main objective of the FCFP is to provide technical assistance which supports countries in building national strategies and systems for REDD. Support by the FCPF is delivered through two mechanisms: the Readiness Fund and the Carbon Fund. US$232 million has been pledged for REDD Readiness FCPF, with US$173 million in committed funds. The FCPF funds activities in 14 countries: Argentina; Cambodia; Colombia; Costa Rica; the Democratic Republic of Congo; Ethiopia; Gabon; Nepal; Guatemala; Kenya; Mexico; Paraguay; Thailand; and Uganda.[147] Each of these countries is matched with a donor country and decisions concerning disbursement of funds and the types of programs to introduce in the host country are made on the basis of approval by both the host and donor countries.

The UN-REDD Program was also launched in 2008 and brings together expertise from the Food and Agriculture Organization of the United Nations (FAO), the United Nations Development Program (UNDP) and the United Nations Environment Program (UNEP).[148] The program supports 14 countries in carrying out national forest program activities, which aim to build capacity to implement REDD+ strategies. US $59.3 million has been invested in national forest programs in Bolivia, Cambodia, Democratic Republic of Congo, Ecuador, Indonesia, Nigeria, Panama, Papua New Guinea, Paraguay, the Philippines, Solomon Islands, Tanzania, Vietnam and Zambia.[149]

5. CONCLUSION

Forests are regulated in a number of ways under the climate change regime. The accounting and reporting methodologies and policies are complex in

---

[146] The website of the Forest Carbon Partnership Facility is available at http://www.forestcarbonpartnership.org/fcp/.

[147] FCPF (2011), 'Forest Carbon Partnership Facility 2011 Annual Report', available at http://www.forestcarbonpartnership.org/fcp/sites/forestcarbonpartnership.org/files/Documents/PDF/Oct2011/FCPF_Carbon_AR_FINAL_10_3.pdf (accessed 8 February 2012).

[148] The website of the UN-REDD program is available at http://www.un-redd.org/.

[149] UN-REDD Program, (2012), 'About the UN REDD Program' available at http://www.un-redd.org/AboutUNREDDProgramme/tabid/583/Default.aspx (accessed 8 February 2012).

nature, reflecting the complexity of what they are attempting to measure. For the forest accounting practices to be more widely used by the parties, the methodologies should be revisited. Ideally, in the future, one standard policy instrument would be created concerning forest sink and source values. This single policy instrument would remove the different formulas currently used within the regime to account for different forest activities. The climate change regime has much stronger political support than the United Nations Forum on Forestry. It also has the ability to deliver finance to implement forest projects. For these reasons, it is anticipated that it will become a dominant player in setting the international forest policy agenda. The current tiered membership within the Kyoto Protocol is a governance feature that should be retained. As this book suggests, the principles of common but differentiated responsibility, the polluter pays principle, and the theory of environmental justice, all should operate to impose heavier obligations on those parties responsible for, and with the capacity to, respond to environmental challenges.

7. Regulation of forest resources by the World Bank

1. INTRODUCTION

A. World Bank Background

This chapter focuses on the role of the World Bank in forest policy, strategy and finance. The World Bank Group and the International Monetary Fund emerged following negotiations at Bretton Woods, New Hampshire, USA in 1944. These global financial institutions were created with the aim of ensuring a sound international economy after World War II. The World Bank Group is an independent specialized agency of the United Nations and works closely with the International Monetary Fund. The World Bank Group operates under the umbrella of the Economic and Social Council of the United Nations.[1]

The founding instrument of the World Bank is the *IBRD Articles of Agreement*. Article 1 broadly provides that the purpose of the Bank is to:

- assist in the reconstruction and development of territories destroyed or disrupted by war;
- promote foreign investment by way of guarantee or loan;
- promote balanced growth of international trade; and
- assist in bringing about a smooth transition from a wartime to a peacetime economy.[2]

The Bank first became involved in the forestry sector in 1949 when it financed forest operations in Finland and the former Yugoslavia.[3] These

---

1 World Bank (2007), *A Guide to the World Bank*, 2nd ed., 43.

2 World Bank Group (1989), 'International Bank for Reconstruction and Development Articles of Agreement, as Amended Effective February 16', available at http://web.worldbank.org/WBSITE/EXTERNAL/EXTABOUTUS/0,,contentMDK:20049557~menuPK:63000601~pagePK:34542~piPK:36600~theSitePK:29708,00.html (accessed 30 July 2012).

3 World Bank (2004), *Sustaining Forests: A Development Strategy*, 19.

projects were funded through the provisions of loans and guarantees. Article III of the *IBRD Articles of Agreement* regulates the authority for these activities. Early forest projects generally financed the purchase of timber processing equipment. Gradually, the Bank's role in financing forest projects evolved from one that focused on timber extraction to trial operations in social forest programs and agro-forestry – and, then, towards an approach that favoured the conservation of remaining forest areas.[4] This evolution was the result of changing forest values within society generally, in combination with pressure from environmental NGOs (which have influenced the Bank's approach to forestry).

B. The Worlds Bank's Role in Forest-related Initiatives

The overarching policy objective of World Bank programs is a reduction in global levels of poverty.[5] Empirical research indicates a strong correlation between areas of high forest coverage and the people living in these areas experiencing high levels of poverty.[6] This link, along with the Bank's interest in investing in socially beneficial projects, has seen the Bank invest in a large number of forest projects aimed at holistically addressing environmental and livelihood issues. Existing research does not indicate that areas with high forest coverage lead to increased poverty. Rather, it indicates there is a high occurrence of marginalized groups living in densely forested areas, and that these are heavily dependent upon forest ecosystem services for their daily survival.[7]

Four reasons have been identified to explain the relationship that exists between high forest coverage and high levels of poverty.[8] First, the evolution from hunting and gathering to shifting cultivation and then to permanent agriculture has improved the living standards of many throughout the world. In some forest areas, this evolution has not taken place and, accordingly, the living standards of these people have remained the same. Secondly, forest communities often lack a voice or representation at the

---

4 Ibid.

5 Ibid.

6 Sunderlin, William D, Sonya Dewi and Atie Puntodewo (2007), 'Poverty and Forests: Multi-country Analysis of Spatial Association and Proposed Policy Solutions' (Occasional Paper No 47, Centre for International Forestry Research). Also, see the World Bank report: Chomitz, Kenneth (2007), *At Loggerheads? Agricultural Expansion, Poverty Reduction, and Environment in the Tropical Forests*, Washington DC: The International Bank for Reconstruction and Development/ The World Bank.

7 Ibid. 1–2.

8 Sunderlin, above n 6, 20 –1.

national level, and this has been identified as an indicator of poverty. This lack of power is associated with the isolation of these people from national political and economic structures. Thirdly, forests areas attract migrants because they are not heavily patrolled by governments. In some circumstances, forests serve as a refuge for people fleeing persecution, conflict and war. And fourthly, forest areas tend to be less densely populated than non-forest areas and, as such, will attract less public and private investment.

Forests are thought to provide a 'safety net' for forest dwellers and forest communities by providing building materials, foodstuffs and other goods during times of hardship.[9] Deforestation and forest degradation can have serious negative consequences for them through the loss of essential services. Forests provide numerous services to humans, directly and indirectly. They directly provide timber and material for housing, firewood for heating and cooking, food, medicine, fodder for livestock, and a variety of sources of income from the sale of forest products. They also act as purifiers of water, serve as buffers to extreme changes in temperature, maintain soil health and integrity, and absorb carbon dioxide – all of which indirectly affect the quality of life experienced by forest communities.[10]

The Millennium Development Goals represent a shared vision of a range of public intergovernmental institutions including the World Bank, the United Nations Development Program and the United Nations Environment Program among many others.[11] At the goals' inception, it was envisaged that they would be implemented by 2015. The Millennium Development goals are to

1. Eradicate extreme poverty and hunger
2. Achieve universal primary education

---

[9] Angelsen, Arild and Sven Wunder (eds) (2003), *Exploring the Forest-Poverty Link: Key Concepts, Issues and Research Implications*, Bogor, Indonesia: Centre for International Forestry Research. See also: Vedeld, Paul, Arild Angelsen, Jan Bojo, Espens Sjaastad, Gertrude Kobugabe Berg (2007), 'Forest Environmental Incomes and the Rural Poor', *Forest Policy and Economics*, **9** (7), 869–879, which distinguishes between three different functions of forest income in rural livelihoods: acting as safety nets; support of current consumption; and a pathway out of poverty.

[10] Program on Forests, United Nations Development Program (1999), *Forest Management to Support Sustainable Livelihoods: Framework for the UNDP Programme on Forests*, iv.

[11] For a complete list of all the public intergovernmental institutions involved please refer to all bodies listed at http://www.un.org/millenniumgoals/.

3. Promote gender equality and empower women
4. Reduce child mortality
5. Improve maternal health
6. Combat HIV/AIDS, malaria and other diseases
7. Ensure environmental sustainability
8. Develop a global partnership for development.

Forest policies and strategies are seen as a mechanism that can make a direct contribution to implementing Goal 1 (eradicating extreme poverty) and Goal 7 (implementing environmental sustainability). World Bank statistics suggest that forest resources directly contribute to the livelihoods of 90 per cent of the 1.2 billion people living in extreme poverty.[12] Sustainably managed forest areas provide forest inhabitants with access to essential forest services and, in some circumstances, may allow forest communities to engage in income-generating forest activities.[13] Forests are also central to maintaining terrestrial biological diversity. Nearly 90 per cent of all terrestrial biodiversity is found within forest areas, with a disproportionate share found in the forests of developing nations.[14]

2. THE INSTITUTIONAL STRUCTURE OF THE WORLD BANK

A. The World Bank Group Institutions

The World Bank Group comprises five bodies that operate collectively to implement the overarching goal of poverty reduction

1. The International Bank for Reconstruction and Development
2. The International Development Association
3. The International Finance Corporation
4. The Multilateral Investment Guarantee Agency
5. The International Centre for Settlement of Investment Disputes.

---

[12] International Bank for Reconstruction and Development (ed.) (2004), *Sustaining Forests: A Development Strategy*, Washington D.C.: The World Bank, 1.

[13] See the following for a discussion about forest income: Vedeld, Paul, Arild Angelsen, Jan Bojo, Espens Sjaastad, Gertrude Kobugabe Berg (2007), 'Forest Environmental Incomes and the Rural Poor', *Forest Policy and Economics*, **9** (7), 869–879. This study finds that forest-related products supply 22 per cent of stakeholders' income in the populations sampled.

[14] International Bank for Reconstruction and Development, above n 12, 1.

The term 'World Bank Group' refers to all five institutions, while the term 'World Bank' refers specifically to the International Bank for Reconstruction and Development and the International Development Association. Each of these bodies plays a distinct role in the mission to fight poverty and improve living standards for people in developing countries.

The International Bank for Reconstruction and Development (IBRD) and the International Development Association (IDA), collectively the World Bank, support a range of initiatives aimed at reducing poverty and improving living standards in developing countries. This is actioned through the provision of loans, guarantees, policy advice and technical assistance.[15] These two bodies share the same employees and headquarters, and use the same standards when determining a project's approval. The IBRD is responsible for providing assistance to middle-income and creditworthy poorer countries, whilst the IDA focuses exclusively on the world's poorest countries.[16] The IDA provides credits and grants on more lenient borrowing terms than the other World Bank Group institutions to assist the poorest countries to reduce poverty.[17]

The International Finance Corporation (IFC) is responsible for promoting private enterprise in developing countries. While the World Bank can only lend to governments of member countries, the IFC is able to lend money directly to private business interests. To be eligible for IFC funding, the project must be profitable for investors, must benefit the economy of the host country, and must comply with environmental and social standards.[18]

The Multilateral Investment Guarantee Agency encourages foreign investment in developing countries by providing political risk insurance for foreign investment, technical assistance, and dispute mediation services.[19] The Agency seeks to build confidence in private enterprises to encourage investment in infrastructure development, investment in conflict-affected countries, investment between developing countries, and

---

15 World Bank, above n 1, 11.

16 For further information on the classification between middle income countries and poor countries see World Bank, above n 2, 15–19. In 2005, countries that had an income of less than $1025 per capita were classified as poor countries and were eligible to apply to the International Development Association.

17 Credits are loans at zero interest, with a 10-year grace period before repayment of the principal begins at maturities of 20, 34 or 40 years. This type of lending is referred to as concessional lending. For further information: World Bank, above n 2, 17.

18 Ibid. 22.

19 Ibid. 23.

investment in frontier markets (high-risk or low-income country markets). The Agency provides investment confidence, due to its position within the World Bank Group. The World Bank Group uses its position to deter activities that may adversely affect foreign investments.[20]

The International Centre for Settlement of Investment Disputes (ICSID) encourages foreign investment by providing international facilities for the conciliation and arbitration of investment disputes. This is believed to foster an atmosphere of mutual confidence between states and foreign investors. The ICSID was established under the Convention on the Settlement of Investment Disputes between States and Nationals of Other States and came into force in 1966. Parties must consent in writing to be bound by decisions of the Centre.[21]

The five institutions of the World Bank Group are owned by their member countries. To obtain membership of the IBRD, states must first join the International Monetary Fund. Membership of the IDA, the IFC, and the Multilateral Investment Guarantee Agency is contingent upon obtaining membership of the IBRD.[22] To become a member of any of these institutions, states must buy shares in the institution – a process known as capital subscription. Figures from the April 2007 record membership levels are:[23]

- 185 members in the International Bank for Reconstruction and Development;
- 166 members in the International Development Association;
- 179 members in the International Finance Corporation;
- 171 members of the Multilateral Investment Guarantee Agency; and
- 143 members of the International Centre for Settlement of Investment Disputes.

Membership of other United Nations' institutions is generally not dependent upon capital subscriptions. This membership arrangement is unique to

20 Ibid. 25.

21 See Article 25 of the Convention at http://icsid.worldbank.org/ICSID/StaticFiles/basicdoc/CRR_English-final.pdf (accessed 26 March 2012).

22 World Bank Group (1989), 'International Bank for Reconstruction and Development Articles of Agreement, as amended effective February 16', available at http://web.worldbank.org/WBSITE/EXTERNAL/EXTABOUTUS/0,,contentMDK:20049557~menuPK:58863~pagePK:43912~piPK:44037~theSitePK:29708,00.html (accessed 1 September 2008), Art. 11.

23 World Bank, above n 1, 94.

global financial institutions. Voting power is determined by the value of the capital subscriptions held by the country member. Thus, members with the highest levels of capital subscriptions receive the highest percentage of 'weighted votes' and acquire greater control over Bank operations.[24] In the IBRD, the following five countries retain the highest capital subscriptions: the United States of America, 16.39 per cent; Japan, 7.87 per cent; Germany, 4.49 per cent; France, 4.3 per cent; and the United Kingdom, 4.3 per cent.

Such voting arrangements are not consistent with the principles of good governance, which requires transparency, accountability, efficiency, fairness, participation and ownership.[25] Originally, all members were allocated a quota of 250 'basic votes'. This allocation was made to reinforce the Bank's role as an intergovernmental public institution.[26] In 1946, basic votes accounted for 11.3 per cent of the total vote while, 60 years later, basic votes account for only 3 per cent of the total vote.[27]

Initially, members of the Bank had two interests in the institution: first, every member had an interest in economic growth and stability. Secondly, major contributors had an interest in ensuring the institution's resources were properly used. However, these interests were based on the Bank's original purpose (to facilitate post war reconstruction in Europe and further development in other countries). Presently, the Bank's operations occur almost exclusively in developing countries and in transitional economies. The Bank provides resources to these members subject to the implementation of its policy and guidelines, and this reduces the sovereign rights of the beneficiary government. Additionally, developing country and transitional economy governments generally hold fewer voting rights at the Bank level, meaning that funding in these countries is dependent upon policies created by the most powerful stakeholders at the World Bank. Voting power based on the percentage of shares owned is characteristic of private enterprise proceedings. The World Bank, as a public intergovernmental institution, should ensure that all participants within

---

[24] Although voting in the World Bank is based on consensus, and not the tallying of individual votes, informally, the percentage of votes held affects a countries' political standing within the Bank structure. See Ngaire Woods, 'The Challenge of Good Governance for the IMF and the World Bank Themselves', (2000), **28** (5), *World Development*, 823, 829.

[25] Woods, Ngaire (2000), 'The Challenge of Good Governance for the IMF and the World Bank Themselves', *World Development*, **28** (5), 823–841.

[26] Gianaris, William (1991), 'Weighted Voting in the International Monetary Fund and the World Bank', *Fordham International Law Journal*, **14** (4), 910–945.

[27] Woods, above n 24, 828.

the system are given more equitable voting rights to address governance concerns.

B. Partnerships with Other Intergovernmental Forest Institutions

The World Bank has formed a number of strategic partnerships to deliver better global forest outcomes. Working collaboratively allows institutions to pool resources and expertise, which should lead to improved outcomes. Four of the Bank's forest partnerships will be examined in this chapter (an alliance with WWF, FLEGT, Program on Forests, and the Forest Carbon Partnership Facility). The World Wide Fund for Nature and World Bank alliance works on increasing the size of protected areas and increasing the percentage of areas governed by independent certification schemes. The Forest Law and Governance Trade Scheme works on addressing issues related to combating the illegal timber market and corruption within producer timber countries. The Program on Forests carries out analytic work that aims to identify barriers to implementation and tools for enhanced implementation of forest policies, whilst the Forest Carbon Partnership Facility provides support for REDD+ demonstration and readiness policies and measures.

World Wide Fund for Nature and World Bank alliance

The World Wide Fund for Nature and the World Bank formed the Global Forest Alliance in 1998 and established three time-bound quantitative global forest conservation goals. Such an approach can be contrasted with other intergovernmental forest initiatives, which tend to set vague non-target driven goals or objectives. The three goals of the Alliance seek to significantly reduce the rate of loss and degradation of all forest types[28]

1. Creating 50 million hectares of new forest protected areas by 2005
2. Ensuring that 50 million hectares of existing highly threatened protected areas are secured under effective management
3. Ensuring 200 million hectares of production forests are bought under independently certified sustainable management.

The Alliance achieved the following outcomes by the end of 2005

---

[28] Leape, James and Paul Wolfowitz (2005), *WWF/World Bank Global Forest Alliance: Alliance Annual Report*.

- Target one was met (in fact, the original target was exceeded with the creation of 55 million hectares of new protected areas of forest)
- Target two was not met (with only 40 million hectares of forest being secured under effective management, a shortfall of 10 million hectares on the original goal)
- Target three fell well short of the original goal (with only 25 million hectares of forest bought under independent certification programs).

Overall, it can be concluded that the Alliance produced an example of successful engagement between a public international institution and a non-governmental organization while also achieving some promising results. A number of lessons can be drawn from this exercise. First, from a governance perspective, the setting of targets and reporting back on the success rate of these targets increases the accountability and transparency of the program. The Alliance was the first international partnership to do this. Even though not all the targets were met in this instance, the setting of the targets and the publication of the results led to outcomes and lessons that otherwise may not have been available.

Secondly, the partnering of the World Bank with an NGO allowed for the pooling of resources and expertise, and led to improved communication between the two groups. Literature in the area is generally favourable towards the inclusion of NGOs in environmental norm-setting and implementation.[29] It is believed that NGOs can: assist with representing the views of wide range of stakeholders; operate as effective monitors of governments, private and international institutions; provide technical assistance in some instances; and offer experience in disseminating information quickly to large numbers of people. A major challenge for the current international forestry framework is the effective engagement, communication and improved interaction between various forestry institutions. It has been suggested that the most valuable contribution of partnerships for forest biodiversity conservation has been filling the gap when governments were neither willing nor able to regulate.[30] On the surface, institutional

---

[29] Tarlock, Dan (1992), 'The Role of Non-Governmental Organizations in the Development of International Environmental Law', *Chicago Kent Law Review*, **68** (1), 61–76; Al Jurf, Saladin (1999), 'Participatory Development and NGOs: A look at the World Bank', *Transnational Law and Contemporary Problems*, **9** (1), 175–185; Bluemel, Erik (2005), 'Overcoming NGO Accountability Concerns in International Governance', *Brooklyn Journal of International Law*, **31** (1), 139–206; and Gerdung, Anja (2004), 'Global Environmental Governance and the Role of Civil Society Groups', *New Zealand Journal of Environmental Law*, **8**, 55–98.

[30] Visseren-Hamakers, Ingrid, and Peter Glasbergen (2007), 'Partnerships in Forest Governance', *Global Environmental Change*, **17** (3–4), 408–417.

engagement exists among forestry institutions. All major forest institutions are given participation and lobbying rights during negotiations that amend or create new forest instruments. They are also given rights to participate and engage at the annual meetings of each other's proceedings. There is, however, a lack of coordination for action among these groups. Apart from 'motherhood' statements (urging increased integration and interaction), little engagement occurs in practice. The alliance between the World Bank and World Wildlife Fund for Nature is a practical example of policy integration and implementation by seemingly different institutions. Given their differences, nevertheless, they were able to identify common interests and formulate a plan for action that addressed the forest priorities of both parties.

Thirdly, the partnership provided valuable lessons for future implementation of sustainable forest management. In relation to the third target (increasing the level of forests under independently verified certification operations), the Alliance identified a number of barriers that exist to the adoption of certification programs. During the implementation of this goal, it was realized that certification is dependent upon a robust, accountable and transparent domestic regulatory regime. The Alliance produced a capacity-building tool entitled the *Forest Certification Assessment Guide* to assist in the pre-certification project period.[31] The Alliance reflected on its experiences by identifying a number of other lessons that it could use to improve its future operations. The other lessons included: the value of visionary targets; the need to focus on forest issues in a broader landscape approach; the difficulty of linking the Alliance program with other international development agendas; and the importance of engaging with the private sector for both additional resources and for outreach programs that influence the management of forestry resources.[32]

Fourthly, a number of highly successful individual projects emerged from the Alliance. The best-known project involved the tripling of existing protected reserves and parks within the Amazon River basin. Some 13.5 million hectares of new, strict nature reserves were created and 7.5 million hectares of new sustainable-use areas were created to benefit local communities. An additional 8 million hectares of parks are receiving support to strengthen their status. The other major project was the creation of the Congo Basin Forest Partnership. This bought governments together from Cameroon, the Central African Republic and the Republic of Congo to establish the 3.6 million hectare Sangha Tri-National Park. This is the

31 Leape and Wolfowitz, above n 29, 5.

32 Ibid. 26.

first multilateral organization to receive verification from an independent certification agency.[33]

Finally, the success of the original partnership led to the renewal of the Alliance, which, again, adopted a target-driven approach for the second phase 2006–2010. The following new goals were created[34]

1. A reduction by 10 per cent of the global levels of deforestation by 2010
2. 25 million hectares of new protected forest areas
3. 75 million hectares of existing forest protected areas under improved management
4. 300 million hectares of forest outside strict protected areas under improved forest management comprising of three sub-targets
   a. 100 million hectares of forest independently certified consistent with the standards outlined in the World Bank's Operational Policy of Forests (OP 4.36);
   b. 100 million hectares of forest progressing towards such independent certification; and
   c. 100 million hectares of forest land under community-based forest management agreements that improve local livelihoods.

Forest Law Enforcement, Governance and Trade Scheme

The Forest Law Enforcement, Governance and Trade Scheme emerged at the World Summit on Sustainable Development, where the European Commission expressed its intention to combat illegal logging and the associated trade in illegally harvested timber.[35] This is a major issue for many timber-producing countries with weak forest law governance arrangements. Illegal logging causes a host of problems: environmental damage; billions of lost revenue dollars; it promotes corruption and undermines the rule of law; and, in certain circumstances, funds the purchase of weapons.[36]

Consumer countries contribute to these problems by importing timber and wood products without ensuring that they are legally sourced. The Forest Law Enforcement, Governance and Trade Scheme recognizes

---

[33] Ibid. 2.

[34] Humphreys, David (ed.) (2006), *Logjam: Deforestation and the Crisis of Global Governance*, London: Earthscan Publications, 188. There is no published information available as yet concerning the implementation of the 2006–2010 goals.

[35] FLEGT Expert Group of the European Commission, European Commission (2007), 'What is FLEGT?', *FLEGT Briefing Note Number 01*, 1.

[36] Ibid.

the connection between consumer countries' timber demands and producer countries' role in meeting this demand. Illegal logging is a result of consumer countries wanting large quantities of cheap timber at low cost without regard to the source of the timber. It therefore seems reasonable for consumer countries to lead the way in addressing illegal logging and the associated trade. Greater responsibility should be placed on those countries responsible for the high consumption of illegally sourced timber. Such responsibility is justifiable on the basis that high consuming timber countries create a market for illegally sourced timber and, as such, encourage illegally timber harvesting.

The original Forest Law Enforcement, Governance and Trade Scheme was formed under a council regulation of the European Union.[37] Three other regional schemes have since been established: one in South East Asia[38]; one in Africa[39]; and one covering Europe and North Asia[40]. The World Bank acts as a coordinator of all of these processes using its convening power and capacity to mobilize financial resources to assist in the implementation of all Forest Law Enforcement, Governance and Trade Scheme initiatives. The World Bank Forest Strategy and the Scheme's policy objectives are compatible and closely aligned.

These schemes organize three types of sequential activities to implement their objectives.[41] First, research is carried out to establish the necessary changes that need to be made within the forest policy and forest institutions. Secondly, ministerial regional meetings take place to raise awareness and secure political commitment to stronger action. A tangible outcome of this activity is the creation of Ministerial Declarations, which commit governments to combat illegal logging and to support the implementation of the scheme at a regional level. Thirdly, partnerships are created among trading partners, institutions of civil society, and industry to improve forest governance. These schemes seek better forest governance at the regional level through existing forest institutions. Wisely, these schemes

---

[37] Regulation (EC) No 2173/2005 of 20 December 2005 on the Establishment of a FLEGT Licensing Scheme for Imports of Timber into the European Community [2005] OJ L 347/1.

[38] Forest Law Enforcement and Governance, East Asia Ministerial Conference, Bali, Indonesia, 11–13 September 2001, Ministerial Declaration.

[39] Africa Forest Law Enforcement and Governance, Ministerial Conference, Yaounde, Cameroon, 16 October 2003, Ministerial Declaration.

[40] St Petersburg Declaration, Forest Law Enforcement and Governance, Ministerial Conference 22-25 November 2005, St Petersburg, Russia, Ministerial Declaration.

[41] Hermosilla, Arnoldo Contreras, and Markku Simula, The World Bank (2007), *The World Bank Forest Strategy: Review of Implementation*, 83.

use existing institutions and work with them to change their approaches to forest governance.[42]

The European Unions' scheme is the most advanced and, as such, will be examined here in further detail. Involvement with the European Union Forest Law Enforcement, Governance and Trade Scheme is on a voluntary basis for timber producing countries. The objective of the scheme is to ensure that all timber imported into the European Union is legally harvested and its authenticity assured. The European Union Timber Regulation commences on 3 March 2013 and will require all operators within the EU to have in place systems that assure that the timber is of a legal origin, with failure to do so resulting in sanctions.[43]

Assurance comes from the operation of the scheme's checks and balances,[44] designed to guarantee that timber traded within the European Union is 'legally produced timber'. Timber is considered to be legally harvested if it is either produced from legally harvested domestic timber, or if the timber is legally imported into a partner country.[45] As the definition of legal forest practices varies from country to country, it has been suggested that a more robust definition of legally harvested timber be provided.[46] The definition of legality is left to the producer country but it must also be acceptable to other stakeholders in the scheme. It is recommended that the definition of legally procured timber incorporate the three pillars of sustainability (economic, environmental and social considerations). The inclusion of sustainability criteria will require laws which:[47]

---

42 Ibid.

43 Regulation EU No 995/2010 of the European Parliament and of the Council of 20 October 2010: Laying down the Obligations of Operators who Place Timber and Timber Products on the Market [2010] OJ L 295/23.

44 Ibid. 4.

45 In furtherance of this objective, a number of European Union community members have implemented green procurement policies that require timber products to be from legal and sustainable sources. These countries are: Belgium; Denmark; France; Germany; and the United Kingdom. See: European Union (2007), *Forest Law Governance,* Europa, http://ec.europa.eu/environment/forests/flegt.htm, accessed on 24 September 2008. Additionally, some European Union members have enacted codes of conduct that aim to remove illegally harvested timber from their supply chains. These countries are: Finland; France; Netherlands; Spain; and the United Kingdom. See: European Union (2007), *Forest Law Governance,* Europa, http://ec.europa.eu/environment/forests/flegt.htm, accessed on 24 September 2008 for further information.

46 European Commission (2006), *Public Consultation on Additional Options to Combat Illegal Logging Analysis and Report*.

47 This list is based on suggestions from the FLEGT Expert Group of the

- provide clear guidance on the legal tenure holding arrangements over forest areas;
- define the process for receiving approval to harvest within these forest areas;
- define the requirements of sustainable forest management; and
- define the taxes, import and export duties, royalties and fees directly related to timber harvesting and timber trade.

The scheme is implemented through the creation of licences,[48] which are attached to timber products certifying that they were legally produced. There are four elements in the legality assurance system.[49] First, (as discussed above), a definition of legally produced timber must exist which clearly sets out the criteria and indicators that can be used as a methodology to test this definition. Secondly, there must be control of the supply chain (which requires systems to exist to trace wood products from harvest to export). Thirdly, verification must be carried out by either the government, market participants, or by a thirty-party organization. This body must have adequate resources and management systems to avoid conflicts of interests.[50] This body verifies compliance with the agreed definition of legality in addition to verifying the supply chain process. Fourthly, the issuing of licenses must be carried out according to a transparent procedure that defines who issues licences and the method to be followed. A legality assurance system can be implemented through the issue of either shipment-based licences or operator-based licences. Shipment-based licences confirm that the individual consignment has complied with the relevant procedure; operator-based licences confirm that a particular operation supplies timber that meets the scheme's requirements.[51]

The legality system discussed above is dependent on a Voluntary Partnership Agreement being in place. Voluntary Partnership Agreements are binding agreements between the European Union and a partner

---

European Commission, (2007), 'What is Legal Timber?', *FLEGT Briefing Note Number 02*, 1.

[48] Regulation (EC) No 2173/2005 of 20 December 2005 on the establishment of a FLEGT licensing scheme for imports of timber into the European Community [2005] OJ L 347/1, art. 3.

[49] FLEGT Expert Group of the European Commission, European, (2007), 'A Timber Legality Assurance System', *FLEGT Briefing Note Number 03*, 1.

[50] FLEGT Expert Group of the European Commission (2007), 'Legality Assurance Systems: Requirements for Verification', *FLEGT Briefing Note Number 05*, 1.

[51] FLEGT Expert Group of the European Commission (2007), 'A Timber Legality Assurance System', *FLEGT Briefing Note Number 03*, 2.

country, whereby both parties undertake to work together to support the aims of the scheme. These agreements are designed, ultimately, to eliminate illegally produced timber from a partner country's international and domestic trade.[52] All agreements require the development of credible legal, administrative and technical systems to verify that timber is produced in accordance with national laws.[53] These partnership agreements should be created with sufficient stakeholder consultation. They should also seek to minimize adverse impacts on local communities and on poor people (who will be affected by the operation of the agreement).

There are currently six countries with Voluntary Partnership Agreements in place: Cameroon; Central Africa Republic; Ghana; Indonesia; Liberia; Republic of Congo (Brazzaville). There are a further four countries[54] with Voluntary Partnership Agreements in the negotiation stage and 15 other countries in the pre-negotiation phase[55]. Meaningful implementation of the forest law enforcement, governance and trade scheme will be dependent upon the implementation of many Voluntary Partnership Agreements. The World Bank's role in the future implementation of the scheme will continue as a coordinating and networking role – linking timber producing countries with timber consuming countries.[56]

Program on Forests

The Program on Forests is a multi-donor partnership formed to enhance the contribution forests make to poverty reduction, sustainable development, and the protection of environmental services through analytical work. The Program on Forests funds research that is likely to lead to an output (either instrument, toolkit or policy) that will assist in the implementation of sustainable forest management. The analytical work is carried out across four thematic areas[57]

1. A 'livelihoods' approach to poverty reduction – work on this topic concentrates on the contribution that forests can make to poor rural

---

[52] FLEGT Expert Group of the European Commission (2007), 'Voluntary Partnership Agreement', *FLEGT Briefing Note Number 06*, 1.

[53] FLEGT Expert Group of the European Commission (2007), 'Voluntary Partnership Agreement', *FLEGT Briefing Note Number 06*.

[54] Democratic Republic of Congo, Gabon, Malaysia and Vietnam.

[55] Bolivia, Columbia, Ecuador, Guatemala, Guyana, Honduras, Peru, Cambodia, Laos, Papua New Guinea, Solomon Islands, Thailand, Cote d' Ivoire, and Sierra Leone.

[56] From 2010 onwards, the FLEG initiative will sit within the Program on Forests Partnership, though it will continue to operate with its same mandate.

[57] Hermosilla and Simula, above n 42, 64.

people through the provision of employment and income creation. Research is conducted in the community and at individual household levels
2. Forest governance – work within this topic looks at the regulation of forests by both the government and other forest stakeholders. It tries to improve decision-making processes, regulatory, and institutional frameworks to ensure good governance, including better enforcement of regulation, improved incentives and enhanced transparency and accountability
3. Innovative approaches to financing sustainable forest management – this group aims to identify ways of increasing the profitability of sustainable forest management over that of unsustainable practices
4. Cross-sector impacts on forests – this work devises and analyses ways to manage links between the sustainable management of forests and other sectors and macroeconomic policy.

The donors to the program are the Department of International Development of the United Kingdom, the Finnish Department for International Development Cooperation, the Japanese International Forestry Cooperation, the Swiss Development Cooperation, and the German Government (which is an in-kind donor). The United Nations Development Program initially administered the scheme but, in 2002, the Program on Forests was relocated to the World Bank.[58] The Program on Forests works collaboratively with the World Bank to implement the Bank's forest strategy.

The Program on Forests is thought to be a time- and cost-efficient arrangement providing problem-driven knowledge. Because this research is demand-driven, it is also thought that the scheme is highly relevant to current forest challenges.[59] It has been said that the program could have more impact if two suggestions were followed.[60] First, it should make greater use of its existing synergies with the World Bank Group, the World Wide Fund for Nature and the Forest Law Enforcement Governance Trade Scheme. Working more pragmatically with these bodies would mean that these other institutions could use the knowledge gained through the program's activities. Secondly, although the scheme's objectives are consistent with the Bank's strategy, further integration is possible between

---

58 Program on Forests (2007), 'About PROFOR', available at http://www.profor.info/about.html (accessed 29 September 2008).

59 Hermosilla and Simula, above n 42, 65.

60 Ibid. 67.

the two bodies. The program could assist by providing project design and relevant case-study knowledge to assist in the project's implementation. A final observation is that the work of this program is similar to the work of the Centre for International Forestry Research (a research institution funded by many international environmental institutions), and to work carried out by the Collaborative Partnership on Forests. These three forest-related research bodies could have greater impact if there was institutional engagement to share knowledge and lessons.

The Forest Carbon Partnership Facility

The Forest Carbon Partnership Facility (FCPF) assists tropical and subtropical forest countries to develop systems and processes for REDD+ and provides them with performance-based payments for emission reductions. Thirty-seven countries have been selected to participate in the partnership (14 in Africa, 15 in Latin America and the Caribbean and eight in Asia and the Pacific). The FCPF is made up of two funds: the Readiness Fund which provides funding for early REDD+ action, including strategy development, technical preparation work and basic capacity building; and the Carbon Fund which has been designed to promote and test innovative programs designed to deliver emission reductions on a significant scale. The World Bank acts as trustee for the REDD+ Readiness and Carbon Funds. It provides secretariat services, technical support, and conducts due diligence on matters related to fiduciary, social and environmental safeguards. As discussed in Chapter 6, the FCFP is a demonstrating body of the REDD+ mechanism. Demonstrating bodies are responsible for implementing REDD+ activities at the domestic level and feeding information and lessons learnt back to the climate change regime.

The FCPF was established in 2008 and has focused on getting a number of countries 'ready for REDD+'. The Charter of the FCPF defines readiness packages as activities designed to support a country's capacity to participate in a possible future REDD scheme.[61] The FCPF implements the Cancun COP decision concerning REDD+[62] by undertaking the following four activities

1. Readiness, organization and consent
   a. National REDD+ management arrangements
   b. Consultation, participation and outreach

---

[61] International Bank for Reconstruction and Development, Charter Establishing the Forest Carbon Partnership Facility (2008), s. 6.3(b).

[62] See Chapter 6, which discusses the Cancun COP Report on REDD+.

2. REDD strategy and preparation
    a. Assessment of land use and land use change drivers, forest law policy and governance
    b. Strategy options
    c. Implementation frameworks
    d. Social and environment impacts
3. Reference emission level/ forest reference level
    a. Reference emissions/forest level
4. Monitoring systems for forests and safeguards
    a. National forest monitoring system
    b. Information systems for multiple benefits, other impacts, governance and safeguards.[63]

Countries submitting readiness proposals must demonstrate compliance with the nine sub-components listed above. Compliance is measured by testing whether each has 'significantly progressed'.[64] Thirteen countries[65] have submitted Readiness Preparation Proposals. Once due diligence is conducted by the Bank on these proposals, grants of up $3.6 million will be provided to these countries to assist them in carrying out the preparatory work which they have identified within their readiness proposals.

C. Engagement with Global Forest Institutions

The World Bank also engages with international forest institutions as well as being supportive of the work carried out by the United Nations Forum on Forestry and its associated research body (the Collaborative Partnership on Forests). One of these is the Global Environmental Facility, an intergovernmental institution whose original mandate was the

---

[63] These requirements are thought to be compatible with paragraph 70 and 71 of the Cancun COP Decision. See Forest Carbon Partnership Facility (2011), 'FCFP Readiness Fund, Readiness Package Content and Assessment Approval', available at http://www.forestcarbonpartnership.org/fcp/sites/forestcarbonpartnership.org/files/FMT%20Note%202011-14%20Readiness%20Package%20Concept.pdf (accessed 14 February 2012).

[64] Forest Carbon Partnership Facility (2011), 'FCFP Readiness Fund, Readiness Package Content and Assessment Approval', available at http://www.forestcarbonpartnership.org/fcp/sites/forestcarbonpartnership.org/files/FMT%20Note%202011-14%20Readiness%20Package%20Concept.pdf (accessed 14 February 2012), 19.

[65] Argentina, Costa Rica, the Democratic Republic of Congo, Ghana, Guyana, Indonesia, Kenya, Lao PDR, Mexico, Nepal, Panama, Republic of Congo, and Tanzania.

implementation of the conventions agreed upon at the UNCED negotiations. The World Bank has played a leadership role in seeking to address the international governance of forests. The Bank proposed the idea of a 'global forest partnership' and sought feedback from a wide variety of international institutions, governments and other bodies on such an initiative. The role of the World Bank in the Global Environmental Facility and the Global Forest Partnership is examined in greater detail below.

Global Environmental Facility

The Global Environmental Facility provides the largest source of funding for environmental projects in developing countries globally.[66] It acts as the financial mechanism for the Convention on Biological Diversity and the United Nations Framework Convention on Climate Change, and is also assisting in the implementation of the Stockholm Convention on Persistent Organic Pollutants and the United Nations Convention to Combat Desertification.

The idea of an international environmental finance mechanism can be traced to the 1987 World Commission on Environment and Development report, *Our Common Future* (also referred to as the *Brundtland Report*).[67] The *Brundtland Report* calls for 'a significant increase in financial support from international sources' to combat global environmental challenges.[68] The World Bank convened a meeting in Paris in March 1990 that resulted in 17 donor countries affirming their interest in forming a facility to fund global environmental programs.[69]

The World Bank, the United Nations Environment Program, and the United Nations Development Program were charged with the responsibility of establishing the operational framework. The initial framework was set up around a three-year pilot program designed to test the effectiveness of a global environmental facility. The program made funds available for projects in developing countries that focused on addressing one of the four key priorities. These priorities were determined as actions that addressed

---

66 Clemencon, Raymond (2006), 'What Future for the Global Environmental Facility?', *The Journal of Environment and Development*, **15** (1), 50–74.

67 Miller, Alan (2000), 'The Global Environmental Facility and the Search for Financial Strategies to Foster Sustainable Development', *Vermont Law Review*, **24** (4), 1229–1244.

68 United Nations, (1972), *Our Common Future* (*Brundtland Report*), available at http://www.un-documents.net/wced-ocf.htm (accessed 26 March 2012) 336.

69 Reed, David (1993), 'The Global Environmental Facility and Non-Governmental Organizations', *American University Journal of International Law and Policy*, **9** (1), 191–213, 192.

climate change, biodiversity, international water resources, and ozone depletion.[70] Forest-related projects are funded under the biodiversity action plan,[71] and also under land degradation plans.[72] The administration of the program was housed and carried out by the World Bank, while the United Nations Environmental Program ensured that policies were consistent with those evolving at the upcoming UNCED negotiations. The United Nations Development Program provided technical assistance and institutional building assistance.[73]

The role of the World Bank was contentious at this time, as the World Bank was held in poor regard by environmental NGOs. These groups believed that, in many instances, the Bank's investments had led to serious environmental degradation. The NGOs were hesitant to entrust the Bank with a stewardship role in the management of global financial resources with the apparent aim of improving environmental standards. These concerns were addressed in part by the restructuring of the Global Environmental Facility that occurred after the UNCED negotiations. The role of the World Bank, the United Nations Environmental Program, and the United Nations Development Program were preserved, but a new council of 32 countries was created. Non-governmental organizations were also given participation rights in council meetings as observers, which helped to build confidence with these stakeholders.

A controversial issue at the facility's inception, and a continuing one, relates to the budget allocation and transparency of the facility's operations.[74] The facility was originally entrusted with US$2 billion for investment during 1992–94. It was recognized at the time that this investment alone was insufficient to address sustainable development. The UNCED secretariat estimated that the additional financial needs required for implementing Agenda 21 were $125 billion per annum – which, at the time, was more than double the level of all international aid.[75] Therefore,

---

[70] Miller, above n 68, 1234.

[71] *Financing the Stewardship of Global Biodiversity: Global Environment Facility Report 2008*. See Section 3, 'Seeking Triple Benefits Through Sustainable Forest Management in Tropical Forests'. This program supports interventions that simultaneously generate multiple benefits in biodiversity, climate change, sustainable land management and livelihoods. More than $44 million was invested during the first six months of the program.

[72] Humphreys, above n 35, 172. By 2004, GEF has committed $778 million for forest-related funding, which has leveraged an additional $1,995 million from other investors.

[73] Miller, above n 68, 1234.

[74] Clemencon, above n 67, 53.

[75] Millar, above n 68, 1237.

a priority for the facility was the leveraging of additional private investment and the creation of self-supporting environmental markets.

Since 1991, the Global Environmental Facility has given $7.4 billion in grants to developing countries and has generated $28 billion in co-financing for environmental projects in developing countries. In total, 1,950 individual projects have been implemented by the facility. In the 2006–2010 project phase, 32 countries pledged $3.13 billion for projects.[76] The negotiations leading to the 2006–2010 funding level were delicate – because the USA drastically reduced its funding, making Germany, France and Sweden contribute additional funds.[77] Considering its budgetary restrictions and huge mandate, it is commonly accepted that the facility has made solid progress in implementing environmental projects in developing countries.[78]

A number of funding programs are available to developing countries and countries with economies in transition

- The small grants program – this includes grants of up to $50,000, and is made in support of community level initiatives. This program is seen as the most effective funding mechanism because it reaches many stakeholders, and benefits are attained at the local level at low cost[79]
- Enabling activities – finance is provided under this option either for the completion of a plan, strategy or report required to fulfil commitments under global environmental conventions, or for the creation of a national communication or report relevant to a convention
- Medium-sized projects – a maximum of $1 million is available for a broad range of programs aimed at increasing the flexibility and faster implementation of projects for the benefit of a wide range of stakeholders
- Full-sized projects – these are projects over $1 million, and they must satisfy three requirements: i) they must satisfy an eligibility requirement under a convention; ii) they must be a strategic priority of the facility; and iii) they must be either an operational program or a short-term response measure.

The operational strategy of the Global Environmental Facility provides that the purpose of the facility is the provision of:

---

[76] Internet site.

[77] Clemencon, Raymond (2007), 'Funding for Global Environmental Facility Continues to Decline', *The Journal of Environment and Development*, **16** (1), 3–7, 5.

[78] Ibid.

[79] Clemencon above n 67, 53–4.

> . . . new, and additional, grant and concessional funding to meet the agreed incremental costs of measures to achieve agreed global environmental benefits in the areas of biological diversity climate change, international waters and ozone depletion.[80]

Global environmental facility finance is only granted to meet incremental costs. Such costs are defined as those incurred for the protection or enhancement of global environmental benefits. The cost of protecting global ecological values is considered to be beyond the costs of achieving a national development goal, thereby making it eligible for facility funding.[81] This consideration, along with six other strategic considerations,[82] is used in the assessment process of applications made to the facility.

Global forest partnership

In 2007, the World Bank launched an initiative with the purpose of exploring interest in creating a 'global forest partnership'. The Bank commissioned the International Institute for Environment and Development (IIED) to carry out consultations worldwide to review stakeholder interest in the creation of a new global forest institution. The initial survey released by the consulting body was vague and generic, and the purpose of the global forest partnership was difficult to decipher.[83] This posed difficulties for the consultation, as respondents were not sure what issues were under consideration by the review.[84] Despite this difficulty (along with other acknowledged limitations),[85] the IIED released a final report on the consultation in July 2008.

The consultation document was based on the Bank's view that a

80 Global Environment Facility (1995), *Operational Strategy of the Global Environmental Facility – Revised Draft GEF Operational Strategy,* Ch. 1, Pt 1.

81 Ibid. Ch. 1, Pt 5.

82 The other six considerations are that the project must: i) be consistent with national and, where appropriate, regional initiatives; ii) strive to ensure the sustainability of global environmental benefits; iii) reduce the risk caused by uncertainty; iv) facilitate effective responses by other entities to address global environmental issues; v) be environmentally, socially and financially sustainable; and vi) avoid transfer of negative environmental impacts between focal areas.

83 This author participated in the original consultation. I found it difficult to offer constructive feedback as the concept was so loosely defined. I was, however, interested in the World Bank pursuing this idea, as no other international institutions are seeking to address the issue of integration among existing global forest initiatives.

84 International Institute for Environment and Development (2008), *Towards a Global Forest Partnership: Consultation, Assessment and Recommendations.*

85 Ibid. 26.

new approach was urgently needed in relation to the old and emerging challenges evident in the forestry sector. The Bank believed that a new approach must link changing local needs with the global forest agenda, capture the true values of forest areas, reduce incentives for forest degradation, and improve the resilience of forest ecosystems and forest communities.[86] The tentative vision and purpose of the global forest partnership is stated as:[87]

> People's forestry: empowering stakeholders to improve livelihoods and ecosystem services through collaborative local, national and international initiatives and supportive investment.

This vision should be amended to incorporate the essential component of the partnership (namely, the creation of a partnership linking current forest initiatives and stakeholders). It is envisaged that the partnership will carry out the following types of international activities. There will be[88]

- Tiered forest stakeholder forays from national to regional to global levels
- A menu-based mechanism for technical and financial support
- Networks for research, learning, knowledge sharing and monitoring
- Preparation of forest investment portfolios
- Amalgamation of some existing forest partnerships
- Roles in global forest financing mechanisms
- A small secretariat to assist the global forest partnerships work.

The report – entitled *Towards a Global Forest Partnership: Consultation, Assessment and Recommendations (by the IIED) Final Report 2008* – makes recommendations about the essential ingredients that a global forest partnership should include. The overarching recommendation of the report is for a three-year development process which establishes the goals of the proposed partnership. The World Bank is currently exploring the role it will play in such a partnership.[89] It needs to be careful in framing

---

[86] Ibid.
[87] Ibid. 13.
[88] Ibid. 16.
[89] World Bank (2009), 'Exploring a New Approach to Partnerships in Forestry', available at http://web.worldbank.org/WBSITE/EXTERNAL/TOPICS/EXTARD/EXTFORESTS/0,,contentMDK:21840644~pagePK:210058~piPK:210062~theSitePK:985785,00.html (accessed 26 March 2012).

its role because feedback from respondents highlights weaknesses in previous Bank environmental projects.[90] It is commonly accepted, however, that the Bank should take a leading role in the funding of the partnership and its processes. This presents a challenge for the future of the partnership, as the Bank is unlikely to heavily fund an initiative without some degree of control over the objectives and policies of the partnership. The future of this partnership will be dependent upon the World Bank taking a leading role in its formation, because no global institution has expressed an interest in convening such a task.

The future of the global forest partnership is uncertain. The World Bank should be encouraged to lead the creation of this partnership, as it has the potential to deliver something missing from existing forest programs (that is, a system that joins closely linked initiatives and creates partnerships between overlapping programs). This role is that of a meta-governance position, where 'meta-governance' can be defined as 'the strategic steering and coordination of the governance of a system'.[91] If successful, this meta-governance could create clear thematic areas of work for forestry and engage relevant stakeholder institutions in working collaboratively to resolve globally agreed ambitions. This integration of forest priorities might well be the catalyst required for achieving progress in the implementation of sustainable forest management.

3. INTERNATIONAL FOREST INSTRUMENTS CREATED BY THE WORLD BANK

A. The World Bank Forest Strategy

Staff members of the World Bank use a document entitled *Sustaining Forests: A Development Strategy* when assessing loan and development applications that concern forest interests. This is the primary policy document used when assessing projects that will have a causal effect on surrounding forest areas. It provides a framework for the Bank to assess applications when they involve forest areas, or when they will affect forest areas. The strategy contains three key elements[92]

---

90 International Institute for Environment and Development, above n 85, 7.

91 Visseren-Hamakers, Ingrid, and Peter Glasbergen (2007), 'Partnerships in Forest Governance', *Global Environmental Change*, **17** (3–4), 408–417, 409.

92 World Bank, *Sustaining Forests: A Development Strategy* (2004) 2.

1. Harnessing the potential of forests to reduce poverty:
   - strengthening the rights of people, especially marginalized forest groups, and fostering their participation in forest management;
   - promoting sustainable forestry, community forestry and agroforestry
2. Integrating forests in sustainable economic development:
   - improving forest governance, and introducing legal and institutional reforms;
   - encouraging investments that catalyse production of forest products, including environmental services
3. Protecting vital local and global environmental services and values:
   - establishing protected areas;
   - improving forest management in other areas;
   - developing markets and finance for international public goods such as biodiversity and carbon sequestration, and helping governments create national markets for environmental services from forests;
   - addressing cross-sector links that affect environmental values.

The Bank now acknowledges that policies beyond the forest sector (for example, land reforms and planning laws) often have a more severe impact on forests and forest-dependent people than forest sector policies, themselves.[93] As such, non-forest interventions (such as rural development, infrastructure programs, and economic adjustment measures) must be carefully formulated, taking into consideration their impact on forest areas. This integrated approval process attempts to overcome shortcomings in the Bank's earlier forest policies.

The current strategy replaces previous forest policies released in 1978 and 1991.[94] The 1978 forest strategy received criticism for being industry-focused (as opposed to community-focused). This was due to the policy's support of commercial logging.[95] The policy indicated that commercial logging would assist local communities by providing much-needed revenue and investment at the local level. A number of campaigns were waged against the Bank to halt its investments in projects, which led to significant deforestation and forest degradation.[96] Due to the extreme

---

93 Hermosilla, above n 42, 2.

94 See Humphreys above n 35 for commentary on old forest policies.

95 Ibid. 169.

96 In particular, the Polonoroeste project in the Brazilian Amazonia led to substantial deforestation. For further information on Bank investments leading to deforestation see Ibid. 170.

pressure from these campaigns, the Bank was forced to reconsider its forest policy.

This led to the introduction of the 1991 Forest Strategy. This strategy contained a firm commitment that 'the Bank group will not under any circumstances finance commercial logging in primary tropical moist forests'. Additionally, the Bank had decided to address forestry issues in a multi-sectoral context (meaning that it would look at the impact of developments outside the forestry sectors, which impacted upon forest resources). This policy was strongly endorsed by civil society; however, it is now acknowledged that the Bank lacked the mechanisms to implement the objectives of this strategy.[97]

The 2002 strategy departed from earlier ones in a number of ways.[98] First, the focus of the 2002 strategy was on poverty reduction and the use of forest resources to improve the livelihood of forest-dependants. This is a change from the earlier focus on combating deforestation and encouraging the conservation of remaining forest areas. Secondly, the strategy now focuses on all types of forests, as well as forest-rich and forest-poor countries. Earlier forest strategies placed most emphasis on the conservation of tropical forest areas. Thirdly, the strategy focuses on improving the management of productive forest areas. It is now recognized that productive forest areas can be managed sustainably to provide a number of forest services. This recognition meant that the 1991 ban on tropical forest harvesting programs was lifted. Fourthly, the new strategy focuses on ways to enhance the quality of governance, particularly issues related to illegal logging, corruption and forest-related trade. Fifthly, the strategy uses the concept of 'forests' as opposed to 'forestry' to show the conceptual change in understanding of forests' multiple values. And sixthly, the 2002 strategy is to be implemented through the formation of partnership at the global, national and local levels.

The International Finance Corporation and the Multilateral Investment Guarantee Agency are responsible for implementing the forest strategy. The strategy recognizes five global challenges relating to forests[99]

1. Failure of governance, leading to the loss of ecosystems services, livelihood services, and forest capital due to the provision of forest concessions and illegal harvesting
2. The need to integrate forests in the fight against rural poverty.

---

97 Humphreys, above n 35, 171.
98 Hermosilla and Simula, above n 42.
99 World Bank Group, above n 2, 17–18.

In many developing countries, rural poverty cannot be reduced unless forest resources are sustainable, developed, and used more efficiently.

3. The contentious nature of forestry policy and reform. There is a need to deal with the inherent conflict present in resolving forest issues. This requires strong institutional frameworks and an overarching effective legal and regulatory regime. This regimen must recognize and protect the rights of the forest-dependent people, while also allowing for the development of sustainable forest businesses.
4. The failure of markets to capture the true value of forest environmental services. Well-established markets exist for forests timber and fuel wood values. However, other forest values (such as contributions to the environment, biodiversity, and the stability of global climate) are not recognized in terms of financial value and, therefore, go unrecognized in markets.
5. The need for accounting systems, which prescribe the global values of forests. At the global level, forests operate as sinks for carbon, and house the majority of terrestrial biodiversity. These values are, therefore, global forest values (which require global initiatives).

The three pillars of the strategy are implemented with the following four considerations in mind. First, project selectivity is important. Projects that are allocated funding must provide dual benefits, either by building in poverty reduction strategies, or by integrating forest practices with larger cross-sectoral and broad economic programs. Secondly, the Bank views the building of partnerships as a crucial element in the building of an effective global forest monitoring system. Three of these partnerships are discussed in detail above. The Bank is also interested in strengthening its relationship with a number of other institutions.[100] Thirdly, the coordination of forest action and policy across the five institutions of the World Bank Group to ensure consistency is an internal structural adjustment to be fulfilled by the Bank. And fourthly, the creation of financial mechanisms to ensure that adequate funds exist for the implementation of the strategy will require the development of blended financial arrangements from multiple sources to reduce the overall financing cost to client countries.

---

[100] These include the Collaborative Partnership on Forests (global research network), Food and Agricultural Organization of the United Nations, International Tropical Timber Organization (private intergovernmental institution), among a number of others. See World Bank, above n 3, 46–51 for further detail.

The policy creates a concept of 'country ownership'. This concept is used throughout the forest strategy, and is defined as a commitment by the borrower and other major local stakeholders to the objectives and means proposed by the Bank.[101] This concept, essentially, requires that countries abide by Bank policy and strategy to obtain funding for projects. Furthermore, the use of the word 'ownership' is evidence of the Bank attempting to place responsibility on countries to abide by the conditions of the grant. Thus, the provision of Bank funds comes with conditions. Depending on the nature of these, it may suit both the Bank's and recipient country's forest objectives.

B. Operational Policies and Bank Policies

The forest strategy is used in conjunction with operational policies and Bank policies. The Bank has ten environmental policies that are based on environmental assessments, natural habitats, pest management, involuntary resettlement, indigenous people, forests, cultural resources, dam safety, international waterways and disputed areas.[102] The relevant operational and Bank policy for forests is OP 4.36 and BP 4.36. Each project must follow these procedures to ensure that a consistent approach is taken to achieving the Bank's corporate priorities and strategic goals. Operational policies are short statements that establish the parameters for conducting operations, describe the circumstances in which exceptions to policy are admissible, and spell out who can authorize any exceptions.[103] Bank procedures explain how staff members carry out operational policies by describing the procedures and documentation required to ensure consistency and quality across the Bank.[104]

The policy objective of OP 4.36 is to assist borrowers to harness the potential of forests to reduce poverty, integrate forests effectively into sustainable economic development, and protect vital local and global environmental services and forest values.[105] Funding is provided under the policy in three instances: where the project will have or may have impacts

---

101 World Bank Group, above n 3.

102 World Bank Group, above n 1, 56.

103 Ibid. 55.

104 Ibid.

105 World Bank (2002), 'Operational Policy 4.36 – Forests', available at http://web.worldbank.org/WBSITE/EXTERNAL/PROJECTS/EXTPOLICIES/EXTOPMANUAL/0,,contentMDK:20064668~menuPK:64701637~pagePK:64709096~piPK:64709108~theSitePK:502184~isCURL:Y,00.html (accessed 26 March 2012), [2].

on the health and quality of forests;[106] for projects that affect the rights and welfare of people and their level of dependence upon or interaction with forests;[107] or, for projects that aim to bring about changes in the management, protection, or use of natural forests or plantations, whether they are publicly, privately or communally owned.[108] These categories are broad in their approach and, as such, most forest projects would fit easily within one (if not two) of these categories.

The Bank will not finance projects that involve significant conversion or degradation of critical forest areas. 'Critical forest areas' are defined to mean: existing protected areas and proposed protected areas; areas protected by traditional local communities; sites that maintain conditions for the viability of these protected areas; areas with high suitability for biodiversity conservation; and sites that are critical for rare, vulnerable, migratory or endangered species.[109] However, the Bank will finance projects that involve significant conversion or degradation in non-critical forest areas, provided that the development of the project is offset or mitigated.[110] This is a departure from previous Bank policy, which banned the conversion of tropical forests under any circumstances.

This ban was strongly endorsed by civil society groups, who were disappointed with the reversion to policies that allowed for significant disturbance to forest ecosystems.[111] The Bank explains the change of policy with reference to implementation reports on its forest strategy, which demonstrate that the ban did not assist local communities who were dependent upon forests (therefore not meeting policy objectives).[112] Under the policy, conversion of non-critical areas only occurs when there is no feasible alternative and the overall benefits from the project substantially outweigh the environmental costs. These requirements may be met easily, especially when large sums of Bank finance are on offer. The Bank does not appear to have substantial environmental offset or mitigation requirements.

The policy allows the Bank to finance commercial harvesting within areas that are deemed to be non-critical forests and non-critical natural habitats. To obtain Bank finance for these operations, independent forest certification must be obtained. Forest certification groups endorse on the basis that sustainable forest management practices are followed. A

---

106 Ibid. [3a].
107 Ibid. [3b].
108 Ibid. [3c].
109 Ibid. Annex A, definition (c).
110 Ibid [5].
111 Humphreys, above n 35, 178–9.
112 World Bank Group, above n 3.

number of different forest certification bodies exist. No particular forest certification group is specified but the Bank's requirements align closely with the procedures of the Forest Stewardship Council.[113] The financing of commercial harvesting initiatives may not sit comfortably with groups that are unyielding in their approach to forest conservation and protection. On the other hand, it should be noted that the establishment of commercial harvesting plants that are certified to be operating sustainably have the potential to provide timber for the growing global timber market. They would also provide local communities with a sustainable industry that they can use to leverage themselves out of poverty. This, of course, is a vision of what sustainable timber harvesting can achieve. In many instances, and for a wide variety of reasons, harvesting operations will not result in such a successful outcome.

To acquire Bank finance, the borrowing country must provide the following information: the borrower's overall policy framework; national forestry legislation; institutional capabilities; and identification of the poverty, social, economic or environmental issues related to forests.[114] The Bank approves finance on the basis that changes are made to strengthen the fiscal, legal, and institutional capacity of the borrower's regulatory framework to improve the forest project's implementation.

Bank Policy 4.36 outlines the process and instruments required to implement the forest strategy and forest policy. BP 4.36 contains two sections: project preparation; and project implementation and supervision. The preparation stage ensures that the project is examined under the relevant environmental assessment.[115] The task team also prepares an assessment of the adequacy of the land use allocations for the management, conservation, and sustainable development of forests. This assessment takes into consideration the existence of critical forest areas and the impact of the project on local communities. The assessment has additional requirements if the project involves commercial harvesting[116] or community-based

---

[113] Humphreys, above n 35, 179–80.

[114] World Bank (2002), 'Operational Policy 4.36 – Forests', available at http://web.worldbank.org/WBSITE/EXTERNAL/PROJECTS/EXTPOLICIES/EXTOPMANUAL/0,,contentMDK:20064668~menuPK:64701637~pagePK:64709096~piPK:64709108~theSitePK:502184~isCURL:Y,00.html (accessed 26 March 2012) [14].

[115] World Bank (2002), 'Bank Policy 4.36 – Forests', available at http://web.worldbank.org/WBSITE/EXTERNAL/PROJECTS/EXTPOLICIES/EXTOPMANUAL/0,,contentMDK:20141282~menuPK:4564187~pagePK:64709096~piPK:64709108~theSitePK:502184,00.html (accessed 26 March 2012) [4].

[116] Ibid [5].

forest management and development.[117] During the project implementation and supervision, the regional vice-president ensures the availability of resources, and each project is supervised in accordance with Operational Policy 13.05 Project Supervision.

The forest strategy recognizes that internal modifications need to occur so that the Bank's operational policy is consistent with the strategy.[118] Of relevance are: operational policies 8.60 (which governs the implementation of structural adjustment programs); operation provision 4.01 (which deals with the impacts of the proposed activity on the natural environment, human health and safety, and social considerations); and operational policy 4.04 (which requires consideration of critical natural habitats). This internal adjustment is a major theme of the report on the review of the forest strategy (discussed in the next section).

C. Useful Instruments Generated from Partnerships

Two instruments generated from the Bank's partnerships are worth closer examination. The World Bank and World Wide Fund for Nature Alliance produced a governance tool that creates a self-assessment regimen (along with guidance on the how to create forest targets). This instrument was prepared by the International Institution for Environment and Development (on behalf of the Alliance) and is entitled *The Pyramid: A Diagnostic and Planning Tool for Good Forest Governance*.[119] The second instrument is a *Poverty Forestry Toolkit*[120] prepared by the Program on Forests. This toolkit provides a simple methodology that can be used to describe the forest-poverty nexus.

Forest pyramid

As the name suggests, the pyramid creates tiers that are required for the implementation of sustainable forest management. Each tier of the pyramid is accompanied by a questionnaire that countries can use to self-assess their progress within the particular tier. Each questionnaire focuses on an element of good governance and on three questions: 'What's working?', 'What's missing?', and 'What needs to be done?'. The straightforward three-question approach may not be the traditional

---

117 Ibid [6].

118 World Bank Group, above n 3, 21.

119 Mayers, J., S. Bass and D. Macqueen, International Institute for Environment and Development (2005), *The Pyramid: a Diagnostic and Planning Tool for Good Forest Governance*.

120 Program on Forests (2007), *Forest Poverty Toolkit*.

format with which legal policy makers are acquainted but the simplicity of this approach makes self-assessment straightforward. There is, of course, room for improvement within the instrument. Some of the elements within the questionaries require clarification and a report which explains each criteria of the tier may be of assistance to regulators. Despite this, the overall purpose and vision of the pyramid is useful for forest regulators. Each tier of the pyramid will be briefly summarized below.[121]

The foundation tier of the pyramid requires the building blocks of forest management to be established. The first of these is the creation of property and tenure rights over the forest area. Also, the operation of the local constitution needs to be taken into account to determine the level of government responsibility and any overarching constitutional requirements related to forest management. The local market and investment conditions must also be considered, as these relate to forests. The major forest institutions must be identified across all sectors (public, private and local).[122]

The tier for 'roles' requires widespread acceptance that there are multiple valid perspectives and stakeholders within the forest-related sectors. Information needs to be obtained on present and future uses, assets and demands. As well, basic forest institutional architecture, structures and decision-making rights and powers must be agreed upon and put in place.[123]

The tier for 'policies' requires an agreed vision of forest management with the roles and basic institutional architecture of the forest sector recognized in central forest policies and laws. There must also be clear, equitable and legally defensible rights in place to manage forest areas and to extract resources from public forests for full economic compensation. Also, relevant stakeholders must be made aware of their rights (especially holders of customary rights).[124] Within this tier, a definition of sustainable forest management needs to be adopted that contains a methodology against which progress towards sustainable forest management can be assessed.[125]

The 'instruments' tier requires the creation of regulatory instruments that provide for: forest tenure rights within an allocation system; protection of public and intergenerational interests in forests; forest

---

[121] There are a number of requirements for each tier, the critical aspects of the tiers are summarized here but this is not a complete list of all of the requirements of each tier. Please see the report for further information.

[122] Ibid.

[123] Ibid. 19.

[124] Ibid. 20–1.

[125] Ibid. 22.

management and investment conditions and control; market access for stakeholders; anti-corruption provisions; and a revenue system dealing with domestic and international forest product export prices. Capacity to create and carry out these regulatory requirements must also be developed.[126]

The 'extension' tier involves reaching beyond the immediate forest stakeholders and educating the wider community about the importance and place of sustainable forest management. This will include forest consumers, forest producers (such as investors, processor, middlemen, retail and the general public).[127] This tier is important because, without wide support and understanding of sustainable forest management, the implementation and effectiveness of the concept will be undermined. 'Extension' is an area that most developed and developing countries could improve upon.

The 'verification' tier involves the independent assessment of forest industry practices. This assessment can be fulfilled by meeting the requirements of certification schemes or any other reputable audit forest schemes.[128] 'Verification' means that forestry practices must be transparent: it increases the accountability of those in charge of the industry. It also means areas that require improvement are identified so that solutions can be created.

Poverty Forestry Toolkit

The Poverty Forestry Toolkit creates a methodology that can be used to assess linkages between forests and poverty at the local and national level. This is the first methodology that provides 'how to' guidance on examining forest-poverty synergies for policy purposes. The purpose of the toolkit is to: rapidly assess dependence on forests; provide a vehicle for poor people to comment upon forest laws and policies; identify policy problems and opportunities; deliver value where there is a willingness to listen to results from key stakeholders; and deliver data on topics impossible to access quickly through orthodox quantitative methods.[129]

The methodology sets out three stages of investigation. Stage one provides a process for selecting a suitable sample group to observe. Guidance is provided on the type of sites that should be selected (sites in poorer areas

---

126 Ibid. 24.
127 Ibid. 25.
128 Ibid. 26.
129 Program on Forests, above n 121, 8.

of country, sites with a wide range of forests, and sites where a variety of forest tenures exist).[130] During this stage, contact should be made with local and national governments to explain the purpose of study, and an understanding should be gained of the different roles that the various levels of government play in forest and poverty policies.[131] Understanding the role of different governments in regulation is crucial, because this will ensure that results are conveyed to the appropriate place. This may result in further action at the appropriate government level once the assessment is complete and data are delivered.

Stage two involves visiting the selected sites and meeting with village officials. Interestingly, the toolkit emphasizes that because the data gathering will not result in direct benefits to the communities, it is essential to compensate the villagers for their time. It is suggested that providing luxury food items from town (such as rice and cold drink, and a slaughtered animal for roasting)[132] would be appropriate. The acceptability of these as payment serves as a reminder of the level of poverty being confronted.

Stage three outlines the process and key questions of the methodology. In total there are eight tools[133]

1. Wealth ranking tool – the attributes of the rich, average, poor, and very poor are identified. These attributes will vary from location to location[134]
2. Local landscape analysis – understanding the way in which members of the village use local resources
3. Timeline and trends – records a short history of the community so that changes in the use of forest resources, agriculture use, local livelihood strategies, and sources of income can be identified
4. Livelihood analysis – carried out to discover the extent of cash and subsistence reliance on forest resources, and the proportion of the total annual livelihood that comes from them
5. Ranking the importance of tree and forest products – undertaken to determine which forest products are used to generate income, and which are used for subsistence

130 Ibid. 1.
131 Ibid. 2.
132 Ibid. 6.
133 Ibid. 9–43.
134 For example, in Tanzania, animal ownership and the amount of land held were key attributes, while costly housing and consumer items were not regarded as very important. Ibid. Pt 2, 9.

6. User rights, duties, and benefits – carried out to determine all forest stakeholders and their associated rights and uses of the forest
7. Forest Problem and Solution Matrix – conducted to identify and rank the main forest problems (including legal and policy problems)
8. Final plenary – used to present the main finding from the subgroups (so that key emerging issues can be summarized).

Once this process has been followed, the knowledge gained from the process must be disseminated to a number of different groups. The information needs to be prepared at the district level, with an emphasis on identifying issues of special relevance to district authorities.[135] Information also needs to be disseminated at the national level to relevant ministries, so that these bodies can include poverty-forestry linkages within relevant instruments. Poverty Reduction Strategy Processes are a requirement for World Bank funding. Information gained through the use of the toolkit will assist countries in completing these strategies and will clearly identify the poverty-forest relationship at the relevant local level. This toolkit has the potential for a wide-ranging application by a variety of participants (including governments, NGOs, international and intergovernmental institutions, and community groups). It provides a much-needed analytical framework for data collection, and identifies avenues where this data can be used. Policy change and 'on the ground' action cannot occur without knowledge of local conditions. The toolkit has the ability to generate this knowledge.

4. IMPLEMENTATION OF WORLD BANK FOREST INITIATIVES

A. Implementation of the Forest Strategy

The implementation of the World Bank Forest Strategy and associated Forest Partnership measures involves implementation at a number of different levels. First, there is implementation at the international level, which requires internal bank instruments and policies to be consistent and supportive of the forest policy. Implementation at this level also requires consideration of the Bank's capacity to finance the implementation and provide appropriate staffing. Secondly, there is implementation of the international policy at the individual country level. World Bank forest

---

[135] Ibid 22.

policy may influence domestic forest policy or may be directly implemented if bank finance is provided. It has been suggested that the ultimate aim of environmental agreements is to influence the behaviour of those who cause or can ameliorate the problem at hand.[136] Changes in target behaviour are the ultimate goals of international environmental agreements; however, these agreements usually go through several stages, or levels, before they influence the target group.[137] Effective implementation, therefore, requires widespread institutional acceptance – and a widespread understanding by stakeholders of their rights and responsibilities within the framework.

A report is available that reviews the implementation of the 2002 World Bank Forest Strategy.[138] The overarching recommendation of the report is a call for greater incorporation of the Bank's forest strategy into core bank policies and operations. This requires internal understanding of the forest strategy's role and the purpose of the Bank's forest partnership and forest engagement relationships. It appears that there is internal confusion as to the operation of the forest strategy. The fragmented nature of the Bank's forest framework could be used as a justification for the creation of the Global Forest Partnership. As discussed earlier in this chapter, this partnership could create thematic program areas, and this would help to increase the clarity of the Bank's forest regulatory framework.

Specifically, the review of the World Bank Forest Strategy suggests four avenues that could be used to incorporate the forest strategy more meaningfully[139]

- Mainstreaming the forest strategy into key Bank instruments (such as poverty reduction strategies, country assistance, development policy loans and broader investment projects)
- Using the forest strategy to promote investment lending in forest projects
- Enhancing the role of partnerships and linkages to implement the forest strategy
- Building due diligence to ensure that key bank operational policies

136 Victor, David, Kal Raustialia and Eugen Sknolnikoff (eds) (1998), *The Implementation and Effectiveness of International Environmental Commitments: Theory and Practice*, Laxenburg, Austria: International Institute for Applied Systems Analysis; Cambridge, Mass: MIT Press, 4.

137 Ibid.

138 Hermosilla and Simula, above n 42.

139 This list has been created after summarizing the main points from the report.

trigger the application of the forest strategy, and identify avenues for improvement in the operation of the forest strategy.

The review team carried out a survey of poverty review strategies, country assistance strategies, and development policy loans to determine if the three pillars of the forest strategy were incorporated into these operations.[140] Poverty-reduction strategies are used to describe a country's macroeconomic, structural, and social policies as well as programs to promote growth and reduce poverty.[141] Of the 43 sampled strategies, only 28 discussed the relationship that exists between forests and poverty.[142] This demonstrates a lack of recognition within policy instruments about the nexus between forest dwellers and poverty. The World Bank should introduce a mandatory component within poverty-reduction strategies discussing the forest/poverty nexus.

Country assistance strategies are designed to promote collaboration and coordination among developing partners in a country. They are used to identify the key areas where Bank assistance can have the biggest impact on poverty reduction. Of the sampled country assistance strategies, it was found that only 23 out of 53 contained actions to address forest-poverty relationships. Furthermore, it was found that even amongst the 23 that considered this linkage, the analysis required significant improvement. This suggests a need for information and training to explain the relationship between forests and poverty and potential policy solutions.

Development policy lending is aimed at supporting the implementation of policy reform at either the macro or sector level through rapidly disbursing financing.[143] This policy lending can have a significant impact upon forests, and these impacts need to be considered. The review found that relatively few lending policies included assessment of the potential effects of policy reform on forests beyond the statement that there was 'no likely significant effect'. This indicates a lack of integration of the forest strategy into Bank operations.

In relation to investment lending, the review found that since the introduction of the Forest Strategy in 2002: 12 forest projects were implemented; 39 projects with a forest component were implemented; and there were 13

---

[140] The three pillars of the forest strategy are: i) harnessing the potential of forests to reduce poverty; ii) integrating forests in sustainable economic development; and iii) protecting vital local and global environmental values.

[141] Hermosilla and Simula, above n 42, 6.

[142] Ibid. 7.

[143] Ibid. 10.

projects in the pipeline (4 of which were forest projects).[144] Comparing these figures with earlier Bank forest investment figures demonstrated that there had been no expansion in forest lending since the introduction of the Forest Strategy.[145] The review also considered the compatibility of these projects with the forest strategy. Generally on this point, it was found that the projects better addressed pillar two[146] and pillar three[147] of the forest policy. It was found that, in relation to poverty alleviation, only 14 out of 40 sampled considered the poverty-forestry connection.[148] This is despite the stated overarching goal of the Forest Strategy to be one of poverty reduction.

The review also recommended greater involvement by the international finance corporation in the implementation of the forest strategy. This body is responsible for lending finance directly to private business interests. The review believed that investment in sustainable forestry projects could be made more appealing to private investors if the Bank and the international finance corporation jointly used their existing expertise to channel additional funding into forestry projects. Private investment in sustainable forest management is low due to: i) the shortage of sustainable private operations; and, ii) a perceived reputational risk that could arise from investing in forest projects.[149] 'Reputational risks' are associated with investments that take over indigenous peoples' land, displace peasant farmers, marginalize small landholders, lack adequate stakeholder considerations, or fail to carry out adequate impact assessments.[150] The review suggests that these reputational risks can be avoided. This can be achieved by the Bank (with assistance of NGOs) demonstrating that sustainably managed and certified forest operations can generate important social and environmental benefits, and can reduce pressure to convert these lands to other uses.[151]

The review also found that the forest operational policy (OP 4.36) was not applied to all forest projects. The operational policy was applied in 79

---

144 Ibid.

145 Ibid.

146 Pillar two is about integrating forests in sustainable economic development – out of the 40 sampled projects, 27 considered this pillar highly, while 6 projects gave it substantial consideration. See ibid. 21.

147 Pillar three is protecting vital local and global environmental values – out of the 40 sampled projects, 26 considered this pillar highly, while 7 gave it substantial consideration. See ibid.

148 Ibid. 21.

149 Ibid. 30.

150 Ibid.

151 Ibid.

per cent of cases but it is still of some concern that such a key Bank policy was not triggered for use in 21 per cent of forest projects. No reason was found to explain the lack of application, though it was inferred (from Bank staff) that the application of operational policies slows down and increases the cost of projects – this results in Bank staff applying fewer operational policies then might be required.[152] The review recommends that bank staff be educated on the importance of complying with operational policies. These are important, because they reduce project risks and improve project quality – and, in relation to forestry, encourage the implementation of sustainable forest management compatible with the Bank's forest strategy.[153]

5. CONCLUSION

The World Bank's approach to forest policy is innovative in a number of ways. This innovation is linked to the banks integrated approach, internally and externally. Internally, the Bank's main goal of poverty reduction means that its forest policy is aligned with this goal – and this means that the Bank's forest policy is more integrated with social and economic considerations than most other international forest policy.

Externally, the Bank has formed a number of strategic partnerships. These have proved effective in terms of pooling expertise and resource, as well as in removing some of the policy duplication within international forest institutions. Furthermore, these partnerships have, in some instances, involved a target-driven approach. Setting and reporting on targets increases the transparency and accountability of the institution. Targets might also assist in increasing compliance and implementation of forest policy. Parties involved in target-driven approaches will feel more compelled to comply when they must report upon the progress made towards reaching the target. The Bank is the only international institution to have adopted a target-driven approach to forest regulation and other international forest institutions should trial them.

The Bank should also be commended for starting an international process aimed at integrating global forest governance arrangements. International forest governance is exceptionally fragmented and, as a result, there is great deal of overlap and duplication between existing international forest processes. The Bank is the most suitable international

---

152 Ibid. 44.
153 Ibid.

institution to lead governance reform of international forest arrangements because it has practical experience in using integrated policy and partnership approaches. The Bank should capitalize upon this experience in leading the reform of global forest governance arrangements.

PART IV

Regulation by non-state forest institutions

Non-state regulation differs significantly in nature from United Nations or public international regulation. The first significant difference lies is the absence of a requirement that policies be adopted on the basis of consensus and be reflective of all stakeholders' perspectives. While parties engaged in public international regulation often aim to achieve international cooperation over a particular issue, parties involved in non-state regulation tend to be more focused on seeking compliance from those interested in implementing their standards. This provides a clear advantage for non-state regulation, in that the time dedicated to negotiating policy can be significantly reduced compared to negotiating timeframes within public international regimes. Generally, membership of non-state regulatory bodies is open to all interested parties, but confirmed by complying with certain conditions and fulfilling certain requirements.

Secondly, non-state forest regulation is voluntary in nature. The schemes are voluntary in the sense that international political pressure to participate in these schemes is markedly less than those associated with participation through the United Nations' approaches. Engagement with non-state regulatory approaches is, therefore, based upon the perceived benefits associated with participating in the scheme (improved forest outcomes, increased price premium for timber, green credentials). The emergence of new voluntary-based instruments could be linked to responsibilities created by the concept of corporate environmental responsibility. Non-state regulation works best when there is a high value placed on the benefit associated with participating with the scheme.

Another feature of non-state regulation is the use of market mechanisms as a major instrument of enforcing policy and regulation. The two forms of non-state forest regulation examined within this book both involve market mechanisms. The forestry markets explored in Chapter 8 (also known as Payment for Ecosystem Services Schemes, Environmental Banking Schemes, or Environmental Offset Schemes) all trade in ecological forest services. Whilst the other form of private forest regulation (forest certification schemes), which is explored in Chapter 9, relies upon an eco-labelling approach to send messages to consumers of forest products concerning the sustainability of the product.

8. The legal requirements of forestry markets

1. INTRODUCTION

Forest regulation has been significantly influenced by market-based regulation, with economic instruments increasingly playing a more dominant role in public and private forestry institutions.[1] The term 'forest market instrument' is used throughout this chapter to describe transactions in which a forest ecosystem service is traded in return for financial remuneration. Market mechanisms are an appealing tool for regulators, as they promise to deliver both improved environmental outcomes and the necessary finance to do so. There are, however, a number of challenges in designing forest market mechanisms that have the ability to improve forest conditions and management, and deliver the necessary finance to fund such improvements. For forest market mechanisms to be more widely used and applied, the demand for 'forest ecosystem credits' needs to be increased. As this chapter will detail, there are a number of avenues open for increasing the demand of forest ecosystem credits. The majority of forest market mechanisms are in the early stages of project implementation. It is, however, anticipated that the introduction of emission trading schemes will generate a market for forest ecosystem services along with other drivers such as environmental offset or environmental banking policy.

[1] Brand, David, 'Forest Investment and Emerging Environmental Markets' (Paper presented at the second Annual Green Trading Summit, New York, 7–8 April 2003); and Scanlon, Jane (2007), 'An Appraisal of the NSW Biobanking Scheme to Promote the Goal of Sustainable Development in NSW', *Macquarie Journal of International and Environment Law*, **4** (1), 71–133, 89.

2. THE EMERGENCE OF ENVIRONMENTAL ECONOMIC REGULATION

A. Economic Environmental Instrument Evolution

Early environmental regulation was simplistic in nature and design. In essence, it placed restrictions upon certain activities and imposed conditions on the carrying out of other activities.[2] Conventional regulation ('command and control' regulation) is generally believed to promote transparency of procedures, accountability of authorities, and participation by the public, with an overall objective of alignment with the public good.[3] Conventional regulation has, more recently, come under scrutiny for lacking in both efficiency and legitimacy.[4] In terms of efficiency, market proponents argued that market-based instruments are able to achieve the same outcomes at a lower cost. Furthermore, conservationists have long complained of the less than vigilant monitoring and enforcement of command and control regulation and perceive market regulation as offering a potential solution to improving the legitimacy of environmental regulation.[5]

In addition to efficiency and legitimacy concerns, there was also a belief that environmental resources were subject to overregulation.[6] Overregulation is blamed for hindering innovation and preventing the development of new approaches.[7] Highly regulated economies were considered to be at a significant disadvantage when compared with less-regulated ones. This was due to the belief that regulation increased transaction costs for industry, and that such systems failed to provide incentives for those who went above and beyond the minimal requirement.[8]

---

[2] Ibid.

[3] Ibid.

[4] Weale, A. (ed.) (1992), *The New Politics of Pollution*, Manchester; New York: St Martin's Press, Ch. 6.

[5] Eckersley, Robyn (1995), 'Markets, the State and the Environment: An Overview' in Robyn Eckersley (ed.) *Markets, the State and the Environment: Towards Integration*, South Melbourne: Macmillan Education Australia, pp. 8–9.

[6] Reitze Jr, A.W. (1984), 'Environmental Policy – It is Time for a New Beginning', *Columbia Journal of Environmental Law,* **14** (1), 111–156, 112–13.

[7] According to Pennington, government intervention can actively suppress the emergence of private approaches through the continued adherence to environmental policy prescriptions that do not allow markets to develop. Pennington, Mark (2001), 'Environmental Markets vs. Environmental Deliberation: A Hayekian Critique of Green Political Economy', *New Political Economy*, **6** (2), 171–190, 186.

[8] Klaus Bosselmann and Benjamin J. Richardson, 'Introduction: New Challenges for Environmental Law and Policy', in Klaus Bosselmann and

The concerns of efficiency, legitimacy, overregulation, and the lack of progress in achieving ecological sustainable development, led to calls for new regulatory approaches, designed to overcome the failings of traditional regulation.[9] Economic regulatory instruments seemed to be the answer, as these promised to generate incentives for ongoing environmental improvement and technological innovation by providing more flexibility – which would promote a wider range of actions from producers and consumers.[10] These new tools were also praised for their ability to generate income for governments and to directly promote the economically efficient allocation of scarce resources. Economists used the term 'cost-benefit analysis' to describe the process that some regulators follow when electing between alternative courses of action. Cost-benefit analyses have conventionally been used to weigh the benefits that would arise from a government project against the costs associated with it.[11] Economic instruments were, thus, a mechanism to implement this theory.

The original advocates of economic regulatory instruments were those with a self-interest in the creation of more flexible approaches (initially industry).[12] Gradually, governments also started to embrace the use of economic instruments in environmental regulation. Industry's initial enthusiasm waned when it realized that the implementation of such an approach meant the potential imposition of green taxes, environmental charges, or tradeable emission rights.[13] Economic instruments have not (as expected by their proponents) led to a decrease in regulation, but have, instead, involved the re-regulation of the environment through the creation of new institutions and instruments to provide for the operation of taxes, charges, and tradeable rights.[14] As such, economic instruments have suffered the

Benjamin J. Richardson (eds) Environmental Justice and Market Mechanisms: Key Challenges for Environmental Law and Policy, (1999), 1, 3.

9 Stewart, R.B. (1992), 'Models for Environmental Regulation: Central Planning versus Market-based Approaches', *Boston College Environmental Affairs Law Review*, **19** (3), 547–562.

10 Bosselmann and Richardson, above n 8, 7.

11 For an analysis of the benefits and pitfalls of 'gross national product' theory and 'cost-benefit analysis' economic theory, see: Beder, Sharon, (ed.) (2006) *Environmental Principles and Policies: An Interdisciplinary Approach*, Sydney: UNSW Press, Ch. 7.

12 Ibid.

13 Bosselmann and Richardson, above n 8, 7. Also see: Ashcroft, Ross (2008), 'Carbon Capture and Storage: A Need for Re-conceiving Property Interests and Resource Management in the Australian Legal System', *Journal of the Law Association for Asia and the Pacific*, **1** (1), 48–61, where the author discusses industry's response to the costs associated with carbon capture and storage.

14 Bosselmann and Richardson, above n 8, 4.

same fate as their regulatory counterparts: increasingly complex rules and regulations, with minimal environmental gains.

B. The Development of Economic Forest Instruments

Following the *Brundtland Report* (1987) and the Earth Summit (1992), there was consensus that the conservation of tropical forests required an integrated management approach.[15] The *Brundtland Report* identified the many interdependencies between economic and environmental administration. In recognition of these interdependencies, the report urged nations to anticipate and prevent environmental damage by integrating environmental policy with economic, trade, energy and agricultural policy.[16] This approach was thought to be able to circumvent environmental damage by incorporating environmental considerations into all facets of government decision-making processes. The *Brundtland Report* encouraged governments to make more effective use of economic instruments to promote clean production.[17]

At around the same time, debates were taking place over the best means of addressing poverty concerns and associated environmental degradation. There seemed to be a common understanding that the alleviation of poverty was central to ensuring the conservation of tropical forest areas. Again, this belief was founded on the principles of integrated sustainable management. In the forestry context, the two concepts that were viewed as a means of implementing such approaches were integrated conservation and development projects, and sustainable forest management.[18] These two concepts were designed to address problems holistically by applying an integrated regulatory approach. Unfortunately, these approaches did not deliver large-scale benefits for forests or peoples, so new regulatory approaches were again called for. The concept of payments for environmental services emerged. The core idea behind the concept of payments for ecosystems' services is that users of ecosystem services make direct, contractual, and conditional payments to local landholders who, in return, adopt practices that secure ecosystem conservation and restoration.[19]

This chapter focuses on the operation of one form of economic

---

[15] Wunder, Sven, 'Payments for Environmental Services: Some Nuts and Bolts' (Occasional Paper No 42, CIFOR, 2005).

[16] Eckersley, above n 5, 264–6.

[17] World Commission on Environmental and Development, *Our Common Future* (1990), 264–66.

[18] Wunder, above n 15, 1.

[19] Ibid.

instrument – that of forest markets. Forest market mechanisms have been justified on the basis that their inherent flexibility increases the recognition of a wide range of forest goods and services and, thus, leads to improved environmental outcomes.[20] Furthermore, markets mechanisms are seen as a tool that engages the private sector in environmental management by creating economic incentives for participation in various schemes. Forests provide a number of environmental services and markets for the following forest ecosystem services that have started to emerge:

- carbon sequestration;
- biodiversity conversation;
- watershed protection;
- landscape values;
- bundled forest services.[21]

The incidence of climate change has been described as the world's biggest market failure.[22] This is because greenhouse gas emissions have historically been considered as a market externality (meaning that the global rise of greenhouse emissions and the associated social, environmental, and economic implications have not been factored into the economic decision-making process). Proponents of market-based regulation were successful in influencing the design of the global regime addressing rising emission levels. Forest deforestation and degradation is now being seen by some as yet another example of market failure.[23] The degradation of forests has continued, according to market proponents, for two reasons: many forest goods and services are considered to be positive externalities; and forest services are considered to be public goods.[24]

A 'positive externality' is any form of uncompensated benefit.[25] Positive

---

[20] See generally: Pagiola, Stefano, Natasha Landell-Mills and Joshua Bishop (2004), 'Market-based Mechanisms for Forest Conservation and Development', in Stefano Pagiola, Natasha Landell-Mills and Joshua Bishop (eds), *Selling Forest Environmental Services: Market-based Mechanisms for Conservation and Development*, New York: Routledge, pp. 20–32, 261–274.

[21] Landell-Mills, Natasha and Ina Porras, International Institute for Environment and Development, *Silver Bullet or Fools' Gold? A Global Review of Markets for Forest Environmental Services and their Impacts on the Poor* (2002).

[22] Stern, Nicholas (2008), 'The Economics of Climate Change', *American Economic Review: Papers and Proceedings*, **98** (2), 1–37.

[23] Pagiola, above n 20, 2 and Richards, Michael (2000), 'Can Sustainable Tropical Forestry be Made Profitable? The Potential and Limitations of Innovative Incentive Mechanisms', *World Development*, **28** (6), 1001–1016, 1003.

[24] Pagiola, above n 20.

[25] Pigou's seminal work in the area discussed the concept of 'environmental

externalities associated with forest protection include: erosion control; reduced risk of flooding; and water quality maintenance.[26] Forest services are viewed as externalities because prices paid for forest goods do not represent the full cost of forest services. Similarly, when forestland is cleared for agricultural or rural expansion, the loss of forest services is usually unaccounted for. Markets typically fail to compensate those who produce positive externalities due to the absence of property rights, or other legal means to require payment for services rendered.

Forest environmental services can also be characterized as 'public goods'. A public good is a special type of externality, and is distinguished from other categories of externality by their 'non-excludability' and 'non-rivalry' characteristics. Fisher, Turner and Morling define 'rival' by stating 'To be "rival" means that if I use this good, there is less of it for you to use'; and define 'excludable' as a means of keeping others from using the good.[27]

Economists believe that government regulation can be justified when the service in question is a public good.[28] As forest services are mainly public goods, governments have taken direct control over their management and use. This has been achieved through the creation of public ownership rights and frameworks defining extraction and use rights. As the area of land under public ownership expands and the associated costs of maintaining such areas increases, governments are looking for alternative sources of income to manage them. This has contributed to driving the creation of markets in environmental services.

Where a market product is non-excludable and non-rival in nature, market formation will be difficult, because no real incentive exists to pay the supplier of the service. As long as an individual cannot be excluded from using a good, they have little reason to pay for access. Similarly, where goods are non-rival, consumers know that, when someone else pays, that they will also benefit.[29] Those who use services without paying are deemed to 'free-riders', and it is suggested that when everyone adopts a free-riding approach, the willingness to pay for a public goods will be zero and the product will not be supplied.[30]

---

externalities', and this theory has been used to justify the basis for pollution taxes. Pigou, A.C. (ed.) (1932), *The Economics of Welfare*, London: Macmillan and Co. Ltd.

26 Landell-Mills and Porras, above n 21, 7.

27 Fisher, Brendan, R. Kerry Turner and Paul Morling (2009), 'Defining and Classifying Ecosystem Services for Decision Making', *Ecological Economics*, **68** (3), 643–653, 647.

28 Landell-Mills and Porras, above n 21, 8.

29 Landell-Mills and Porras, above n 21, 8.

30 Ibid.

However, this is not entirely accurate in relation to the provision of environmental services. This is because the environment will continue to provide services irrespective of whether payments have been tendered by those using the service. In instances where payments are made to improve the management and ensure the long-term supply, then the user payments are more directly attributable to the production of the service.

To overcome non-excludability and non-rivalry issues, environmental markets require the creation of clear and enforceable property rights over natural resources.[31] The holders of these rights then have an incentive to manage the resource more efficiently. This line of reasoning can be linked to Hardin's 'tragedy of the commons' analysis.[32] This proposed that, when things are left open to the public, they are wasted by over-use or under-use. In such instances, there is minimal investment in property because there is no security associated with ownership. Interests may be lost at any time. Furthermore, unclear ownership rules prevent market exchanges. This ultimately results in users grasping what is available for capture in the short term, leaving behind a wasteland.[33] Hardin's theory has been used to justify the need for innovative property-based interests in natural resources.[34]

3. EMERGING ECONOMIC FOREST INSTRUMENTS

A. The Meaning of Economic Incentive Instruments

Economic instruments may take a variety of forms.[35] Despite this, all types of economic instrument use financial incentives or disincentives as the

---

[31] According to Bosselmann and Richardson, above n 8, 7, Coase's theory can be interpreted to mean that bargaining and compensation among property rights holders is the most efficient way to resolve independent uses of the environment so as to maximize net social welfare.

[32] The phrase 'tragedy of the commons' originates in Hardin, Garrett (1968), 'The Tragedy of the Commons', *Science*, **162**, 1243–1248.

[33] Rose, Carol (1986), 'The Comedy of the Commons: Custom, Commerce and Inherently Public Property', *The University of Chicago Law Review*, **53**(3), 711–781, 712.

[34] There are even instances where the development of property-based interests in natural resources had lead to improved management. For example, in relation to river management, the creation of property rights has led to improvements in the maintenance of fish stocks and the reduction of pollution. See Pennington above n 7, 184.

[35] Ali, Paul, and Kanako, Yano (eds) (2004), *Eco-Finance: The Legal Design and Regulation of Market-Based Environmental Instruments*, The Hague: Kluwer Law International, 7.

motivator for compliance. The Organization for Economic Cooperation and Development (OECD) deems 'economic instruments' to be tools promoting more cost-efficient decision-making, and creating level playing fields for trans-national economic activity. It defines economic instruments as those that affect the costs and benefits of alternative actions open to economic agents, with the effect of influencing behaviour in a way favourable to the environment.[36] An 'economic instrument' is, therefore, a tool which requires that a certain activity be carried out. It either provides an incentive for compliance, or a disincentive for failing to comply.[37] The focus of this chapter will be on the operation of 'market' economic instruments but to clarify the role of markets in regulation, other economic forest instruments will also be briefly examined.

Panayotou[38] classifies economic-incentive instruments into the following categories

- Property right-based systems – this involves the creation of clear property rights over environmental services
- Market creation – this requires creating of property rights, recognized in the form of credits or certificates, that can they be traded among market participants
- Fiscal instruments charge systems – to encourage responsible behaviour through the full or partial cost pricing of consumption or production
- Financial instruments – these aim to mobilize financial revenue for conservation and environmental protection through the use of green funds, favourable interest rates, and soft loans
- Liability instruments – these work by creating economic penalties for failing to comply with regulatory standards
- Performance bonds and deposit refund systems – these require the posting of a deposit, which is refundable only if the parties comply with environmental obligations.

Market-based instruments minimize the aggregate cost of achieving a given level of environmental protection and provide dynamic incentives for the adoption and diffusion of cheaper and better control technolo-

---

[36] Bosselmann and Richardson, above n 8, 7.

[37] See generally: Gunningham, Neil and Peter Grabosky (eds) (1998), *Smart Regulation: Designing Environmental Policy*, Oxford: Clarendon, 70.

[38] Panayotou, Theodore (ed.) (1994), *Economic Instruments for Environmental Management and Sustainable Development*, Nairobi: United Nations Environment Program.

gies.[39] Within the market-based instrument category, there are two broad forms.[40] First, systems based on the 'polluter pays' principle are ones that set a charge to reflect the cost value of damage caused to the environment by the polluter. Thus, the cost of mitigating the negative environmental outcomes is passed onto the parties responsible for depleting or damaging the environmental resource.

The second model involves creating a market that allows for trade in credits or permits. This type of system requires the creation of a baseline that determines the level of activity, which can be carried out within a particular sector. Participants in these schemes are allocated a number of credits and may only engage in the specified activity up to the level of credits they possess. If the participants reduce their level of activity, they can sell the excess credits to another scheme participant. Those participants who fail to modify are forced to pay for additional credits.[41]

B. Recognized Forest Ecosystem Services

> Forests regulate local and global climate, ameliorate weather events, regulate the hydrological cycle, protect watersheds and their vegetation and water flows and soils, and provide a vast store of genetic information much of which yet to be uncovered.[42]

It is recognized that forests provide a number of different services that are beneficial to human existence. Broadly, forest values associated with forest areas can be classified into: ecological and environmental benefits; economic and trade values; and social and cultural values. Forest goods and services include timber, fuel wood, non-timber forest products, genetic information, recreation and tourism, watershed, climate, biodiversity, amenity, and cultural or non-use values.[43] Economic instruments have been developed to recognize and attribute a monetary value to some forest services.

---

[39] Leohane, Nathaniel, Richard Revesz and Robert Stavins (1998), 'The Choice of Regulatory Instruments in Environmental Policy', *Harvard Environmental Law Review*, **22** (2), 313–367, 314.

[40] European Commission Directorate-General for Environment (1994), *Potential Benefits of Integration of Environmental and Economic Policies: An Incentive-based Approach to Policy Integration* 7.

[41] Ibid.

[42] Secretariat of the Convention on Biological Diversity, UNEP, *The Value of Forest Ecosystems*, Convention on Biological Diversity Technical Series No 4 (2001) 1.

[43] Ibid. 46.

To create economic incentives or disincentives, the relevant forest product must be clearly definable. Forest values and services are not always capable of being expressed in clear and definable ways. This is because forest values are often appreciated in an indirect manner and forests are also commonly valued on the basis of their inherent worth.[44] For these reasons, economic instruments have tended to emerge when the forest service is definable and the law has the ability to recognize this service by attaching rights and responsibilities to its use and management. Economic instruments have been created and implemented for the following forest services[45]

- Economic incentives that recognize forests' ability to promote biodiversity;
- Economic incentives that recognize forests' ability to store carbon;
- Economic incentives that recognize forests' ability to provide watershed protection;
- Economic incentives that recognize the value associated with forest landscape beauty.

C. Economic Instruments for Forest Biodiversity Conservation

Forests (and in particular tropical forests) are major locations of biological diversity and environmental services.[46] The ongoing loss of biological diversity across the planet is an issue of serious concern to the international community, as evidenced by the Convention on Biological Diversity and its associated implementing institutions. In 2002, the Conference of the Parties of the Convention on Biological Diversity adopted a strategic plan to achieve (by 2010) a 'significant reduction of the current rate of biodiversity loss at the global, regional and national level as a contribution to poverty alleviation and to the benefit of all life on Earth'.[47] The findings of the Millennium Ecosystem Assessment were that:

> . . . biodiversity loss, and in particular the loss of species diversity and transformation of habitats, is likely to continue for the foreseeable future, and

---

[44] Ibid. 3.

[45] Landell-Mills and Porras, above n 21.

[46] Secretariat of the Convention on Biological Diversity, UNEP, *The Value of Forest Ecosystems*, Convention on Biological Diversity Technical Series No 4 (2001) 1.

[47] UNEP, *Convention on Biological Diversity: Summary of the Second Global Biodiversity Outlook*, 8th mtg, Agenda Items 13 and 20, UN Doc UNEP/CBD/COP/8/12 (15 February 2006) annex [3].

> certainly beyond 2010 . . . Progress towards the 2010 biodiversity target is impossible.[48]

The Millennium Ecosystem Assessment concluded that there has been a failure by the international community to protect and conserve remaining biodiversity. To tackle the ongoing threats to biodiversity, improved compliance and implementation with existing environmental regulatory instruments (command and control, market or otherwise) are needed.[49] Thus, there is an impetus behind the wider implementation of economic instruments aimed at increasing or improving biological diversity within specified areas.

Economic instruments aimed at improving biological diversity have taken a number of different forms. There are many publications detailing individual projects in which economic instruments have been used to achieve better environmental outcomes.[50] The global review of such projects, conducted by Landell-Mills and Porras, noted that, in many instances, credible data demonstrating improved environmental and social outcomes were not included within the individual project analysis: 'The studies examined offered superficial reviews of economic, social and environmental benefits with virtually no assessment of costs'.[51] This suggests that some caution should be exercised when contemplating the development of economic instruments aimed at improving biological diversity.

The economic instruments listed below have been identified as operating in the biodiversity enhancement context[52]

- Biodiversity business shares – these instruments capture the willingness that exists within society to pay for biodiversity protection. Under this approach, companies are established whose main

---

48 Ibid. [20–21].

49 Ibid. [21].

50 For example, see the eleven case studies in: Pagiola, Stefano, Joshua Bishop and Natasha Landell-Mills (eds) (2004), *Selling Forest Environmental Services: Market-based Mechanisms for Conservation and Development*, London: Earthscan Publications. The most commonly touted example is: Chomitz, Kenneth C., Esteban Brenes and Luis Constantion, (1998), 'Financing Environmental Services: The Costa Rica Experience and its Implications', available at http://www.elaw.org/system/files/Financing.Environmental.Services.pdf (accessed 20 February 2009). This example is praised for mobilizing finance for local community members to carry out reforestation, natural forest management, and forest regeneration and forest protection practices.

51 Landell-Mills and Porras, above n 21.

52 Ibid. 29–31.

objective is biodiversity conservation. Such companies sell shares, and the price of the share price determines the value of biodiversity protection and the market demand for such products.

- Biodiversity credits and offsets – these systems are part of regulatory frameworks that prescribe the manner in which biodiversity is to be managed. Within these systems, market values are attributed to biodiversity rights when a demand for these rights exists.[53]
- Biodiversity-friendly products – these systems use existing markets to determine the willingness of consumers to pay a higher premium for products generated in a manner consistent with biological diversity protection.
- Bioprospecting rights – these systems sell a right allowing for collection and testing of genetic material from designated forest areas. Some systems will require a share in the profits generated from the commercial production of any forest goods.
- Conservation easement – governments entice landowners to manage land in a manner consistent with promoting biological diversity by providing economic assistance.
- Debt-for-nature swap – governments purchase discounted developing country debt, which is exchanged for domestic financial resources to invest in conservation.
- Land lease or conservation concession – this occurs when governments provide leases on concessional terms that require the sustainable use of the forest area.

A number of barriers exist that prevent the wider use of economic instruments in the biodiversity protection context. These include: unclear property rights; lack of monitoring and enforcement; complex stakeholder's participation;[54] low capacity for service delivery; and cultural resistance.[55] Furthermore, the form of biodiversity services means that these services are rarely consumed by a clearly identifiable clientele.

---

[53] An example of this type of system can nbe found in New South Wales under the *Threatened Species Conservation Act 1995* (NSW) ('TSC Act'). This act establishes a system for biodiversity banking. For further explanation of such systems, see Maguire, Rowena (2008), 'Environmental Offsets in Queensland: New Mechanisms for Managing Natural Resources?', *Queensland Environmental Practice Reporter*, **13** (61), 160–167.

[54] Sayer, Jeffrey, Unna Chokkalingam and John Poulsen (2004), 'The Restoration of Forest Biodiversity and Ecological Values', *Forest Ecology and Management*, **201** (1), 3–11.

[55] Landell-Mills and Porras, above n 21.

D. Economic Instruments for Forest Carbon

The Framework Convention on Climate Change and the Kyoto Protocol collectively operate to create an international market for carbon.[56] The focus of this regime is a reduction in global greenhouse gas emissions: such forest activities that sequester carbon are valued by the regime. Currently, the framework allows participants to account for the following forestry practices: reforestation; afforestation and certain land use change activities. However, the cumbersome approvals process and uncertainty surrounding the scientific credibility of awarding carbon credits for forestry practices has resulted in few countries implementing the rules created by the international regime.[57]

The REDD+ initiative differs from the existing carbon forest rule by creating incentives to avoid deforestation. The creation of economic instruments that reward participants for *not* carrying out a certain activity is fraught with complications; and this challenge is further compounded by the difficulty associated with calculating emission reductions from forest activities.[58] To speed up the process and overcome a number of these barriers, it has been suggested a new instrument (such as a 'Carbon Conservation Mechanism') could be created.[59]

The Carbon Conservation Mechanism would be a voluntary scheme based on the approach of paying for ecosystem services. This allows the mechanism to consider not only the carbon storage capacity of forests, but also other forest services (such as biodiversity and soil and water quality). This would be achieved by creating a market for verified conservation credits (representing a tonne of carbon dioxide or equivalent). For the credits to become verified, a number of qualitative criteria (linked to promoting other forest services) would need to be met before the credits were issued.[60]

The carbon conservation mechanism would be different from any mechanism for reduced future emissions from deforestation and degradation,

---

[56] See Chapter 6 for an in-depth discussion of this topic.

[57] See: Streck, Charlotte (2008), 'Forests, Carbon Markets, and Avoided Deforestation: Legal Implications', *Carbon and Climate Law Review*, (3), 239–247, 240.

[58] For further analysis on this point, see: O'Sullivan, Robert, and Charlotte Streck (2008), 'Conservation Carbon: A New Voluntary Market Mechanism to Protect Forests', *Carbon and Climate Law Review*, (3), 312–321, 315–16.

[59] See ibid., which outlines the nature, requirements and processes associated with such a mechanism.

[60] Ibid. 316.

because it would allow for credits to be generated from forests that are not in immediate threat of deforestation, so long as that forest ecosystem provided other important ecosystem services. Thus, the carbon conservation mechanism is inclusive of more forest values (as opposed to current carbon forestry rules that focus solely on carbon). To address 'additionality'[61] concerns within the existing carbon market, a mechanism for reduced future emissions from deforestation and degradation would only recognize credits from areas where deforestation is imminent.[62]

O'Sullivan and Streck note that the operation of such a mechanism would be through voluntary engagement. Creating market demand for these types of credits would require consideration. These authors suggest two avenues for voluntary purchases of such credits

1. Existing voluntary retail carbon markets – these markets operate due to corporate responsibility and individual business commitments to sustainability. Airlines, travel agencies, retail chains, banks and the automotive sector have all felt it necessary to engage within existing voluntary carbon markets.[63]
2. Within the greening compliance markets under the European Union Emissions Trading Scheme or the Kyoto Protocol, reform is required to address current environmental and social considerations (criticisms of the scheme). The introduction of verified conservation credits could be used to address these concerns.[64]

E. Economic Instruments for Forest Watershed Services

Forests are considered by many to be important for the provision and health of watersheds. This is because forests are associated with assisting in the following[65]

- Water flow regulation – maintenance of dry season flows and flood control;
- Water quality maintenance – sediment load control, nutrient load

---

[61] 'Additionality' requires that the activity is *additional* to normal practice. Therefore, if an area were not under imminent threat of deforestation, the additionality requirements would not be satisfied.

[62] See O'Sullivan and Streck, above n 58, 317.

[63] Maguire, Rowena (2011), 'Opportunities for Forest Finance: Compliance and Voluntary Markets', *Carbon and Climate Law Review*, (1), 100–112.

[64] For further information on the greening compliance market see ibid. 319.

[65] Landell-Mills and Porras, above n 21, 112.

control (for example phosphorous and nitrogen), chemical load control, and salinity control;

- Erosion and sedimentation control;
- Land salinization reduction and water table regulation;
- Maintenance of aquatic habitats (for example, maintaining water temperature, shading rivers and streams, ensuring adequate woody debris in the water).

The global review conducted by Landell-Mills and Porras found that there was limited scientific basis for these assumptions. The review concluded that the role of forests in watershed health and provision is dependent upon a range of site-specific factors such as climate, terrain, soil, composition, forest management, and access to road and transport systems.[66] The existence of a forest is not unequivocally linked to improved watershed values. Pagiola also makes this point in his assessment of the market for water services in Costa Rica.[67]

> The main remaining weakness, which affects both the present effectiveness of the program and its future sustainability, is the lack of reliable and precise qualitative and quantitative information about the links between forest cover and water services.[68]

The study conducted by Landell-Mills identifies a number of economic instruments to promote watershed values in forests[69]

- Best management practice contracts – contracts negotiated with watershed landholders and downstream beneficiaries setting out detailed 'best management practices' that need to be implemented in return for set payments;
- Salinity-friendly products – where payments for forests' salinity control function are piggybacked onto sales of existing commodities;
- Salinity credits – these operate within the offset regulatory framework

---

66 Landell-Mills and Porras, above n 21, 114.

67 Though the precautionary principle could be used as a basis for arguing that lack of full scientific data on forests and watersheds should not operate to restrict conservation of forest areas containing watershed values.

68 Pagiola, Stefano (2004), 'Paying for Water Services in Central America: Learning from Costa Rica', in Stefano Pagiola, Joshua Bishop and Natasha Landell-Mills (eds), *Selling Forest Environmental Services: Market-based Mechanisms for Conservation and Development*, New York: Routledge, pp. 37–56.

69 Landell-Mills and Porras, above n 21, 116.

requiring polluters to correlate the amounts of credits in their possession with their emission levels;
- Salmon habitat restoration contracts, and salmon habitat credits – where landholders receive incentives to manage land for salmon habitat;
- Salmon-safe products – where payments for a forest salmon habitat protection function are piggybacked onto sales of agricultural produce;
- Transpiration credits – used to commercialize a forest's role in evapotranspiration and water table regulation. Transpiration credits are supplied through tree planting in critical points of catchments;
- Water rights – creates a regulatory system that assigns property rights to water use and management;
- Watershed lease – where land in watersheds is leased by downstream beneficiaries for watershed protection activities.

The biggest driver behind the development of watershed markets is a 'demand' for these services. Demand for watershed markets accounts for 50 per cent of the market's development.[70] Therefore, the perceived linkages between watershed health and forest cover are valued highly by participants within schemes. Despite the lack of clear environmental benefits associated with protecting forest watershed values, a number of social improvements can be identified as a result of watershed markets. These include direct health benefits, such as improved drinking water, and indirect health benefits, such as improved agricultural and fisheries production. Furthermore, the creation of such a market provides an opportunity for community outreach programs that increase environmental awareness and education, and provide training on the ground in land use management practices.

A number of barriers to expanding watershed market development exist, such as: dealing with the large number of stakeholders involved in watershed use; the lack of cost-effective implementing institutions; poorly defined and insecure property rights; lack of a clear and comprehensive regulatory framework; lack of scientific information on forest-water linkages; and lack of stakeholder participation.[71] Before establishing forest watershed markets, the literature suggests that it would be advisable to carry out a study of the effect of forests on watershed values in any par-

---

70 Landell-Mills and Porras, above n 21, 136.
71 Ibid. 145.

ticular region and to identify the expected objectives of schemes (be they solely environmental or otherwise).

F. Economic Instruments for Forest Landscape Beauty

Markets for landscape beauty are not as developed as markets for other forest values.[72] This could be explained by social attitudes that assume a right to appreciate forest landscape amenity without contributing to the cost of preserving and conserving it. However, in certain instances, society has demonstrated a willingness to pay for forest landscape beauty. This is evidenced by the growth and development of the ecotourism market. The global growth of ecotourism-related travel has created a market incentive to protect and enhance certain forest areas. The payment for forest landscape beauty is determined by the willingness of visitors to pay for access to forest areas. There are many examples that demonstrate visitors are willing to pay for certain nature experiences.[73]

In general, private tour operators and service providers (that is, accommodation and meal providers) have benefited the most from the expansion of ecotourism-related travel. Landholders of forest areas have not shared in the increased revenue from eco-tourism sources.[74] This is because visitors to natural areas generally do not pay access rights directly to local communities. Entrance fees are charged to visitors of World Heritage and other protected areas and the revenue generated from these fees is generally collected by the government and used in meeting the cost of maintenance associated with managing such areas. The expansion of economic instruments used in valuing forest landscape beauty is really dependent

---

72 Originally, the US Forest Service (Department of Interior) used the scenic amenity value of forests to convince local communities that endless fir/pine forests, imposed on the community by commercial forestry, was a socially valuable and beautiful thing. Tarrant, Michael, and Ken Cordell (2002), 'Amenity Values of Public and Private Forests: Examining the Value-Attitude Relationship', *Environmental Management*, **30** (5), 692-703.

73 One of the best-known examples is the eco-tourism market based around visiting gorillas in the wild in Rwanda. The habitat of the gorillas (mountain forest regions) was being destroyed as a result of civil conflict. The protection of forest areas was funded by revenue generated from ecotourism initiatives. For further examples, see: Durst, Patrick, Chris Brown, Henrylito D. Tacio and Miyuki Ishikawa (eds) (2005), *In Search of Excellence: Exemplary Forest Management in Asia and the Pacific*, Bangkok: Food and Agricultural Organization of the United Nations and Regional Community Forestry Training Centre for Asia and the Pacific.

74 Landell-Mills and Porras, above n 21, 168.

upon whether we, as a society, are comfortable with being charged to enter public natural areas that are the common property of society as whole.

4. FOREST MARKET DESIGNS AND GOVERNANCE CONSIDERATIONS

A. Suitability of the Market Mechanism to Address Environmental Activity

For any type of market to operate successfully there must be a demand for the service being provided. In the forestry context, creating a demand for forest services is a challenging task. This is because forest services have typically not been incorporated within economic valuation frameworks. The economic value of an item or service is measured by the summation of many individuals' willingness to pay for them. In turn, this willingness to pay reflects individuals' preferences for the item or service in question.[75] Thus, economic valuation in the environmental context is about measuring the preferences of people who will pay to induce an environmentally sound practice, and or to avoid unsustainable or damaging environmental practices.

The lack of economic value being attributed to forest services has resulted in a widely held perception that forest services are of a public nature and, therefore, no cost arises from their use. As users of forest services are not liable to pay for the use of them, as things currently stand, regulator intervention will be required to create a demand for forest services coupled with an acceptance of the need to pay for these services. The most crucial aspect, therefore, is a market demand for the enhancement or preservation of an existing forest service that is backed up by evidence of willingness to pay for this service.

Demand and a willingness to pay for forest services can arise in two instances: under private voluntary investment, or as a result of government intervention. Private voluntary investment is usually made on the premise that the item or service is of economic worth, and that this worth will not decrease in value. There is some evidence of voluntary investment in reforestation and afforestation initiatives.[76] The method that has been

---

[75] Pearce, David W. (ed.) (1993), *Economic Values and the Natural World*, London: Earthscan Publications, 13.

[76] See generally: Harris, Elizabeth, International Institute for Sustainable Development (2007), *The Voluntary Carbon Offsets Market: An Analysis of Market Characteristics and Opportunities for Sustainable Development*.

more widely used in this context for creating a demand and a willingness to pay for forest ecosystem services is regulatory intervention. A government-like institution usually undertakes regulatory intervention.

In Costa Rica, the government intervened to create a market demand for forest ecosystem services by providing finance for activities that enhanced or preserved forest ecosystem services. The government financed these activities by diverting 3.5 per cent of a tax from fossil fuel sales (around $US 10 million a year), together with grants from a number of international institutions to initiate the *Pago por Servicios Ambientales* (PAS) program.[77] Forest Law N° 7575 (enacted in 1996) explicitly recognized four environmental services provided by forest ecosystems: (i) mitigation of greenhouse gas emissions; (ii) hydrological services (including provision of water for human consumption, irrigation and energy production); (iii) biodiversity conservation; and (iv) provision of scenic beauty for recreation and ecotourism. Holders of forest areas, who managed their land according to the management criteria for enhancing these four services, would become eligible for land management payment. The PAS program envisaged that all beneficiaries of environmental services would pay for the services they received, which would ultimately finance the payment made to landholders who enhance the forest services.[78] This example demonstrates one way in which government intervention can be used to create a demand and a willingness to pay for forest ecosystem services.[79]

In Queensland, Australia, the government has taken a different interventionist approach, which attempts to foster a market for forest ecosystem services. The government has created a mandatory offsetting regime, which requires that the effects of certain developments be offset.[80] This

[77] Pagiola, above n 68, 713.

[78] This objective has been only partially fulfilled. Much of the finance to landowners who manage forest areas to provide services is generated from government and other donor funds. See ibid.

[79] For an analysis that examines transferring this type of regulation to Papua New Guinea, see: Thorpe, Amelia, and Kristy Graham, 'A Payment for Ecosystem Services Scheme in Papua New Guinea: Incorporating Lessons from Costa Rica's PSA' (Paper presented at the New Zealand Centre for Environmental Law Conference 2009: Property Rights and Sustainability, Auckland, New Zealand, 16–18 April 2009).

[80] For further details of the legal requirements and operation of this framework, see: Maguire, Rowena (2008), 'Environmental Offsets in Queensland: New Mechanisms for Managing Natural Resources?', *Queensland Environmental Practice Reporter*, **13** (61), 160–167; Maguire, Rowena (2008), 'The Compatibility of Leasehold and Freehold Tenure Arrangements with Environmental Offset and Trading Initiatives', *Queensland Environmental Practice Reporter*, **14** (63), 27–38; and Maguire, Rowena (2008), 'Legal Issues in The Design and Implementation

provides an opportunity for landholders to manage their land to meet certain ecological criteria and be financially rewarded for it. Landholders are financially rewarded when developers buy credits from them to meet their own offset requirements. It should be noted, though, that this approach does not provide for the preservation of existing forest services. Instead, it provides for degraded forestland to be improved to meet offset requirements. Therefore, essentially, a swap occurs so that the quantity of forest ecosystem services does not alter when development take place. This, of course, raises questions about environmental effectiveness.

Both of these interventionist government approaches artificially contribute to the creation of a demand and willingness to pay for the preservation and enhancement of forest ecosystem services. Analysis of these two regimes demonstrates

1. That government support for the creation of the environmental market can be important for establishing a demand, and a willingness to pay, for forest ecosystem services. If sustainable forest management is accepted as a goal of society, then a mixed or integrated policy approach (including regulatory intervention and institutional reform providing for the operation of a market mechanism) will be necessary to achieve that goal.[81] The determination of environmental and social objectives should not be left to the market. It is important to understand the distinction between general community interests and the private interests of market players. In some instances, their interests may be complementary but, in instances where the interests do not coincide, the community interest must prevail.[82] In these cases, the government is usually the most appropriate regulator to ensure that broader public interests prevail.
2. That these schemes will probably not be entirely self-funded and will require additional government start-up finance (and perhaps even donor assistance).
3. The government will require the capacity to implement and monitor this type of scheme.[83]

---

of Environmental Offset and Environmental Trading Frameworks, *Queensland Environmental Practice Reporter*, **14** (64), 91–100.

[81] Kinrade, Peter (1995), 'Towards Ecologically Sustainable Development: The Role and Shortcomings of Markets' in Robyn Eckersley (ed.) *Markets, the State and the Environment: Towards Integration*, South Melbourne: Macmillan Education Australia, pp. 86–99.

[82] Ibid. 105.

[83] In the Costa Rica context, institutional capacity has been credited as

4. Community participation in the formulation and implementation of policy will be vital to the attainment of sustainable forest management.

No policy approach, regardless of how well conceived and implemented, will achieve the anticipated environmental outcomes without community support and value shifts. Neither government nor markets can induce those value shifts; at best, they can merely help to facilitate them.[84] These are the sorts of considerations that should be applied by regulators when examining whether a market mechanism would contribute to addressing unsustainable or degrading forest practices.

As a final consideration, governments might consider the use of a different economic instrument, such as an environmental tax, to achieve the desired environmental objective. The imposition of environmental taxes is believed to produce better environmental outcomes than those produced by market instruments.[85] In theory, the choice between carrots (tradeable rights), and sticks (taxes) is mainly a matter of ethical and distributional judgement (rather than of efficiency and pollution control).[86] When a tax is imposed, the government places a charge on every unit of pollution or other activity from the outset (for example, carbon dioxide).[87] Those who wish to limit their economic liability will be motivated to change their practices. When tradeable rights are used, the government, again, determines the level of activity permitted, and then allocates rights to participants to continue this activity to a certain level.[88] Once this level is reached, participants must trade these rights amongst themselves. Again, those who wish to limit their economic liability will reduce their actions and sell their rights to other participants. In theory, therefore, both of these approaches should result in the same outcome.

In practice, though, there are many issues associated with the implementation of tradeable rights schemes, namely additionality (paying for activities that would have been conducted in any event) and leakage

one of key contributing pillars to the success of the PSA. Pagiola, above n 68, 713.

84 Kinrade, above n 81, 105.

85 Ibid. 86. and Engel, Stefanie, Stefano Pagiola and Sven Wunder (2008), 'Designing Payments of Environmental Services in Theory and Practice: An Overview of the Issues', *Ecological Economics*, **65** (4), 663–674, 668–69.

86 Moran, Alan (1995), 'Tools of Environmental Policy: Market Instruments versus Command-and-Control' in Robyn Eckersley (ed.) *Markets, the State and the Environment: Towards Integration*, South Melbourne: Macmillan Education Australia, pp. 73–76.

87 Ibid. 76.

88 Ibid.

(shifting environmentally-damaging activities elsewhere in space).[89] The implementation of environmental taxes, although less palatable to industry, will involve less cumbersome regulation processes and will not be as susceptible to leakage and additionality concerns.

B. Design Considerations for Forest Market Mechanisms

Clear environment objectives

For environmental regulation to achieve desired changes in human behaviour towards the environment, the instrument must establish clear and measurable objectives. The innovative nature of the regulation must not be the most celebrated feature. Fine-tuning of the mechanism should not draw attention away from overarching considerations as to whether the scheme is achieving its objective. To date, the fine-tuning of both traditional and market-based instruments has drawn attention away from the primary purpose of the instrument. It is commonplace within bureaucratic society to have processes to improve processes. These allow for focus on micro-issues but, ultimately, result in loss of focus on the 'big picture' issues. Similarly, the creation of complex requirements makes compliance with the process more complicated.

Rather, the focus should be squarely placed on ensuring that there are improvements in the forest condition. The forest objective should be optimistic, yet realistic, in nature. Perhaps more challenging is the aim that the objectives should be both consistent with international forest law concepts and inclusive of local interests associated with the forest area.

To improve the effectiveness of the scheme, the objective (as already mentioned) must be clearly defined. In addition to this, the forest commodity[90] to be traded, and the method by which it will be measured, needs to be clearly defined.[91] This requires the use of terminology that is gener-

---

[89] These issues will be discussed in more detail in section 7.E. Also, see: Engel, Pagiola and Wunder, above n 85, 669.

[90] A major issue for the inclusion of forest-generated carbon credits was based on conceptual challenges. Standard definitions did not exist for forest, deforestation, afforestation, and reforestation. To incorporate forest-generated carbon credits into the trading regime, these definitional issues had to be addressed. See Graichen, Patrick (2005), 'Can Forestry Gain from Emissions Trading? Rules Governing Sinks Projects Under the UNFCCC and the EU Emissions Trading System', *Review of the European Community and International Environmental Law*, **14** (1), 11–18.

[91] In Costa Rica, the forest commodity was defined as a bundle of services that included forest watershed values, forest biodiversity values, forest carbon values, and forest landscape values. See Pagiola, above n 68, 715–16.

ally understood, and the avoidance of terms with multiple meanings (for example, the term 'sustainable'). Thus, a suitable objective may be one that requires a measurable increase in forest cover and health.

A clear overarching objective will define the scope and ambit of the scheme's existence.[92] Markets that sell forest ecosystem services will generally be established with the end-goal of an improvement in forest ecosystem health, or an increase in the provision of forest ecosystems' services. Promoting improvement in forest ecosystem health is the preferable goal, because this will mean that there is a 'functional lift' in the quality and quantity of the relevant forest service. Functional lift is the preferable indicator because it measures not only increased activity, but also focuses on the quality of the increased activity. By comparison, if the purpose of the scheme were to only ensure an increase in the production of the forest eco-system service, then this would not be as desirable, because the environmental integrity of the scheme could be called in question.

Environmental and economic effectiveness

Environmental effectiveness requires that the forest market leads to an actual increase in environmental services compared to the situation without it.[93] Markets that are established on the basis of providing a particular forest service must demonstrate their environmental effectiveness in creating a marketable forest product.[94] For the market to operate successfully, the forest service must be capable of differentiation from competing forest services. Therefore, the provision of the forest service through the market must be an *improved* forest service in order to attract an economic value. This is normally achieved by introducing management regimes that require that the health of the particular ecosystem service be improved.

Economic efficiency requires that the cost of improving the forest service is compatible with the market cost of the forest ecosystem

---

[92] There are many examples of schemes that are politically driven to achieve both economic and environmental benefits. Many of these fail to identify clear objectives and, as such, fail to deliver economic benefits to local communities or environmental benefits to the global community. For further detail, see: Sayer, Jeffrey, Unna Chokkalingam and John Poulsen (2004), 'The Restoration of Forest Biodiversity and Ecological Values', *Forest Ecology and Management*, **201** (1), 3–11.

[93] Engel, Stefanie, and Charles Palmer (2008), 'Payments for Environmental Services as an Alternative to Logging under Weak Property Rights: The Case of Indonesia', *Ecological Economics*, **65** (4), 799–809, 800.

[94] Duraiappah, Anantha Kumar, International Institute for Sustainable Development (2006), *Markets for Ecosystem Services: A Potential Tool for Multilateral Environmental Agreements* 13.

service,[95] thereby avoiding a situation where the cost of restoring or improving the service is higher than the cost at which it is sold.[96] Economic efficiency must consider the alternative income available if the forest area is used for other purposes (for example, income from timber). Ideally, forest markets would create monetary incentives that are approximately equivalent to the income generated from past activities. If old activities remain more profitable, then participation in forest markets schemes is only likely if the more profitable forest activity continues.[97]

C. Good Governance Requirements for Forest Market Mechanisms

For markets to operate effectively for forest resources, a number of governance-related considerations must be examined. 'Governance' is a term broadly defining the decision-making procedures and behavioural conventions within established institutions.[98] Environmental governance not only refers to government regulation and law enforcement for conservation, but also involves the political, organizational, and cultural frameworks through which diverse interests in natural and cultural resources are coordinated and controlled.[99] Good governance requires evidence of transparency, accountability, and the engagement of participants within the scheme.[100] In some instances, governance reform may be required, particularly in developing countries.

It is crucial that forest-based trading regimes are transparent. Transparency requires that the procedures on which decisions are based are as open and clear as possible.[101] This will require making all information about the envi-

---

95 Ibid.

96 For further analysis of economic models, see: Hunt, Colin (2007), 'Economy and Ecology of Emerging Markets and Credits for Bio-sequestered Carbon on Private Land in Tropical Australia', *Ecological Economics*, **66** (2–3), 309–318.

97 Engel and Palmer, above n 93, 800. In this instance, the authors explored the variance in income earned from timber profits in Indonesia. The authors found that lower prices were paid to those with fewer legal rights, or those who did not have the capacity to enforce their legal rights.

98 Goldsmith, Arthur (2007), 'Is Governance Reform a Catalyst for Development?', *Governance: An International Journal of Policy, Administration, and Institutions*, **20** (2), 165 –186.

99 Cronkleton, Peter, Peter L. Taylor, Deborah Barry, Samantha Stone-Jovicich, Marianne Schmink (2008), 'Environmental Governance and the Emergence of Forest-Based Social Movements', Occasional Paper No 49, Centre for International Forestry Research, 1.

100 Ibid.

101 Fisher, Douglas (2006), 'Markets, Water Rights and Sustainable Development', *Environmental and Planning Law Journal*, **23** (2), 100 –112, 104.

ronmental offset process publicly available. The types of information that should be made available include: information and data used by regulators in the formation of the scheme; information about individual commodity rights and responsibilities; information about the practical operation of the whole scheme; and general information about forest market-based regimes.[102] These will ensure that market-based forest regimes are perceived as legitimate and accountable by participants and by the general public.

Accountability requires that the institution, and the individuals within the institution, are answerable for the activities carried out by it. Liability to answer may also involve reporting to a higher authority, or to those with a vested interest in the institution. The notion of accountability also involves the setting of goals and the reporting and publication of results and data.[103] Accountability within market-based forest regimes will require the institution responsible for providing the forest product to create regular reports about the trade, and ensure they are made available for higher authorities or stakeholders to access.

For market-based forest regimes to enjoy widespread take-up and implementation, there must be appropriate stakeholder engagement.[104] The stakeholders involved in forest trading schemes may include[105]

- Local communities, indigenous groups and landholders who will be responsible for carrying out the work associated with improving the quality of the forest service;
- Local communities, indigenous groups and landholders whose individual interests would be affected by the implementation of a forest-trading regime (for example, access and use rights may be varied as a result of the trading regime);
- Private industry that would buy the forest service;
- Government agencies interested in promoting a particular forest value;

---

102 Fisher, Douglas (ed.) (2006), *Implementing the National Water Initiative: A Generic Set of Arrangements for Managing Interests in Water*, Kelvin Grove, Qld: Queensland University of Technology, 19.

103 Hodge, Graeme, and Ken Coghill (2007), 'Accountability in the Privatized State', *Governance: An International Journal of Policy, Administration, and Institutions*, **20** (4), 675–702, 676.

104 Colfer, C. and R. Wadley, Centre for International Forestry Research (1996), *Assessing Participation in Forest Management*.

105 See, generally Pagiola, Stefano, Natasha Landell-Mills and Joshua Bishop (2004), 'Market-based Mechanisms for Forest Conservation and Development', in Stefano Pagiola, Natasha Landell-Mills and Joshua Bishop (eds), *Selling Forest Environmental Services: Market-based Mechanisms for Conservation and Development*, New York: Routledge, pp. 20–32, 261–274.

- Parties obligated to participate in a forest-trading scheme (for example, developers or farmers may be required to provide an offset for their activities);
- Parties responsible for managing the forest area to meet specified objectives and standards;
- Agencies that create policies concerning forest use and management;
- Independent verification and monitoring bodies that ensure that requirements are met;
- Government bodies responsible for registering the interests that environmental offsets create.

The success of forest market-based regimes will be dependent upon stakeholder engagement and participation.[106] Changes to the management and use of forest areas will impact upon current access and use rights held by local communities and other stakeholder groups. It is essential that those stakeholders whose interests are affected by the imposition of a forest-trading regime be actively consulted to determine their ongoing interests in the forest area.[107] Every effort should be made to ensure that their interests are not overlooked, and that their interests are incorporated into the regulatory framework.[108]

Without adequate consultation with local interest groups, forestry trading regimes will not be able to deliver on their stated objectives because these groups will not understand the potential benefits and incentives that market-based regulation offers. Unless clear incentives are provided for stakeholders, it is unlikely that a change of behaviour will occur. The Pareto optimality criterion, used in neo-classical economics, requires that no one person is worse off than they were before the intervention.[109]

---

[106] See, generally: Sayer, Chokkalingam and Poulsen, above n 54, 7.

[107] Often, marginalized and powerless groups are left out of the stakeholder consultation process, and this undermines the equity and legitimacy of market-based approaches. For further information on this topic, see: Corbera, Esteve, Katrina Brown and W. Neil Adger (2007), 'The Equity and Legitimacy of Markets for Ecosystem Service', *Development and Change*, **38** (4), 587–613.

[108] For further information on environmental markets and their contribution to equity and legitimacy, see ibid. 'Legitimacy' is defined as the way in which outcomes are negotiated, administered and accepted by stakeholders. It encompasses issues such as the recognition of stakeholders, the acknowledgment and hearing of their concerns, the participation of stakeholders in decision-making, and the distribution of decision-making power.

[109] Duraiappah, Anantha Kumar, International Institute for Sustainable Development, (2006), *Markets for Ecosystem Services: A Potential Tool for Multilateral Environmental Agreements* 9.

If a change of practice is required, groups need to be presented with one of the following incentives to encourage compliance

- Be financially rewarded for changes to management and use practices;
- Be provided with an alternative supply of the forest service relied upon;
- See improvement in the productivity or health of the forest service product relied upon through the introduction of management policy, which, in turn, provides an incentive for changed practice.

When examining forest market-based trading regimes, the case studies indicate that the support of the local community and stakeholders is essential to a scheme's success.[110] However, engaging the sector that will pay for the improved forest service is also critical. The private sector may become involved in market-based forest trading regimes with a number of motivations: the belief that such investment will, in the long term, deliver economic gain; a sense of a duty to contribute to the local community and environmental health; to meet an obligation required by law. Despite the origin of the motivation, the success of market-based schemes relies on actively engaging resources from the private sector.[111]

The role of the government will also be important depending upon the nature of the forest-trading regime. This may be designed so that it is compatible with existing forest governmental policy and, furthermore, government policy may, in particular instances, obligate parties to participate in a forest-trading regime.[112] In this situation, the government may act as a broker, generating business for the regime. Pagiola, Landell-Mills

---

[110] Brown, Chris, Patrick Durst and Thomas Enters (2005), 'Perception of Excellence: Ingredients of Good Forest Management', in Durst, Patrick, Chris Brown, Henrylito D. Tacio and Miyuki Ishikawa (eds), *In Search of Excellence: Exemplary Forest Management in Asia and the Pacific*, Regional Community Forestry Training Centre for Asia and the Pacific, Bangkok: Food and Agriculture Organization of the United Nations, pp. 7–28, 9. This report identifies a number of excellent forest-based projects, and finds that people's participation in schemes is a common characteristic among successful forest projects.

[111] For a discussion related to the four conditions that allow markets to function, see: Ashford, Nicholas and Charles Caldart (eds) (2008), *Environmental Law, Policy and Economics: Reclaim the Environmental Agenda*, Cambridge, Mass: MIT Press.

[112] An example of this is government policy that requires development activity be offset. For an examination of the Queensland Government's approach in Australia, see Maguire, Rowena (2008), 'Environmental Offsets in Queensland: New Mechanisms for Managing Natural Resources?', *Queensland Environmental Practice Reporter*, **13** (61), 160–167.

and Bishop suggest that the government may play a number of roles in forest-trading regimes such as: the buyer of the forest product; the seller of the forest product; or as an intermediary.[113] Depending upon the structure of the particular forest-trading regime, the level and role of government involvement will vary. Despite the variance of government involvement, forest-trading regimes should seek to ensure that its objectives and requirements are compatible with existing government policy.

In the context of developing countries, much attention is directed towards improving governance processes; however, debate exists as to the effectiveness of the governance reform in these countries. While there is mounting evidence to suggest that governance reform rarely delivers on its stated objectives, this has not altered the agendas of international aid agencies or other development assistance bodies.[114] There is often great difficulty in targeting local governance reform, due to limited targeting of partners, inappropriate communications, language barriers, resource constraints, cultural differences, and capacity limitations.[115] Often, agencies underestimate the time and political effort required to change governance practices and the economic impacts of such reform are overestimated.[116]

It has been suggested that the overall benefits attained by governance reform packages are linked to improved transparency, accountability and participation as opposed to increased economic development.[117] Thus, any reforms directed towards improving governance and capacity in developing countries should acknowledge the inherent challenges, and the time and resource commitments, required to secure long-term transformations.

5. FOREST MARKET LEGAL ISSUES

A. Land Tenure

Secure, stable land tenure is perhaps the most fundamental requirement for forest market mechanisms. This is because it allows the holder of the

---

113 Pagiola, Landell-Mills and Bishop, above n 20, 57.

114 Colchester, Marcus, Tejaswini Apte, Michel Laforge, Alois Mandondo and Neema Pathak (2003), 'Bridging the Gap: Communities, Forests and International Networks' (Occasional Paper No 41, Centre for International Forestry Research).

115 Ibid. 15.

116 Goldsmith, Arthur (2007), 'Is Governance Reform a Catalyst for Development?', *Governance: An International Journal of Policy, Administration, and Institutions*, **20** (2), 165–186.

117 Ibid. 181.

land to commit to, and have the capacity to, participate in forest market regulation. Without secure rights, forest users have few incentives – and often lack legal status – for investment in managing and protecting their forest resources. Generally speaking, in developed countries, holders of land have a higher sense of tenure security than they do in developing countries.[118] Thus, developing and developed countries face different challenges in relation to tenure security.

B. Developing Countries

In the developing country context, a major barrier to the effective implementation of forest market mechanisms is the lack of secure tenure rights.[119] Many international development and environment institutions that regularly implement initiatives to improve tenure security have acknowledged this.[120] The lack of secure tenure rights in developing countries is a significant barrier to the implementation of forest market frameworks. Countries where states fail to specify even the most fundamental property rights to land, building sites, and other arable resources, usually experience poor economic growth and persistent poverty.[121]

This relationship is explained by a number of factors. First, if no secure tenure rights exist over the forest area that generates the forest market product, there is no guarantee that the service will continue in perpetuity. In this instance, there is no environmental effectiveness, because any benefits obtained by the imposition of a forest market framework may be temporary in nature.[122] Secondly, the lack of secure tenure rights will act as a major disincentive to attracting participants to the scheme. Forest dwellers are unlikely to participate in a scheme aimed at improving the health of the forest ecosystem unless they have a stake in its future. Similarly, it will be more difficult to generate market demand for the forest ecosystem service,

---

118 White, Andy, and Alejandra Martin, Forest Trends (2002), *Who Owns the World's Forests? Forest Tenure and Public Forests in Transition.*

119 Wunder, above n 15.

120 The Rights and Resources Initiative is a group of 10 international bodies that have banded together to address the cross-cutting issues associated with forests, conflict, and climate change. They examine the role of land rights as being part of the solution. For further information on the initiative see the website http://www.rightsandresources.org/index.php .

121 Sandberg, Audun (2007), 'Property Rights and Ecosystem Properties', *Land Use Policy*, **24** (4), 613–623, 619.

122 This concern is linked to two concepts, that of 'permanence' and 'additionality', both of which will be discussed in greater detail further on in this section.

as the purchasers of services will view the investment as neither secure nor environmentally viable without evidence of secure tenure rights.

In developing countries, economic growth increases development – which, in turn, increases the demand for land. As a result, forest areas are cleared to make way for infrastructure and agricultural development, resulting in rapid deforestation and forest degradation. However, it is suggested that at a certain point of economic growth, a change sweeps through society that results in demands for better use and management of environmental resources.[123] This is referred to as an inverted U-shaped 'environmental Kuznets curve' between income and deforestation.[124] The environmental Kuznets curve has been taken to imply that economic growth will eventually redress the environmental impacts of the early stages of development, and that growth will lead to further environmental improvements in these circumstances.[125]

This environmental Kuznets curve theory was applied in a study across 66 countries in Latin America, Africa and Asia.[126] It was found that to reduce deforestation, the role of institutional structures and governance processes was crucial in adapting to more environmentally compatible activities. Of relevance to this section was the finding that the rule of law and the protection of personal property rights, among other political rights, are related to improving environmental quality.[127] Furthermore, it has also been found that institutions that secure property rights can reduce deforestation without adversely affecting economic growth (thus flattening out the Kuznets curve[128]). The environmental Kuznets curve theory demonstrates that property rights are linked to decreasing deforestation and highlights the significance of property rights in addressing deforestation.

---

[123] For further detail on this concept, see: Kant, Promode and Wu Shuirong (2008), 'Reducing Deforestation and Degradation through Post-colonial Settlement of Land Rights: A Case Study in India', *Carbon and Climate Law Review*, (3), 300–305.

[124] The 'environmental Kuznets curve' is named after Simon Kuznets who hypothesized that the relationship between a measure of inequality in the distribution of income and the level of income is an inverted U-shape curve. See: Kuznets, S. (1955), 'Economic Growth and Income Inequality', *American Economic Review*, **45** (1), 1–28.

[125] Stern, David (1998), 'Progress on the Environmental Kuznets Curve', *Environmental and Development Economics*, **3** (2), 173 –196.

[126] Bhattari, Madhusudan, and Micheal Hammig (2001), 'Institutions and the Environmental Kuznets Curve for Deforestation: A Cross-Country Analysis for Latin America, Africa and Asia', *World Development*, **29** (6), 995–1010.

[127] Ibid. 999.

[128] Culas, R.J. (2007), 'Deforestation and the Environmental Kuznets Curve: An Institutional Perspective', *Ecological Economics*, **61** (2–3), 429–437.

C. Developed Countries

In the context of developed countries (where property rights institutions are better established and provide landholders with legally recognized and enforceable rights), tenure security is not a major barrier to implementing forest market mechanisms. There are, however, other issues related to land tenure that become relevant to the operation of forest market mechanisms.[129] When examining land tenure in the countries with established property right institutions and instruments, two questions still remain relevant:

1. What is the form of the tenure holding over the relevant forest area?
2. Are the requirements of the forest market mechanism compatible with the form of the tenure holding?

In relation to the first question, the most common forms of land tenure in common law countries are freehold title and leasehold title. Those who hold a freehold or leasehold interest over a forest area will need to ensure that their land interest is inclusive of all forest eco-system services. Therefore, landholders will need to check if their interest in land includes:

- rights to the trees on the forest land;
- rights to biodiversity on the forest land;
- rights to carbon stored in the trees on the forest land;
- rights to watershed values within the forest land.

Once this is established, landholders will then be able to judge if they have the appropriate right to participate in a forest market regulatory framework.

Secondly, the landholder must ensure that the management requirements prescribed by the forest market regulatory framework are consistent with their tenure use rights. This will usually require that the holder of the land has the right to manage the forest area in a particular way so as to achieve specified environmental standards. Therefore, in the developed country context, tenure issues are linked to form and use considerations.

Market-based approaches to environmental management require the

---

[129] For an in-depth analysis of an environmental offset trading regime on freehold and leasehold tenure in Queensland, see: Maguire, Rowena (2008), 'The Compatibility of Leasehold and Freehold Tenure Arrangements with Environmental Offset and Trading Initiatives', *Queensland Environmental Practice Reporter*, **14** (63), 27–38.

development of property rights in environmental resources. This is because markets are premised on the economic definition that 'property' means 'ownership'. According to this reasoning, once ownership and/or property rights are established, incentives are created to manage and care for the environmental value to which the property rights attach. Environmental degradation could be viewed as a market failure. If this is accepted, then it can be suggested that the creation of property rights in environmental resources results in an economic value being attributed to these environmental values This, in turn, means that an incentive exists to manage and look after this environmental value.[130] Property rights, therefore, provide the holder of the rights with a sense of security, which, in turn, acts as incentive for better management or protection of the interest.

It should be noted that securing property rights alone will not lead to improved environmental outcomes,[131] since secure rights alone do not prevent a landholder from engaging in unsustainable forest practices.[132] Rather, they must accompany traditional forms of regulation requiring certain standards to be met (or the use of a market instrument that provides an incentive for certain standards to be met) to assure improved forest outcomes. Regulatory standards alone have not been able to promote a certain level of sustainable forest management. In many instances, this is likely to be attributable to the voluntary nature of sustainable forest management obligations.

D. Legal Recognition of Tenure and Property Rights

Form of recognition

To have secure tenure and property rights, there must be a system for legal recognition and registration of such interests in land. Within environmental market frameworks, there are two general approaches to recognition and registration of forest property-based interests.

There are frameworks that use existing legal mechanisms to enforce the legal rights and obligations associated with offset creation. Under

---

[130] Passero, Michelle (2008), 'The Nature of the Right of Interest Created by a Market for Forest Carbon', *Carbon and Climate Law Review*, (3), 248 – 253, 250.

[131] 'Furthermore, a private property rights approach to environmental protection without regulation, as favoured by free market economists, will almost certainly lead to a trade-off of species and ecosystems for other objectives, although some free-market economists resist this interpretation.' See Kinrade above n81; also see Bosselmann and Richardson, above n8.

[132] Richards, Michael (2000), 'Can Sustainable Tropical Forestry be Made Profitable? The Potential and Limitations of Innovative Incentive Mechanisms', *World Development*, **28** (6), 1001–1016, 1010.

this approach, contract law is used to create rights and obligations for all parties involved in the transaction. Unless additional steps are taken, the participant's rights will be based upon the scope of the contract between the parties. Parties may wish to increase their legal rights by obtaining recognition of their interests via a property-based mechanism. This may involve registering their interest on the land title registry.[133] Once an interest is registered on the register, legislative protection will be afforded to the right, and the right is enforceable against the current landholders and all future landholders. The level of protection afforded depends upon the nature of the registration.

There are also models that create a central registry where all forest property rights associated with the program are registered. This registry may or may not be linked to other registries, such as the land registry. An example of this approach is the Biodiversity Banking Scheme in New South Wales.[134] This scheme has been created through legislative reform,[135] and provides participants with legal rights to enforce all participants' obligations and rights. The scheme allows participants to register their biodiversity offsets on the land registry in addition to the biodiversity banking registry.[136]

E. Legal Market Concepts

There are four legal market concepts that must be addressed within forest market regulation:

- additionality;
- permanence;
- leakage;
- double-counting.

'Additionality' is a principle within market frameworks that requires that something has taken place that otherwise would not have taken place.

---

[133] For example, wetland mitigation credits in the USA are protected by registering an easement over the bank (offset) site. In Victoria, a Forest Property Agreement can be registered on the land registry. For further information, see: Australian Greenhouse Office (2005), *Planning Forest Sink Projects: A Guide to Legal, Taxation and Contractual Issues.*

[134] See *Threatened Species Conservation Act* 1995 (NSW); see Part 7A Biodiversity Banking, Division 9 Registers.

[135] See *Threatened Species Conservation Amendment (Biodiversity Banking) Act* 2006 No 125 (NSW).

[136] *Threatened Species Conservation Act* 1995 (NSW) s127ZZB.

Specifically, in the forestry context, the enhancement or preservation of the forest service must be additional to standard practices.[137] Additionality, therefore, requires that the forest practice go above and beyond business as usual. The theory underlying this is that parties should not be rewarded for doing the bare minimum. Rewards or incentives should be reserved for those parties who surpass current obligations.[138]

'Permanence' requires continuance in the same state or place.[139] There are two separate permanence issues

1. Natural environmental change – the environment will not remain perpetually in the same state or condition
2. Forest interest security – once a forest interest has been created, the interest must be recognized by the law to prevent future incompatible land use.

The provision of environmental services can never be stable or fixed in nature. The environment has a dynamic nature that brings change over time, and that is susceptible to such changes. During different periods, the quantity and quality of the provision of environmental services will vary. At the international level, the changing nature of the environment has been vigorously debated in relation to the creation of forestry carbon credits.[140] Carbon credits (certified emission reduction credits) are issued on the basis that the activity for which they have been issued will contrib-

---

[137] 'Additionality' in the carbon context involves a number of different types: i) *program additionality*, which requires that emission reductions are additional to emission reductions required by law or government; ii) *financial additionality* is the requirement that funding for the implementation of projects must not from come from overseas development or environmental assistance funds; and *investment additionality*, where a project may justify additionality by showing that the creation of carbon offsets will involve costs that would not be incurred in a business as usual scenario. See: Minany, Peter, Hans Bressers, Margaret Skutsh and Michael McCall (2007), 'National Forest Policy as a Platform for Biosphere Carbon Management: the Case of Community Forestry in Cameroon', *Environmental Science and Policy*, **10** (3), 204 –218, 206.

[138] This is sometimes called *Payment for Ecosystem Service*; see: Asquity, Nigel, Maria Vargas and Sven Wunder (2008), 'Selling Two Environmental Services: in Kind Payments for Bird Habitat and Watershed Protection in Los Negros, Bolivia', *Ecological Economics*, **65** (4), 675–684; see additionally discussion at 680.

[139] Webster Dictionary (2008), 'Definition of Permanence', available at http://www.webster dictionary.net/definition/permanence (accessed 8 May 2008).

[140] Subsidiary Body for Scientific and Technological Advice, *Reducing Emissions from Deforestation in Developing Countries: Approaches to Stimulate Action*, 11th sess, Agenda item 6, FCCC/SBSTA/2009/L.9 (11 November 2005).

ute to the reduction of carbon and other substances in the atmosphere. The natural processes associated with forest carbon sequestration present a challenge for their inclusion in emission-reduction trading frameworks. This is because, during certain periods, forests act as sequesters of carbon but, during other periods, they act as emitters of carbon.[141]

The forest sector is seen as particularly vulnerable to the risk of impermanence. These may be caused by people through changes in land ownership, unclear land ownership, deforestation, conversion and unsustainable harvests; or they may be naturally caused disturbances (like fires, pests and disease).[142] The challenge is not so significant for environmental offset programs, because these seek to replicate the natural cycles of environmental services. Difficulties may still arise where there are unexpected events (such as severe fire, drought, or storm conditions), as these may have a substantial impact upon the provision of environmental services.[143]

'Leakage' in the environmental market context means avoiding a situation whereby a damaging environmental practice is moved to another location.[144] In the forest context, the concept seeks to avoid instances where forest standards are improved in one area, and which then only intensifies undesirable forest practices in another. For example, the implementation of a forest-market mechanism may prevent deforestation in a particular region but if the driver of the deforestation is a demand for timber sales, it is more than likely that the deforestation will simply occur in another. Leakage is perhaps one of the biggest challenges to all environmental market mechanisms, because it is virtually impossible to be assured that the implementation of the mechanism is not responsible for increasing undesirable forest activities in other locations.

Issues related to 'double-counting' arise in two circumstances

1. *Where competing schemes exist for carrying out the forest activity* – a landholder may attempt to get recognition under both schemes, and

---

141 Schulp, Catharina, Gert-Jan Nabuurs and Peter Verbug (2008), 'Future Carbon Sequestration in Europe – Effects of Land Use Change', *Agriculture, Ecosystems and Environment*, **127** (3–4), 251–264.

142 Passero, Michelle (2008), 'The Nature of the Right of Interest Created by a Market for Forest Carbon', *Carbon and Climate Law Review*, (3), 248–253, 249.

143 It has been suggested that, to deal with intentional or unforseen losses of credits, either an insurance pool is created by the offset framework or private insurance is taken out by the participants in the offset scheme. See: Curnow, Paul, and Louisa Fitz-Gerald (2006), 'Biobanking in New South Wales: Legal issues in the design and implementation of a biodiversity offsets and banking scheme', *Environmental and Planning Law Journal*, **23** (4), 298–308.

144 Engel, Pagiola and Wunder, above n 85, 668.

this may prevent an overall gain in forest services. The party could receive a number of incentives or cash payments for carrying out only the one activity. To overcome this situation, participants should be required to sign an acknowledgment form that would require the parties to state that the forest practice has not already been accounted for elsewhere. In the event that a party is found to have gained recognition from two environmental offset schemes, a penalty and/or sanction should be imposed upon the participant.[145]

2. *Where one parcel of land may be used to generate many different types of forest interests* – for example, if a forest area is established it would provide many environmental services, such as biodiversity services, carbon storage, water purification services, and soil health improvement. The question, then, is whether the landowner should be able to create individual credits for each individual environmental service provided for by the establishment of the forest. The answer depends on the scientific soundness of rewarding these competing environmental services (most likely recognized and measured in forms of functional lift).[146] Some frameworks have overcome this concern by creating a bundled forest interest, which recognises all the interests within a forest.[147]

6. CRITIQUE OF MODERN ENVIRONMENTAL MARKETS

> A better approach would seem to be to start our analysis with a situation approximating that which actually exists, to examine the effects of a proposed policy change and to attempt to decide whether the new situation would be, in total, better or worse than the original one. In this way conclusions for policy would have some relevance to the actual situation.[148]

---

145 See the discussion of double counting issues relating to an international carbon market under the clean development mechanism: United Nations Framework Convention on Climate Change (2005), 'Input to UNFCCC on Double Counting and Methodological Issues', available at http://cdm.unfccc.int/public_inputs/meth_double_counting/DC4_Inputs_DoubleCounting_MA_SS.pdf (accessed 27 June 2008).

146 Asquity and Wunder, above n 138, for a discussion of the benefits of creating two environmental services on the one area of land.

147 Pagiola, Stefano (2008), 'Payments for Environmental Services in Costa Rica', *Ecological Economics*, **65** (4), 712–724.

148 Coase, R.H. (1960), 'The Problem of Social Cost', *The Journal of Law and Economics*, **3** (1), 1–44.

This quotation comes from Ronald Coase, whose work is associated with creating the underlying theory used to justify the existence of modern environmental market frameworks. Environmental markets apply an aspect of Coase's work that suggests that the external effects of an individual's actions can, in some instances, be addressed through private negotiation between affected parties.[149] Coase suggests that the exercising of one individual's right usually compromises another's rights. A change within existing rights' structures may lead to improvements in some instances but this must be tempered with the reality that improvements to one interest may lead to a worsening of conditions for other interests. Fundamentally, Coase believes that, in devising and choosing between social arrangements, one should have regard to the total effect of the change.[150] Therefore, Coase is an advocate of private negotiations between parties to settle disputes (in essence, a supporter of mediation as opposed to legislation to solve disputes). Ultimately, Coase believes that an action should only be undertaken once all possible ramifications associated with implementing the action are understood.

Environmental markets have been praised for their innovativeness in addressing environmental issues. This is because markets are believed to promote efficiency and flexibility.[151] Environmental markets are in reality neither flexible nor efficient,[152] due to the complex myriad of institutions and instruments that have been developed to implement them.[153] The increasing complexity associated with creating new interests in the environment, regulating and controlling the use of these interests with existing interests, and the stringent monitoring, verification and other guidelines created by regimes is time-consuming and laborious. Following such processes, and meeting the requirements of such regimes, is often at the forefront of participants' minds. Likewise, the regulatory bodies and their employees now specialize in identifying flaws in current policy design, as

---

149 Engel, Pagiola, and Wunder, above n 85, 665.

150 Coase above n 148, 44.

151 Bosselmann and Richardson, above n 8, 3.

152 In support of this, see Scanlon above n 1, 88. Participants were interviewed about the Biodiversity Banking Scheme in New South Wales. One stated that banking schemes 'especially if they are added to, rather than integrated into a planning system, further complicate an already complex system'. This means that stakeholders, including local communities and government, lose understanding of and faith in both the system and administrators of biodiversity protection.

153 The efficiency, flexibility, and accompanying complexity of international emissions trading market is persuasive on this front. See, for example, the plethora of legal analysis of existing arrangements in the journal of *Carbon and Climate Law Review*.

opposed to actually changing society's relationship with the environment. This draws attention away from the fundamental purpose of the scheme, which is likely to be preventing the further degradation of an identified ecosystem service.

There is also a valid concern that markets based upon the theory of economic rationalism fail to distribute resources in way that is just and equitable.[154] In addition, the legitimacy of markets can be called into question, because participation within market frameworks is skewed in favour of the wealthy and powerful.[155] Markets that work on the basis of directing resources to the highest bidder run an extreme risk of increasing the cost to use what has traditionally been considered a public good – while still continuing practices that are ecologically unsustainable. Conversely, it could be suggested that governments have been unable to cover the cost of effective environmental maintenance, and that transferring responsibility to the private sector results in broad public gains.

Modern environmental problems have also spawned the creation of new employment opportunities in the sphere of environmental regulation. Interestingly, many people engaged in environmental-based work are not driven by a desire to improve the use and management of our environment, but instead see issues surrounding sustainability, climate change, and the degradation of ecosystems services as business opportunities. With increasing numbers of individuals driven by personal interests to work on environmental regulation (as opposed to individuals in pursuit of broader public interests) there looms a weakening of environmental goals. The credibility of such schemes is at high risk.

Market-based regulation is appealing, as it appears to be able to recognize and provide for all the interests associated with the environment. Indeed, there are examples of successful environmental market initiatives[156] – although it must also be noted that a number of environmental markets have failed to deliver on their stated objectives.[157] Environmental markets seem to be developing at an exponential rate. There appears to be a widely-

---

154 Corbera, Brown, Adger, above n 107, 590–1.

155 Lyster, Rosemary (2002), '(De)regulating the Rural Environment', *Environmental and Planning Law Journal,* **19** (1), 445–468.

156 One of the most celebrated environmental markets is the one created for financing forest protection in Costa Rica. Carbon sequestration rates were improved upon the land by protecting forest, water, and biodiversity. Currently, 10 per cent of the land in Costa Rica is regulated by environmental markets. See Pagiola, above n 147.

157 For further information, see the following work that discusses the dismal environmental outcomes of tradeable fishing rights, water trading, and wetland mitigation banking schemes. Beder, above n 11.

held belief that environmental market regulation is the answer to all conflicts connected with environmental use and management.[158] This chapter identifies the major regulatory issues faced by environmental markets, focusing on markets associated with forest-related ecosystem services. It seeks to demonstrate that a cautious approach should be taken before assuming that a market mechanism will solve the problem.[159] Market regimes are not necessarily the most appropriate and effective mechanisms for achieving desired outcomes – regulators must consider this when designing schemes to achieve environmental objectives.[160]

This is not to suggest that market regulation is fundamentally flawed.[161] However, markets operate on the premise that a service will continue to be available and that current consumption levels are appropriate.[162] In relation to natural resources, it is now well understood that environmental services will not exist indefinitely and that the planet cannot support current consumption levels.[163] Current market design contributes to perpetrating the belief that current consumption and development levels are appropriate.

---

158 As suggested by Bayon: 'Markets are tools, tools that are inappropriate for some situations, and tools that, when used incorrectly can backfire. As the saying goes, to a man with a hammer, everything starts to look like a nail'. Bayon, Ricardo, Forest Trends, *Making Environmental Markets Work: Lessons from Early Experience with Sulphur, Carbon, Wetlands, and other Related Markets* (2004).

159 In support of this, see: Lyster, Rosemary (2002), '(De)regulating the Rural Environment', *Environmental and Planning Law Journal,* **19** (1), 445–68; 'It is not suggested that the search for alternative methods of regulation be abandoned. However, policy makers in Australia need to ensure that before rushing to the market, they are confident that the services that are traded are properly evaluated and accounted for'.

160 Keohane, Nathaniel, Richard Revesz and Robert Stavins (1998), 'The Choice of Regulatory Instruments in Environmental Policy', *Harvard Environmental Law Review*, **22** (2), 313–367.

161 For an analysis of market failure and government failure see: Datta-Chaudhuri, Mrinal (1990), 'Market Failure and Government Failure', *The Journal of Economic Perspectives*, **4** (3), 25–39.

162 See generally: Pearce, David W. (ed.) (1993), *Economic Values and the Natural World*, London: Earthscan Publications, 1 – where it is stated 'that if the earth's resources were available in infinite quantities, and if they could be deployed at zero cost, there would be no economic problem'.

163 Myers, Norman (1997), 'Consumption in Relation to Population, Environment and Development', *The Environmentalist*, **17** (1), 33–44; and de Sherbinin, Alex, David Carr, Susan Cassels and Leiwen Jiang (2007),'The Annual Review of Environmental Resources: Population and Environment', available at http://arjournals.annualreviews.org/doi/pdf/10.1146/annurev.energy.32.041306.100243?cookieSet=1 (accessed 6 May 2008).

To radically alter the human relationship with the environment, society's expectations concerning quality of life need to be transformed. There are simply not enough natural resources for current consumption and development to continue. As stated by Gandhi in 1947: 'The Earth has enough for everyone's need but not for everyone's greed'.[164] Within society, we need to manage our relationship with the earth so that human and other biological populations have a future. Markets will need to play a role in this transformation due to their ability to provide both incentives and disincentives to encourage specified outcomes.[165]

7. CONCLUSION

Market mechanisms have emerged as a solution to concerns about sustainability, flexibility and the economic cost of environmental management. At this stage, forest market mechanisms, like conventional forest regulation, have been unable to avoid complex frameworks that undermine the goals of sustainability, flexibility, and economic effectiveness. From a legal perspective, the increased use of forest market mechanisms is dependent upon resolving three issues. First, the environmental objective of the forest mechanism must be clear.[166] There must be a clear goal or purpose behind creating new interests in the environment. Secondly, the legal nature of the forest interest being traded within such schemes must be clarified. As discussed in this chapter, forest values may be protected under contract, legislation, or property rights. And thirdly, the governance process of trade in forest values must be designed so that it is clear and simple. If these issues can be resolved, forest market mechanisms may be of great assistance in implementing sustainable forest management due to their ability to provide incentives encouraging compliance.

---

164 Myers, above n 163.

165 See Bosselmann and Richardson, above n 8 which discusses the role of markets as a carrot or stick mechanism.

166 Crook, Carolyn, and Roger Clapp (1998), 'Is Market-Oriented Forest Conservation a Contradiction in Terms?', *Environmental Conservation*, **25** (2), 131–142.

9. Forest certification schemes

1. INTRODUCTION

Certification is a mechanism that seeks to implement sustainable forest management through the labelling of consumer products. Such systems are premised on the view that consumers will seek out forest products that are reputably certified.[1] Consumer demand for sustainably harvested and produced forest products then creates a market demand for them. This, in turn, leads to the better management and use of global forest resources. The potential economic benefits of forest certification depend primarily upon certification programs being able to generate access to eco-sensitive markets that differentiate and favour certified forest products.[2] Access to such eco-sensitive markets provides an economic incentive in the form of a premium price for certified timber. Certification schemes have been used to promote other social and environmental objectives (for example, fair trade and organic schemes). For certification schemes to flourish, there must be sufficient consumer demand, coupled with a willingness by consumers to pay any price premiums associated with the production of the item. Certification programs foster constructive discussions about forest governance and destructive logging practices by increasing awareness of sustainable management alternatives, and enhancing dialogue and cooperation among environmentalists, timber producers, buyers and governments.[3]

Forest certification schemes, thus, offer a new avenue for the implementation of sustainable forest management. This avenue taps into consumer demand for sustainable timber, and requires the producing industries to change their practices (to preserve their market share). For wood-based certification schemes to be compatible with World Trade Organization

[1] Shanley, Patricia, Alan Pierce, Sarah Laird and Dawn Robinson, Rights and Resources Initiatives (2008), *Beyond Timber: Certification and Management of Non-Timber Forest Products,* 1.

[2] Ebeling, Johannes, and Mai Yasue (2009), 'The Effectiveness of Market-based Conservation in the Tropics: Forest Certification in Ecuador and Bolivia', *Journal of Environmental Management*, **90** (2), 1145–1153, 1146.

[3] Ibid. 1151.

(WTO) or General Agreement on Tariffs and Trade (GATT) guidelines, the framework must be voluntary.[4] This is also the case for voluntary governmental certification programs, as long as they conform to the rules set forth by the WTO (such as non-discriminatory treatment and transparency). Non-governmental bodies are not subject to WTO jurisdiction, although the Technical Barriers to Trade Agreement contains provisions for certification systems for non-governmental bodies.[5] A mandatory preference for sustainably harvested timber could be seen as a barrier to trade (in contravention of international trade principles). This imposed voluntary requirement, thus, makes market-driven, or non-state governance, the most appropriate method for the implementation of forest certification schemes. The development of the Forest Law Enforcement Governance Program,[6] and the advent of forest certification schemes, suggests that unsustainable timber harvesting is becoming a barrier to trade.

When first conceived, there was great optimism that forest certification schemes would improve silviculture and harvesting processes worldwide but, to date, they have not been as widely implemented as first thought. This is due to a wide range of barriers that will be discussed in more detail throughout this chapter. Such barriers include: fostering consumer demand and a willingness to pay a premium price for certified timber products; improving consumer understanding of forest certification logos and processes; increasing incentives for those bodies that comply with certification standards; making forest certification more accessible to small-scale forest producers; and significantly increasing the capacity within tropical forest countries to participate in certification processes.

If such barriers can be overcome, forest certification could significantly alter the use, management, and consumption patterns of forests and their products. It has been stated:

> Forest certification caters for many different peers and their respective interests. For industry and trade it is an instrument for environmental marketing and market access. For buyers and consumers, it provides information on the impacts of products they purchase. For forest owners and managers, it is a tool for market access or gaining market advantage. For governments, it is a soft policy instrument to promote sustainable forest management and sustain-

---

[4] Rametsteiner, Ewald (2002), 'The Role of Governments in Forest Certification – a Normative Analysis Based on New Institutional Economics Theories', *Forest Policy and Economics*, **4** (3), 163–173, 167.

[5] Ibid.

[6] Discussed in more detail in Chapter 7. The program from the European Union requires a commitment from both producing and consuming countries to combat illegal logging.

able consumption patterns. For the environmental movement, it is a means to influence how forests are managed – to promote, inter alia, biodiversity maintenance.[7]

Certification can be defined as:

> [an] evaluation system used to verify compliance with a set of standards that ensure product quality, consistency or safety.[8]

Certification schemes require the creation of standards. These are agreed-upon measures that prescribe specific requirements, benchmarks, or protocols. Parties seeking certification of their processes must meet the requirements established by the standards. To determine if a certification applicant has complied with the necessary standards, a verification process is followed. This usually includes an audit or assessment where practices and instrumentation are reviewed to evaluate the applicant's compliance with the standard. Verification can be carried out in three different ways. The first involves first-party assessment (otherwise known as an internal audit process). Parties using this method carry out the process within their own organization. The second method involves second-party assessment, and is carried out by an organization with a relationship to the institution being assessed (for example, supplier audit). The final method is third-party certification, which is carried out by an independent organization. Third-party verification processes tend to be more rigorous than first- and second-party processes (and, as such, are more accountable and transparent).

In order for verification bodies to carry out assessments they, themselves, must be accredited. Accreditation is, thus, an evaluation process which verification bodies complete. The aim is to ensure that organizations undertaking certification assessments adhere to established standards and provide competent, independent, and credible results. Certification can refer to one component of forest practices (for example, the growing and harvesting of timber) or it may relate to the entire timber production process. 'Chain of custody' certification verifies the successive links in the supply chain of products from their origin (in the forest) through the different stages of processing, transport and distribution.

---

[7] Rametsteiner, Ewald and Markku Simula (2003), 'Forest Certification – an Instrument to Promote Sustainable Forest Management?', *Journal of Environmental Management*, **67** (1), 87–98, 88.

[8] Shanley, above n 1, 7.

2. THE EMERGENCE OF A FOREST CERTIFICATION MECHANISM

A. The Impact of the Earth Summit Negotiations on Forest Regulation

International forest regulation differs substantially in form from other international environmental regulation. Most environmental resources, or environment-related challenges, have one legally binding international instrument, and an accompanying institution that oversees the implementation of the agreement. Examples of such regimes include the biological diversity regimen, the wetland regimen, the climate change regimen, and the desertification regimen. By contrast, forests are not regulated by a single entity, and forest use and management is regulated under a number of voluntary international arrangements. It is widely accepted that the failure at the Earth Summit 1992 to create a legally binding forest instrument led to the current unique style and form of international forest regulation.[9] Concerns that prevented the creation of a forest convention were those connected with state sovereignty and finance. The latter required compensation, technology transfer, and capacity-building assistance to implement obligations arising from such an agreement. As documented in earlier chapters, the public international forest regimen has continued to be plagued by sovereignty and funding issues.

Cashore classified two distinct paths of international forest regulation that emerged after the Earth Summit negotiations in 1992, and suggested that both of these paths sidestepped state sovereignty issues connected with forest resources.[10] The first path involved a process to define sustainable forest management. The 'criteria and indicator' approach to achieving sustainable forest management is the primary example of a defining international forest process. This method does not prescribe standards to be met, nor does it require evidence of implementation. Instead, it concentrates on establishing definitions, practices and ways to measure progress towards a prescribed goal (sustainable forest management).[11] Domestic

---

9 See generally, Humphreys, David (ed.) (2006), *Logjam: Deforestation and the Crisis of Global Governance*, London: Earthscan Publications, 22.

10 Cashore, Benjamin, Graeme Auld, Steven Bernstein and Constance McDermott (2007), 'Can non-State Governance "Ratchet up" Global Environmental Standards? Lessons from the Forest Sector', *Review of European Community and International Environmental Law*, **16** (2), 158–172.

11 See Rametsteiner and Simula, above n7, who compare the similarities and differences between criteria and indicator process and certification. See

forestry agencies, intergovernmental negotiators, and public international agencies are the institutions that have participated in defining approaches to forest management.[12]

The second approach favoured market-driven approaches to address worsening global forest conditions. The two main market-driven approaches concerning forest use and management are the development of markets for forest ecosystem services and forest certification schemes. While both of these are market-based in style, they seek to promote different forest values. Markets for forest ecosystem services are aimed at enhancing and preserving the protective values associated with forest areas, while forest certifications are more focused on improving harvesting and production. As such, these approaches are complementary. Institutions that have favoured market-driven approaches include environmental groups, forest companies and retailers, governmental and aid agencies, and the World Bank.[13]

B. Conditions Allowing for the Development of Forest Certification Schemes

After the Earth Summit, there was growing dissatisfaction among environmental non-governmental organizations about international forest regulation. Such groups felt that governments would be unable to agree quickly on a legally binding international forest instrument. NGOs decided to capitalize on earlier trials of forest certification and labelling schemes.[14] The World Wide Fund for Nature (WWF) started certifying forest landowners and forest companies that complied with WWF's sustainable forestry definition and requirements. Similarly, Friends of the Earth created 'a good wood buyers guide'.[15] Such approaches were a departure from earlier environmental NGOs'

---

also: Rametsteiner, Ewald and Markku Simula (2003), 'Forest Certification – an Instrument to Promote Sustainable Forest Management?', *Journal of Environmental Management*, **67** (1), 87–98, 91.

12 Cashore, above n 10.

13 Ibid.

14 Friends of the Earth first started a timber certification scheme in 1985; the scheme was withdrawn after several years of experimentation due to concerns over the rigour of the system. See Counsell, Simon, and Kim Terje Lorras (eds) (2002), *Trading in Credibility: the Myth and the Reality of the Forest Stewardship Council*, London: Rainforest Foundation, 12.

15 Klooster, Dan (2005), 'Environmental Certification of Forests: The Evolution of Environmental Governance in a Commodity Network', *Journal of Rural Studies*, **21** (4), 403–417, 405.

activities, which had previously favoured boycotts and awareness-raising campaigns.[16]

The certification of forest practices meant that NGOs had something with which they could start to negotiate with the timber industry. The timber industry in return was interested in the potential advantages offered by having 'green' credentials.[17] The more traditional boycotting and educational campaigns of environmental NGOs have previously only resulted in increased hostility between the two groups, while this new certification mechanism provided new communication opportunities between both sides. Environmental NGO's were also willing to work with the timber industry because of their increasing doubts about the effectiveness of timber boycotts (since only a small percentage of tropical wood appears in export markets).[18]

During the 1990s, political leaders and others began to give increased attention to a promising set of new policies that recognized the potential role of market forces in achieving sustained environmental progress.[19] Parallel to this broader trend in environmental market regulation, there was increasing concern about the lack of progress in relation to tropical deforestation rates.[20] Unsustainable forest practices in the tropics were associated with market, policy and institutional failure.[21] Market failure occurs when forest goods and services are undervalued or not valued at all. According to Richards, the sources of market failure include[22]

- Externalities in which the effect of an action on another party is not taken into account by the perpetrator;
- Missing markets for environmental services and other open-access public goods;
- Market imperfections (such as a lack of knowledge, which causes uncertainty).

---

16 Cashore, above n 10.

17 See in general on this point: Klooster, above n 15, 405.

18 Ibid. 406.

19 Hahn, Robert, and Robert Stavins (1991), 'Incentive-Based Environmental Regulation: A New Era from an Old Idea', *Ecology Law Quarterly*, **18** (1), 1–42, 18.

20 Cashore, Benjamin, Graeme Auld and Deanna Newsom (2003), 'Forest Certification (Eco-labelling) Programs and their Policy-Making Authority: Explaining Divergence among North American and European Case Studies', *Forest Policy and Economics*, **5** (3), 225–247, 226.

21 Barbier, E., J. Burgess and C. Folke (eds) (1994), *Paradise Lost? The Ecological Economics of Biodiversity*, London: Earthscan Publications.

22 Richards, Michael (2000), 'Can Sustainable Tropical Forestry be Made Profitable? The Potential and Limitations of Innovative Incentive Mechanisms', *World Development*, **28** (6), 100 –116, 1003.

Richards defined policy failure as state actions or policies that provide disincentives (either directly or indirectly) to sustainable forest management. Such actions or policies may be related to land tenure, the weak enforcement of laws, or subsidies that allow unsustainable timber practices to continue.[23] Richards argues that institutional failure takes place when institutions are poorly designed and do not coordinate their activities well with other relevant institutions. Because governments are largely responsible for policy and institutional failures, non-state stakeholders are unable to intervene in such matters. However, market failure is an area that allows a role for non-state participants. Within the forestry context, markets were introduced to fill in missing markets (payment for ecosystem services), and also to address market imperfections (forest certification schemes).

C. Theories Which Support Certification Regulation

Vogel's work on consumer and environmental regulation identified two trends that can occur within market-based regulation: the 'Delaware effect' and the 'California effect'.[24] The 'Delaware effect' (which is sometime referred to as a 'race to the bottom') describes the phenomenon of businesses fleeing to less regulated markets.[25] Markets that operate in such conditions place price considerations above all others and, as such, usually correlate highly with unsustainable and environmentally degrading practices. An example of the 'Delaware effect' is the operation of foreign timber companies in Indonesia. These companies are able to reduce their overheads and operating costs substantially by conducting their business in a forest sector marketplace where there is low enforcement of regulation and high levels of corruption.[26]

The opposite of the 'Delaware effect' is the 'California effect' (sometimes referred to as a 'race to the top'). The 'California effect' is a phenomenon where firms in highly regulated markets are able to influence the practices of those in less-regulated markets by imposing their higher standards upon

---

[23] Ibid.

[24] Vogel, David (ed.) (1995), *Trading Up: Consumer and Environmental Regulation in a Global Economy*, Cambridge, Mass: Harvard University Press.

[25] Cashore, Benjamin, 'Governing Through Markets' (Paper presented at International Regime, Avoided Deforestation and the Evolution of Public and Private Forest Policies in the South, Paris, France, 21–23 April 2007).

[26] Vogel, above n 24, 259.

their less-regulated competitors.[27] Vogel defines the 'California effect' as having three components[28]

1. The market advantage offered by stricter regulation (competitive advantage)
2. Rich nations creating domestic 'green product standards', forcing foreign producers to raise their product standards to maintain market access
3. Agreements reducing trade barriers that provide rich and green nations with the opportunity to pressure other nations into adopting stricter product and production standards.[29]

Forest certification is an example of the 'California effect', whereby higher standards of timber producing and harvesting are imposed on countries whose own sustainable forestry standards are not enforced, or do not exist. As outlined above, global forest regulation shows evidence of both the 'Delaware' and 'California' effects.

Cashore's work builds on Vogler's studies by creating the term 'internationalization'. Cashore finds an increasing trend towards domestic policy being scrutinized by transitional actors and international rules and measures. The influence of such scrutiny sometimes results in a reversal of the 'downward' effect of globalization; this process is defined as internationalization.[30] The concept of internationalization provided lessons to environmental non-governmental organizations about the power of using market forces to shape policy responses (where efforts to change were often easier than attempting to influence domestic and international business dominated policy networks).[31] Many tropical countries lack the resources and governance capacity to enforce forestry and land use regulations effectively and to provide secure land tenure. As a result, in the last two decades, market-based instruments involving non-state people (such as forest certification) have been promoted as an economically attractive

---

27 Cashore, above n 10, 159.

28 Vogel, above n 24, 25–60.

29 Ibid.

30 Bernstein, Steven and Benjamin Cashore (2000), 'Globalization, Four Paths of Internationalization and Domestic Policy Change: The Case of Eco-forestry Policy Change in British Columbia, Canada', *Canadian Journal of Political Science*, **33** (1), 67–99.

31 Cashore, Benjamin, Graeme Auld and Deanna Newsom (2003), 'Forest Certification (Eco-labelling) Programs and their Policy-making Authority: Explaining Divergence among North American and European case studies', *Forest Policy and Economics*, **5** (3), 225–247.

alternative that is less dependent on public resources and governance capacity (and, therefore, potentially more effective in tropical developing countries).[32]

D. Forest Certification and Free Trade

The International Tropical Timber Organization (ITTO) is the international body that deals with trade in forest products. This body has two primary instruments: the *International Tropical Timber Agreement 2006* (ITTA 2006); and the *ITTO Action Plan 2008–2011*. The ITTA 2006 is a foundation document outlining objectives, definitions, organizational and administrative guidelines, and other miscellaneous articles. The objective of the instrument is to promote the expansion and diversification of international trade in tropical timber from sustainably managed and legally harvested forests.[33] This objective clearly indicates that the focus of this institution is on protecting the resource base for the international tropical timber industry.

The ITTA 2006 contains a number of definitions but the concept of 'sustainable forest management' is not given a definition by the agreement.[34] The agreement creates two forms of membership: producer countries, and consumer countries.[35] The agreement gives equal voting power to consumer and producer companies, assigning each group 1000 votes.[36] These votes are broken down among the producer countries in Annex A to the agreement: the scale of their tropical forest timber holdings determines countries' votes. Developed countries that have tropical forest areas (such as Australia) are considered to be consumer countries by this instrument.

The agreement acknowledges the need for cooperation and coordination with other international institutions although the only body specially mentioned is the United Nations Conference on Trade and Development (UNCTAD), suggesting the focus of the agreement is squarely on the trade and production of tropical timber products.[37] Signatories to the agreement

---

[32] Gulbrandsen, Lars (2004), 'Overlapping Public and Private Governance: Can Forest Certification Fill the Gaps in the Global Forest Regime?', *Global Environmental Politics*, **4** (2), 75–99.

[33] United Nations, *International Tropical Timber Agreement 2006*, 4th pt, Agenda Item 7, UN Doc TD/TIMBER.3/12 (1 February 2006) art 1.

[34] See ibid. Art. 2 (1), which provides that sustainable forest management will be understood according to the organization's relevant policy documents and technical guidelines.

[35] Ibid. Art. 4.

[36] Ibid. Art. 10.

[37] Ibid. Art. 15.

are required to make contributions to the administrative account (which covers the cost of the body's operations).[38] Other funds (such as the project funds which finance projects to improve sustainable forest management outcomes) are made available on a voluntary basis only.[39]

The other particularly interesting clause in the agreement is found at Article 34, which provides:

> Nothing in this agreement authorizes the use of measures to restrict or ban international trade in, and in particular as they concern imports of and utilization of, timber and timber products.

The ITTO is a firm believer in the operation of free markets. The website of the ITTO comments:

> Market access refers to the extent to which products and services can be freely traded into export markets. In practice, the quest for complete market access has been motivated by a desire to gain maximum benefits from an expanding and liberalising global trend. However, improving market access has been a laborious and contentious exercise, and governments continue to resort to a host of protective measures (in the form of tariff barriers, non-tariff barriers and other market impediments) in order to restrict trade (often to protect domestic industries).[40]

The ITTO believes that policies – such as encouraging domestic processing, promoting products from sustainably managed forests, restricting imports of unsustainably or illegally produced timber – are barriers to free trade. This stance is inconsistent with many other international forest processes. Forest certification schemes work by labelling products from sustainably managed forests so that consumers have the opportunity to source products made from sustainably harvested timber. The Forest Law Enforcement and Governance Program (FLEG)[41] operates to prevent the importation of timber that is sourced from unsustainable sources. The recent amendment to the Lacey Act in the United States of America now allows the government to fine, and even jail, parties who traffic in illegally harvested timber.[42] Furthermore, the non-discrimination provision

---

[38] Ibid. Art. 18 (1) (a).

[39] Ibid. Art. 18 (1) (b).

[40] International Tropical Timber Organization (2004–2009), 'Market Access', available at http://www.itto.int/en/feature11/ (accessed 11 June 2009).

[41] Discussed in more detail in World Bank chapters.

[42] For further information on the operation of this Act, see Gregg, R. Juge, and Amelia Porges (2008), 'Amendment to the US Lacey Act: Implications for Chinese Forest Products Exporters', available at http://www.forest-trends.org/

of the ITTO is in direct contrast to the operation of the *Convention on International Trade in Endangered Species of Wild Fauna and Flora* (which seeks to ban trade in endangered fauna and flora, including forest fauna and flora). It is suggested that the ITTO's firm commitment to 'non-discrimination' should be reconsidered in the light of the other international forest processes that seek to improve the management of forest areas through market intervention.

The *ITTO Action Plan 2008–2011* identifies a number of trends in the global trade in tropical forest products.[43] These indicate that:

- the incidence of poverty in forest-dependent communities remains high;
- tropical forest areas and associated ecosystem services continue to be lost at an unsustainable rate;
- deforestation contributes 20 per cent to global greenhouse gas emissions;
- forest certification schemes are having an impact on the international timber market;
- in many countries, the responsibility for the management of forest resources is increasingly being transferred to the private sector.[44]

Recognizing that forest certification schemes have impacted on the international timber market could be interpreted as suggesting that the ITTO may eventually abandon the non-discrimination principle.

The thematic programs of ITTO for the 2008 to 2011 period are: Forest Law Enforcement, Governance and Trade; Community Forest Management and Enterprise; Trade and Market Transparency; Industry Development; and Efficiency and Reducing Deforestation; and Forest Degradation and Enhancing Environmental Services in Tropical Forests.[45] The final program related to REDD is a new addition – and appears to be a reflection of the current global consensus that protection of the

---

documents/publications/Lacey%20Act%20Brief_China_EN.pdf (accessed 11 June 2009).

[43] ITTO Working Group, International Tropical Timbers, *ITTO Action Plan 2008–2011: ITTO Policy Development Series No 18* (2008) 3.

[44] For example, as a result of the global economic downturn, the Queensland government in Australia decided to outsource management of forest reserves to the private sector. See: ABC News Online (2009), 'Asset Sell-off Only Decided Last Week: Bligh', available at http://www.abc.net.au/news/stories/2009/06/03/2587961.htm (accessed 11 June 2009).

[45] ITTO Working Group, International Tropical Timber Organization, *ITTO Action Plan 2008–2011: ITTO Policy Development Series No 18* (2008) 5.

remaining tropical forest areas is crucial in the mitigation component of addressing climate change.

The trade and market transparency program identifies the following two outcomes as evidence of its work: the promotion of public awareness about progress made in implementing SFM; and the increased availability of tropical timber from sustainably managed and legally harvested sources.[46] This objective of public awareness is also present in the ITTO *Yokohama Action Plan 2002–2006*.[47] The purpose of this is to increase public awareness about products available from sustainably managed sources. At the international level, there is a lack of attention to changing individual consumer behaviour. Eco-labelling programs have not been marketed in a way that allows individual consumers of forest products to know they have the option of purchasing timber products made from sustainable timber sources. The ITTO seems to have recognized this lack of effective marketing and communication, and is seeking to redress it. This action lends support to the view that the ITTO is moving away from the absolute principle of non-discrimination, perhaps at least in the sale of timber from tropical forests.

3. FOREST CERTIFICATION INSTITUTIONS

A. Engaging with Forest Certifying Bodies

There are a number of forest certification institutions. These institutions can be divided into two main groups: those that exist at the international level and require independent, third-party verification processes; and those that exist at the domestic level and involve internal verification. At the international level, the Forest Stewardship Council and the Pan European Forest Certification initiatives provide forest certification accreditation. Some country-based certification schemes have also emerged (such as the Sustainable Forestry Initiative in the United States of America, and the Canadian Standards Association, which has a forest certification arm).[48]

---

[46] ITTO Working Group, International Tropical Timber Organization, *ITTO Action Plan 2008–2011: ITTO Policy Development Series No 18* (2008) Outcome 4 (B), 8.

[47] ITTO Working Group, International Tropical Timber Organization, *ITTO Yokohama Action Plan 2002–2006, ITTO Policy Development Series No 11* (2002) Goal 2 Action 1, 10.

[48] For an analysis of the differences between each certification scheme, see: Cashore, Benjamin, G.C. van Kooten, Ilan Vertinsky, Graeme Auld, Julia

The development of alternative forestry certification schemes (particularly industry certification schemes) has raised legitimacy concerns about the value and use of forest certification status.[49]

Legitimacy can be defined as a generalized perception (or assumption) that the actions of an entity are desirable, proper, or appropriate within some socially constructed system of measures, values, beliefs and definitions.[50] The legitimacy of forest certification schemes can be considered from a number of standpoints, such as those of the timber producer, the timber retailer, consumer, and independent environmental organization. Work has been carried out by Cashore to examine why forest harvesters and retailers of forest products engage in forest certification processes.

According to Cashore, there are two main forms of legitimacy – pragmatic, and moral.[51] Pragmatic legitimacy rests on the self-interested calculation of an organization's most immediate audiences (in which the material well-being of the grantor of legitimacy is enhanced). Moral legitimacy reflects a positive normative evaluation of an organization and its activities. It rests not on judgements about whether a given activity promotes the goals of the evaluator, but rather on whether the activity is the correct thing to do. Cashore found pragmatic legitimacy played a larger role in explaining why firms may decide to engage in a voluntary regulatory forest certification scheme.[52] Van Kooten, Nelson and Vertinksy's work examines the reasons why firms might engage in certification practices. They find that companies primarily do so in pursuit of economic interests and social capital (which is consistent with Cashore's findings).[53]

There has also been work to examine why timber retailers engage with

Affolderbach (2005), 'Private or Self Regulation? A Comparative Study of Forest Certification Choices in Canada, the United States and Germany', *Forest Policy and Economics*, **7** (1), 53–69, 57.

49 Counsell, Simon, and Kim Terje Lorras (eds) (2002), *Trading in Credibility: the Myth and the Reality of the Forest Stewardship Council*, London: Rainforest Foundation, 36–7.

50 Suchman, M.C. (1995), 'Managing Legitimacy: Strategic and Institutional Approaches', *Academic Management Review*, **20** (3), 571–610.

51 Cashore, Benjamin, Graeme Auld and Deanna Newsom (2003), 'Forest Certification (Eco-Labelling) Programs and their Policy-making Authority: Explaining Divergence Among North American and European Case Studies', *Forest Policy and Economics*, **5** (3), 225–247, 228.

52 Ibid.

53 van Kooten, G. Cornelis, Harry Nelson and Ilan Vertinsky (2005), 'Certification of Sustainable Forest Management Practices: A Global Perspective on Why Countries Certify', *Forest Policy and Economics*, **7** (6) , 857–867.

forest certification programs. Again, pragmatic legitimacy can explain their involvement. This is a particularly telling quote from a manager of a large homewares retailer in the UK (B&Q):

> We weren't ever going to have customers demanding sustainable timber in our stores. But we knew that if our name [B&Q] was associated with the destruction of tropical forests or even temperate forests, our brand name would be damaged.[54]

So it appears that those interested in the economic value of forests (timber producers and retailers) are enticed into engaging with certification schemes from a practical business standpoint. Consumer's knowledge of forest certification schemes, their labels and their standards is largely non-existent.[55] This lack of consumer knowledge is discussed further in the chapter, where it is proposed that a fundamental flaw in forest certification schemes is the lack of motivation and engagement of the consumers of timber products.

Some NGOs (for example, the Rainforest Foundation) have concerns about the legitimacy of forest certification labels. Primarily, they are concerned that the label does not necessarily ensure that sustainable practices have been followed. They are also concerned that different standards and interpretations are used when granting certification status.[56] The World Wide Fund for Nature (WWFN) previously partnered the World Bank to implement the ambitious goal of securing 200 million hectares of the world's production forests (independently certified as sustainably managed) by the year 2005. The WWFN has frequently stated that it considers the Forest Stewardship Council as the only credible certification scheme currently available. The WWFN and World Bank alliance has significant potential to affect forest management in the Bank's client countries.[57]

The remainder of this chapter will focus on the operation of the Forest

---

[54] Taylor, Peter Leigh (2005), 'In the Market but not of it: Fair Trade Coffee and Forest Stewardship Council Certification as Market-Based Social Change', *World Development*, **33** (1), 129–147, 142.

[55] Research has been carried out in USA to determine consumers' knowledge about forest certification schemes. Teisl, Mario (2003), 'What We May Have is a Failure to Communicate: Labelling Environmentally Certified Forest Products', *Forest Science*, **49** (5), 668–680.

[56] Counsell and Lorras, above n 49.

[57] Dingwerth, Klaus (2008), 'North-South Parity in Global Governance: The Affirmative Procedures of the Forest Stewardship Council', *Global Governance*, **14** (1), 53–71, 59.

Stewardship Council (FSC) as a case study for forest certification regulation. The FSC was selected as a case study for a number of reasons

- Support from the World Bank and the WWFN for its operations and objectives.
- The FSC was the first institution to establish forest labelling and standards processes and, to date, has certified timber practices in more than 82 countries.[58]
- The FSC is the only internationally focused forest certification regimen.[59]
- The unique institutional structure of the FSC.

B. Governance of the Forest Stewardship Council

In 1990, the Forest Stewardship Council (FSC) was created when timber users, traders and representatives of environmental and human rights organizations met in California to discuss the need for a credible system to identify products from well-managed forests. Immediately before the Earth Summit, the FSC held consultations in 10 countries to seek support for its proposals for a global certification and accreditation system for forests and forest products. When the Earth Summit failed to deliver a legally binding forest instrument, the FSC decided to pursue its agenda independently from existing public international institutions. The Founding Assembly of the Forest Stewardship Council was held in Toronto, Canada, in 1993, with 130 participants from 26 countries attending.[60] Originally, the FSC was funded through the World Wide Fund for Nature and B&Q (Home Improvement Company).[61] Today, the institution is self-funded through evaluation fees, accreditation fees, fees generated from use of trademarks, grants and donations, membership dues, and returns on investment.[62]

---

58 Forest Stewardship Council (2009), 'Global FSC Certificates: Type and Distribution', available at http://www.fsc.org/fileadmin/web-data/public/document_center/powerpoints_graphs/facts_figures/09-06-15_Global_FSC_certificates_-_type_and_distribution_-_FINAL.pdf (accessed 16 September 2009).

59 The Pan European Forest Certification initiative is focused on Europe, but does allow participation from other countries.

60 See the Forest Stewardship Council (2008), 'History', available at http://www.fsc.org/history.html (accessed 4 September 2009).

61 Eden, Sally (2009), 'The Work of Environmental Governance Networks: Traceability, Credibility and Certification by the Forest Stewardship Council, *Geoforum*, **40** (3), 383–394, 385.

62 Forest Stewardship Council, *Forest Stewardship Council A.C. By-Laws* (Ratified September 1994) cl. 10 Finance.

The Forest Stewardship Council Statue provides that the purpose of the association is to:[63]

- promote adequate management of forests and provide assistance to achieve environmentally appropriate and economically viable exploitation of natural resources;
- promote sustainable forestry principles through a voluntary accreditation program;
- conduct educational activities aimed at increasing the awareness of the importance of improving forest management;
- provide guidance and assistance to forest policy makers;
- provide assistance and collaborate with all kinds of entities and agencies around the world, public or private, dedicated to forests;
- promote the development of all kinds of activities addressed to the preservation and maintenance of forests;
- establish offices for research, information, and analysis of forest matters.

The FSC is currently promoted in more then 80 countries worldwide through the use of forest management certification and chain of custody certification. Around 10 per cent of the world's managed forests have been certified to FSC Standards.[64] The Council has, thus, played a significant role in shaping and implementing forest harvesting and production policy across the globe. During the implementation of certification processes, a wide array of stakeholders became involved in Council processes, including people and organizations involved in supplying wood and fibre to the worlds markets, landowners, forest managers, community groups, manufactures, traders, and retailers. In broad terms, the FSC has been successful in developing and implementing international legal measures for the production and harvesting of timber.[65] There is, of course, still much scope for expanding the percentage of the world's forests managed according to such criteria.

The institutional structure of the Council is worthy of examination due to its innovative design. The General Assembly is the highest organ of the FSC. Its by-laws require that the General Assembly consist of three cham-

---

63 Forest Stewardship Council, *Statute* (Revised August 2000, November 2002, June 2005, February 2009) cl. 4.

64 FSC Board of Directors and Staff, Forest Stewardship Council, *Strengthening Forest Conservation, Communities and Markets: The Global Strategy of the Forest Stewardship Council* (2007) 1.

65 Dingwerth, above n 57, 58–9.

bers: a social and indigenous chamber, an environmental chamber, and an economic chamber.[66] Each of these is given 33.3 per cent of voting power, and each must contain Northern (high-income countries) and Southern (low-income countries) representatives. Bodies eligible for membership within the social and indigenous chambers must be indigenous and social movements seeking to promote environmentally appropriate, socially beneficial, and economically viable forest management.[67] Members of the environment chamber must be either a not for profit, non-governmental organization, or an appointed individual with a demonstrated commitment to environmentally appropriate, socially beneficial, and economically viable forest management.[68] Those eligible for membership of the economic chamber are: forest production companies; certification bodies; industry associations; wholesalers; retailers; traders, end-users; and consulting companies.[69]

The structure of the General Assembly ensures that all relevant stakeholders are represented and, furthermore, have voting rights. As a result, stakeholders of the regime feel higher levels of ownership of the policy and procedures. It also means that while one chamber may not get everything that it wants, it will be at least happy with the compromise reached during the process. In terms of voting arrangements, the by-laws require all decisions to be made by consensus; in the event that this is not possible, 66.6 per cent of the vote is required to make a change.[70] The voluntary acceptance of the Forest Stewardship Principles can, in part, be explained by this representation of interests within the founding organ. Pragmatists who believe that all forest stakeholders must be able to live with the policy would support such approaches.

While the governance structure of the Council recognizes the importance of economic, environmental and social representation, it has been suggested that recognition alone will not be sufficient to ensure that the interests of different parties are met. In particular, Dingwerth identifies barriers to the recognition of Southern interests.[71] There are a number of issues associated with selecting the most suitable people to represent Southern interests. First, there needs to be recognition of the great diversity within Southern interests and there need to be representatives from a

[66] Forest Stewardship Council, *Forest Stewardship Council A.C. By-Laws* (Ratified September 1994) cl. 11–3.
[67] Ibid. cl. 31.
[68] Ibid.
[69] Ibid cl. 28.
[70] Ibid cl. 15.
[71] Dingwerth, above n 57, 53.

variety of Southern countries. Currently, within the FSC, Latin America is well represented whilst stakeholders from African and Asian countries play a lesser role in the governance process.[72] This lack of representation may also explain why certification rates are lower in these countries. Secondly, the people selected must be involved at the local level to ensure that representation is fulfilling its purpose (that is, local people and their issues are understood by the organization).

In total, 28 national or regional standards have been created by the FSC. Of these, eight are in Southern countries, and all of these regional standards have been created within Latin America.[73] There has, thus, been no regional development of Council policies in Africa or Asia. The lack of regional policies in these two regions means that local forest issues and conditions there are not currently addressed by FSC operating modalities. This suggests that the Council needs to strategically expand its operations into those countries in Asia and Africa responsible for producing and harvesting global timber supplies.

In a similar vein, it has been suggested that the economic chamber dominates within the Forest Stewardship governance structure. While each chamber is given equal voting rights, the membership levels of each chamber are not equal. The economic chamber accounts for 46 per cent of membership, the environmental chamber 36 per cent and the social chamber holds the remaining 18 per cent.[74]

It has been suggested that, although NGO groups initially conceived the FSC, large-scale commercial interests are increasingly gaining control over its operations. Two particular issues concerning the social chamber and its participation were identified by an NGO-commissioned environmental report on the operations of the FSC.[75] First, there was concern that in many instances local forest stakeholders were not adequately consulted during certification processes. In some instances, certification was seen as being illegitimately imposed on traditional land tenure regimes.[76] Secondly, there was evidence that FSC operations at the national level

---

[72] For a breakdown of membership and board membership, see: Dingwerth, Klaus (2008), 'North-South Parity in Global Governance: The Affirmative Procedures of the Forest Stewardship Council', *Global Governance*, **14** (1), 53–71, 54–65.

[73] Dingwerth, above n 57, 66.

[74] Counsell, Simon, and Kim Terje Lorras (eds) (2002), *Trading in Credibility: the Myth and the Reality of the Forest Stewardship Council*, London: Rainforest Foundation, 32.

[75] Ibid. Pt 3.

[76] Ibid. 32.

did not actively ensure that each chamber was adequately represented during negotiations.[77] While voting rights are divided according to the three chambers, over-representation of the economic chamber at meetings means that its influence and interests are able to permeate council decisions. This accords with Dingwerth's view that recognition of the three interest groups within the chamber does not necessarily align with the three chambers gaining equal control over the development and implementation of policies.[78] In essence, there is concern that the timber industry is gaining increasing control over the principles, measures, and standards of the organization. There is, however, evidence to suggest that the Council has recognized these criticisms, and has developed new guidelines for regional group certification, and has channelled more support to the social chamber.[79]

C. Non-State Governance

The FSC has been described as providing an example of governance without government.[80] This form of governance has also been referred to as non-state governance arrangements, private systems of regulation, private governance and non-governmental market-based regulation.[81] The FSC is an often-cited example of a new environmental governance model that operates across science, policy and business sectors.[82] Importantly, these new governance networks operate largely outside governmental frameworks and being based on voluntary take-up must compete for

---

77 For example, at the UK FSC national meeting, there were 30 members present, with only three representing social interests: ibid. 33.

78 Dingwerth, above n 57, 66.

79 Bass, Stephen, 'Global Forest Governance: Emerging Impacts of the Forest Stewardship Council' (Paper presented at the International SUSTRA Workshop 'Architecture of the Global System of Governance of Trade and Sustainable', Berlin, 9–10 December 2002); and Forest Stewardship Council International Centre, Forest Stewardship Council (2003), *FSC Social Strategy: Building and Implementing a Social Agenda*. The social strategy was created to address a number of issues raised by social constituencies, such as: respecting the rights of people involved in, or impacted by, forest management; addressing issues of scale and intensity; and building capacity in the FSC system. See also the Global Strategy of the Forest Stewardship Council that identifies the need to improve social criteria.

80 Bell, Stephen and Andrew Hindmoor (2011), 'Governance with Government? The Case of the Forestry Stewardship Council', *Public Administration*, 1–6.

81 Eden, above n 61, 384.

82 See Rametsteiner, above n 4, and Taylor, Peter Leigh (2005), 'In the Market but not of it: Fair Trade Coffee and Forest Stewardship Council Certification as Market-Based Social Change', *World Development*, **33** (1), 129–147.

authority.[83] Perhaps this competitive process makes such institutions more responsive, adaptive, and inclusive of relevant stakeholder's inputs (because firms must establish their right to exist and govern).

Some features of non-state, market-driven governance include[84]

- Governments do not create or require adherence to the rules;
- An institution and its processes are more open and transparent and involve a wide range of stakeholders;
- Profit-maximizing firms are required to undertake costly reforms that they would not otherwise pursue;
- Authority is granted through the market's supply chain (consumer demand);
- The existence of verification procedures designed to ensure that the regulated entity actually meets the stated standards.

The Forest Stewardship Council has been described as being a transnational regime. This type of regime emerges when international regulation is either absent or relatively weak.[85] Within these regimes (as already discussed), it is the non-state participants that create the principles, measures, rules, and decision-making procedures. Bass comments:

> It is notable how FSC has been able to develop norms of behaviour, develop procedures for compliance, tackle the issue of multifactor coherence, and ensure a case-based approach to judgement and appeal. In other words – to develop 'law'.[86]

Dingwerth defines the status of the Council operations as being quasi-legislative in nature. FSC processes, aimed at developing specific standards for private and public behaviour, are one of the most advanced cases of non-state-driven rule-making dynamics in the environmental field globally.[87] Transnational regimes have thus created for themselves power to create forest law, entirely independent of the traditional sources of power associated with sovereignty.

While governments are not the most authoritative party within tran-

---

83 Eden, above n 61, 383.

84 Cashore, above n 10, 161–2.

85 Dingwerth, above n 58, 54.

86 Bass, Stephen, 'Global Forest Governance: Emerging Impacts of the Forest Stewardship Council' (Paper presented at the International SUSTRA Workshop, Architecture of the Global System of Governance of Trade and Sustainable, Berlin, 9–10 December 2002) 7.

87 Dingwerth, above n 58, 59.

snational regimes, they still play an important role in either providing or withholding their support for the regime. Government support for forest certification programs may include setting up governmental certification systems, or providing incentives or finance for forest certification, and ensuring the compatibility of regional codes.[88] In instances where the government owns large reserves of forest resources, the government can provide significant support to certifying bodies by requiring compliance with certifying standards on all government forest land.

Clause 8 of the FSC by-laws provides that the institution is a non-profit, non-governmental organization that is to remain independent of control by commercial interests, government, and multilateral and bilateral agencies. Contemporary circumstances have created opportunities for NGOs to play a greater role in the development and implementation of international law.[89] This is has been explained, in part, by a shift in international law from focusing on state interests to focusing on individual interests.[90] Eden suggests that NGOs have the potential to better link scientific knowledge with policy and instrument requirements. This is because NGOs are often regarded as being less formalized, more reactive, and less bureaucratic than traditional policy agencies.[91] There is now wide acceptance by many scholars of the positive influence that NGOs can play in contemporary international law in areas related to human beings, rights, environmental protection, sustainable development, indigenous rights, non-violent conflict resolution, participatory democracy, social diversity, and social and economic justice.[92]

NGOs are not without their critics. There are those who argue that NGOs are undemocratic because they are narrowly elected bodies that pursue particular-interest agendas. Other critics suggest that the policies of NGOs are based on emotional and moral responses (as opposed to being grounded in objective and scientific criteria).[93] Furthermore, there are some concerns that NGOs lack the capacity to carry out the wide range

---

88 Rametsteiner, above n 4.

89 For example, the International Committee of the Red Cross is charged with monitoring compliance with international humanitarian law as it relates to prisoners of war.

90 Ku, Charlotte, and Paul Diehl (2006), 'Filling in the Gaps: Extra Systemic Mechanisms for Addressing Imbalances between the International Legal Operating System and the Normative System', *Global Governance*, **12** (1), 161–183, 167.

91 Eden above n 61, 384.

92 Ku and Diehl, above n 90, 171.

93 Eden, above n 61, 384.

of tasks assigned, or self-assigned, to them.[94] Transnational regulation has legitimized the role of NGOs in environmental regulation and while these criticisms have not completely vanished, they have, in more recent times diminished as the unique contribution of NGOs has become increasingly valued within international rule making institutions.

4. FOREST CERTIFICATION INSTRUMENTS

The key instrument of the Forest Stewardship Council is the *FSC International Standard*;[95] which contains 10 principles and associated criteria for meeting these principles. These apply to all tropical, temperate, and boreal forests. More detailed standards for particular forest types and conditions may be prepared at the national and local level (by FSC regional bodies).[96] The standards outline the FSC-approved principles and practices. In response to criticism over the costs associated with certification for small forest holders, an additional standard has been created for use in small and low-intensity managed forests.[97] This standard is used in conjunction with other instruments. The certification process involves the creation of accredited bodies to carry out assessments of certification compliance. Such a process necessitates the creation of policies on procedures for obtaining accreditation and guidelines for certifiers.

A. The Forest Stewardship Council Process

Applicants have a choice of three different types of certification. The first is forest management certification. This is designed for those who want to provide forest operations that are socially beneficial, and managed in an environmentally appropriate and economically viable manner. The second option is 'chain of custody' certification for companies that manufacture, process, or trade in timber and want to demonstrate to their customers that they use responsibly produced raw materials. The third option is controlled wood certification, which is used by companies to control the non-

---

94 Ku and Diehl, above n 90, 171.

95 Forest Stewardship Council (1996), *FSC International Standard: FSC Principles and Criteria for Forest Stewardship.*

96 Forest Stewardship Council Australia has an objective of creating regional principles and criteria but, at this stage, uses the international standard.

97 Forest Stewardship Council (2004), *FSC Standard: SLIMF Eligibility Criteria.*

certified wood in their products. There are five stages to an application for FSC certification status[98]

1. The applicant approaches a certification body, which will estimate the costs and time associated with forest certification on their lot. Details of certification bodies are available on the FSC international and regional websites.
2. After consulting with one or several certification bodies, the applicant decides which certification body to work with, and then enters into an agreement with it.
3. A certification audit takes place to assess the applicant's suitability for certification status. During this process, the certifier visits the forest site and requires certain documentation from the applicant.
4. The data collected at the audit is used to compile an auditor's report – the auditor uses this to make a decision about whether to grant certification status.
5. If the certification status is positive, the applicant will receive FSC certification and will be able to use its logo on their products. The audit report may reveal that the applicant needs to modify some practices to achieve certification status. The applicant can then modify its practices and undertake a further audit to demonstrate compliance with the FSC requirements. Once a forest certification status has been given, it is valid for five years.

Holders of certification status are required to undergo an annual audit assessment. The certification body that verifies if there is continued compliance with FSC requirements carries this out.

B. The FSC International Standard

The FSC International Standard is an instrument designed to outline the objectives and requirements of the FSC. The Council does not insist on perfection in satisfying the principles and criteria contained within the standard but major deficiencies in an individual principle will normally disqualify a candidate from certification (or lead to de-certification). Flexibility is required in interpreting the standard to take into account local forest conditions and local forest practices.

Principle 1 requires compliance with the prevailing law and with FSC

---

[98] Forest Stewardship Council (2009), '5 Steps Towards FSC Certification', available at http://www.fsc.org/106.html (accessed 11 September 2009).

principles. Essentially, this requires that forest areas be certified with respect to all international, national, and local laws, and to administrative requirements. If local laws require a higher standard of forest certification status, the applicant would be obliged to meet that standard. In the event of conflict between local laws and FSC criteria, discretion is given to the certifying body to resolve the issue.[99]

Principle 2 deals with issues connected to tenure and land use rights and responsibilities. For certification status to be granted, an applicant must show evidence of long-term forest use rights to the land. Evidence may be in the form of land title recognition, evidence of customary rights over land, or a valid lease agreement where the duration is compatible with granting certification status. Disputes of substantial magnitude involving a significant number of interests will normally disqualify an applicant from being certified.[100] In many developing countries, proving tenure security and use rights may prove difficult for some applicants.[101] It has been suggested that the FSC could undertake some capacity-building work aimed at improving tenure security for applicants in these areas to assist them with meeting this requirement.

Principle 3 requires that areas under FSC certification recognize and respect any indigenous holders' rights over the land to be certified. In particular, it is stated that forest management should not threaten or diminish (either directly or indirectly) the resources or tenure rights of indigenous people.[102] The inclusion of indigenous considerations on the face of it appears praiseworthy. However, there is some evidence to suggest that certification has been issued over areas where indigenous land use rights have not been adequately recognized or respected.[103]

Principle 4 deals with community relations and workers' rights. The principle requires forest management operations to maintain or enhance the long-term social and economic well-being of forest workers and local communities. Forest certification does not currently involve a fair trade component. Such a component would require that forest workers were paid a reasonable wage. There has been some criticism directed at

---

99 Forest Stewardship Council (1996), *FSC International Standard: FSC Principles and Criteria for Forest Stewardship*, principle 1.4.

100 Ibid. principle 2.3.

101 Shanley, Patricia, Alan Pierce, Sarah Laird and Dawn Robinson, Rights and Resources Initiatives (2008), *Beyond Timber: Certification and Management of Non-Timber Forest Products.*

102 Forest Stewardship Council (1996), *FSC International Standard: FSC Principles and Criteria for Forest Stewardship*, principle 3.2.

103 Counsell and Lorras, above n 49, 32.

forest certification schemes on this point, and it has been suggested that the schemes could learn from fair trade initiatives.[104] The Council has responded by starting a FSC and fair trade dual certification pilot project. This is an 18-month project aiming to develop an affordable dual certification system for communities that will, ultimately, be incorporated into the existing FSC and fair trade certification systems.

Principle 5 requires forest management to encourage the efficient use of forests' multiplicity of products and services to ensure economic viability as well as a wide range of environmental and social benefits. It encourages diversification of the local economy by avoiding dependence on a single forest product.[105]

Principle 6 deals with the environmental impacts associated with the operations of harvested forests. It obliges forest managers to conserve biological diversity and its associated values (such as water resources, soils, and unique and fragile ecosystems). Representative samples of existing ecosystems within landscapes are to be protected in their natural state and recorded on maps.[106] Also, genetically modified organisms are prohibited on certified forest lands.[107]

Principle 7 deals with the requirements of a management plan. The plan must, among other requirements, contain: management objectives; a description of forest resources; environmental limitations; land use and ownership statutes; socio-economic conditions; a profile of the adjacent land; a description of silvicultural management systems; a rationale for the rate of annual harvest; and provisions for monitoring forest growth.[108]

Principle 8 deals with monitoring and assessment obligations. These are imposed upon the certification holder and are determined by reference to the scale and intensity of forest management operations, as well as the relative complexity and fragility of the affected environment.[109] Monitoring, at a minimum, must include data on: the yield of forest products harvested; growth and regeneration rates; composition and observed changes in the flora and fauna; environmental and social impacts of harvesting and other operations and costs; and the productivity and efficiency of forest management.[110]

---

104 Leigh Taylor, above n 54.

105 Forest Stewardship Council (1996), *FSC International Standard: FSC Principles and Criteria for Forest Stewardship*, principle 5.4.

106 Ibid. principle 6.4.

107 Ibid. principle 6.8.

108 Ibid. principle 7.1(a–f).

109 Ibid. principle 8.1.

110 Ibid. principle 8.2.

Principle 9 requires areas containing conservation attributes to be identified during the certification process. Management activities in areas of identified high conservation value should maintain or enhance such attributes. Decisions regarding high conservation value forests shall always be considered in the context of the precautionary approach.[111]

Principle 10 deals with the requirements for establishing and managing a plantation certified forest. Some of the conditions imposed include wildlife corridors, streamside zones, and a mosaic of strands of different forest ages and rotation periods.[112]

5. IMPLEMENTATION OF FOREST CERTIFICATION SCHEMES

A. Implementation of Forest Stewardship Council Certification to Date

To date, about 5 per cent of the world's productive forests have received forest stewardship certification status. This equates to around 100 million hectares worldwide, distributed across 82 countries. The value of timber from Forest Stewardship Council sales is estimated to be over US$20 billion.[113] The Council breaks down certification status into six regions (highest to lowest of total certified areas): Europe, 45.21 per cent; North America, 34.5 per cent; South America and the Caribbean, 10.15 per cent; Africa, 6.18 per cent; Asia, 2.57 per cent; and Oceania, 1.40 per cent.

The FSC is deservedly proud of bringing 5 per cent of the world's productive forest estate under their regime. This accomplishment is significant in the light of the international and national forestry-related political, economic, social and environmental issues that have plagued progress in other international forestry regimes. No other international forest institution can publish figures demonstrating an impact on the ground above the 5 per cent global level achieved by the FSC. It should be noted in this context, though, that the Council figures refer only to the world's productive forest estate, not the entire global forest estate (that is, productive and protective forest areas). Whilst 5 per cent is significant, there is clearly a need for greater expansion of the areas under Council certification. There is still much progress to make before forest certification schemes operate over the majority of the world's productive forest estate.

---

111 Ibid. principle 9.

112 Ibid. principle 10.2.

113 Forest Stewardship Council (2009), 'FSC facts and figures', available at http://www.fsc.org/facts-figures.html (accessed 16 September 2009).

On examining the regions with the highest certification levels, it becomes clear that Europe (45.21 per cent) and North America (34.5 per cent) dominate. It has been suggested by a number of commentators that certification within these regions is, perhaps, not as necessary as certification in tropical and or developing countries (where the bulk of unsustainable and illegal timber harvesting takes place).[114] It could be suggested that the majority of forestry practices in Europe and North America have only had to make minimal changes to their forestry operations, because existing 'command and control' type domestic legislation already requires compliance with sustainable forest practices. The certification process will, nevertheless, have improved existing forest practices. Following on from this, it could be suggested that the Council has not actively changed the implementation of sustainable timber practices in 5 per cent of the world's productive forest estate but, rather, influenced the practices of sustainable timber harvesting in 5 per cent of it.

There is a need for this form of non-state, market-driven governance regimen to expand in countries where compliance with sustainable timber practices is weak or non-existent. This is because domestic regulatory approaches indicate a lack of capacity or unwillingness to implement sustainable timber practices. Such regulation stands more chance of being implemented in these regions if there are appropriate incentives and support at government and community level. Incentives may come in the form of increased market access, potential price premiums for sustainably harvested timber (this is something that forest certification schemes have not yet been able to offer their participants), or improved community engagement and control over forest estates. The expansion of forest certification schemes is, thus, reliant upon the FSC improving opportunities and incentives for forestry operations in developing countries to become certified. As noted by one authority in the area of forest certification:

> The original goal of certification programmes was to protect tropical forests. However, today certified logging operations exist mostly in boreal and temperate environments; at present only 17% of FSC-certified forests are located in the tropics.[115]

Given the significantly higher need for conservation in tropical forests, it is crucial to understand the factors limiting the adoption of certified

---

114 Rametsteiner and Simula, above n 7.
115 Cashore, above n 10, 7.

forest management in the relevant countries.[116] The identification of these 'barriers' to forest certification will assist in the prediction of the likely success rate of certification in different tropical forest countries. Ebeling and Yasue suggest that once these certification rates have been established, the realistic potential of forest certification mechanism as a global forest conservation tool for tropical forests will be better understood. The focus could then be directed towards improving capacity for participation within certified forestry initiatives in targeted tropical forestry countries. Alternatively, certification could be promoted in tropical forestry countries where existing policy is most compatible and supportive of certification requirements.[117] Such a strategy would allow the FSC to expand its operations in tropical forestry counties and increase its influence over the tropical forest timber industry overall.

B. Barriers to Further Implementation of Forest Certification Initiatives

There are a number of barriers that need to be overcome for forestry certification schemes to be implemented more broadly. Some of the barriers preventing their further expansion are the same ones faced by other international forestry regimes (for example, land and tenure security). Other barriers, like consumer awareness, are unique to forest certification regimes. The general barriers (such as costs and government support) will be examined, followed by analysis of particular barriers that arise in the developing country context (such as the standard gap and tenure security issues), and concluding with an examination of issues related to consumer awareness of certification schemes.

The costs/benefits associated with certification and government support or compatibility with certification requirements are barriers to forest certification, which are applicable in both developing countries and in timber-consuming countries. Because forestry certification is a voluntary process, it is assumed that those who engage in the process perceive there to be some form of benefit associated with meeting and complying with the requirements.[118] The cost of participating in voluntary initiatives is a relevant factor at the forefront of participants' minds. It has been identified within existing literature that the cost of forest

---

116 Ebeling and Yasue, above n 2, 1146.

117 Ibid.

118 Also see: Overdevest, Christine and Mark Rickenbach (2006), 'Forest Certification and Institutional Governance: An Empirical Study of Forest Stewardship Council Certificate Holders in the United States', *Forest Policy and Economics*, **9** (1), 93–102.

certification is a significant barrier to the expansion of forest certification programs.[119] Forest managers must cover the costs of audit, certification fees and corrective action requests. In Mexico, the average evaluation and monitoring costs are $US36,000 over 5 years and, if corrective action requests were included, the cost would reach $60,000 over 5 years.[120]

Linked to concerns surrounding the cost of certification is the issue that forest certification schemes have, as yet, been unable to deliver higher returns to participants in the certification scheme.[121] Although demand for certified wood is growing, price premiums have rarely materialized.[122] This means that the cost of certification cannot currently be recouped through improved or higher premiums for sustainable timber. A cost/benefit analysis of whether to certify or not would go against certification from a solely financial perspective. This suggests that the direct costs of auditing, verification and certification fees may need to be reduced (at least until the FSC can provide participants with access to eco-markets where consumers are willing to pay a premium). Klooster states:

> Some analysts doubt that forest certification will ever generate price premiums able to cover the cost of evaluations, audits, licensing and the management improvements that it usually requires. Retailers are the most powerful actors in wood commodity chains, and they have little interest in either increasing the cost of the products to consumers or in passing any increased revenue back to their certified suppliers.[123]

The Council has recognized that the costs associated with certification may limit the ability of small forest landholders or community groups to participate in certification processes. It is easier to find economies of scale in larger managed forest areas in relation to certification costs. To reduce the cost of certification for small- and low-intensity managed forests, the FSC permits certification bodies to evaluate and report on these forests using streamlined certificated procedures. A small forest management unit is defined as any lot up to 1,000 hectares.[124] It has been suggested that a streamlining of certification procedures alone would not help

---

119 Leigh Taylor, above n 54.

120 Klooster, above n 15, 408.

121 Ibid.

122 Ebeling and Yasue, above n 2, 1146.

123 Klooster, above n 15, 412.

124 Forest Stewardship Council (2004), *FSC Standard: SLIMF Eligibility Criteria*, 2.2.

significantly. The main costs associated with certification regimes are for documentation and monitoring.[125]

Government support for forest certification schemes could have a significant impact on the take-up of certification schemes. If domestic forest laws are compatible with certification requirements, this will assist in increasing participation as only minimal change (if any) to existing practices will be required.[126] Furthermore, if the government provides incentives for forest certification (for example, tax benefits,[127] or preferential treatment for state leasehold land), this will also impact upon the number of stakeholders engaging in forestry certification schemes. A study examining forest certification rates in Bolivia and Ecuador highlights the role that governments can play in encouraging forest certification. Bolivia, a poorer country, had much higher rates of certification due to strong government support (evidenced through compatible domestic forestry laws and tax incentives to offset certification fees). In addition, limited corruption within the Bolivian forestry department also assisted in increasing FSC certification.[128]

In developing countries, barriers to the expansion of forest certification include 'the standard gap' between domestic requirements and certification requirements, and issues connected with tenure security. In many industrialized countries, effective enforcement of forestry regulations leads to a small gap between actual forestry practices and certification standards. In contrast, poor environmental law enforcement in most developing countries – the original impetus for non-state market-driven forest certification – arguably creates a hurdle to certification.[129]

The introduction of certifiable forest management practices can entail considerable direct and indirect costs. Logging operations in tropical developing countries are often poorly planned and do not follow legal requirements due to weak enforcement of forestry laws.[130] The implementation of sustainable forestry criteria will likely result in increased costs associated with the planting and harvesting of trees, and may also result in harvesting being restricted in certain areas. Such changes would make for-

---

[125] Molnar, A., Forest Trends (2003), *Forest Certification and Communities: Looking Forward to the Next Decade*.

[126] Ebeling and Yasue, above n 2.

[127] In Bolivia, forest certificate holders enjoy tax benefits of 14–28 per cent, which roughly offsets direct certification costs. Certified producers are also exempted from government audits, saving them time and money. Ibid. 1149.

[128] Ibid.

[129] Ibid. 1151.

[130] Ibid. 1146.

estry operations less lucrative in the short term (although the returns may well be greater in the long term if it is assumed that the area previously logged was to remain barren). Those with long-term interests in the land would have more motivation to follow sustainable practices than those with no long-term formal rights to the land.

For FSC certification the following tenure arrangements must be:

> Long-term tenure and use rights to the land and forest resources shall be clearly defined, documented and legally established.[131]

Ensuring that the area certified has secure tenure arrangements and there is appropriate recognition of community land use rights to the forest remains a challenge for certification schemes. Many of the regions where forest certification schemes would be of assistance are the same regions that lack tenure security and land use rights. This creates a situation where forest users lack incentives to comply with certification guidelines. The FSC could seek to identify large-scale timber harvesting areas that have uncertain tenure status. It could then work with local groups to pursue tenure security on the basis that such support requires the parties to comply with the sustainable forestry certification practices. This would act as an incentive for forest users because they would obtain long-term access to an area, and improve their long-term livelihood through the practice of sustainable harvesting techniques.

In consumer timber countries, the single biggest barrier to increasing the demand for certified timber is the lack of consumer knowledge and understanding about the role of forestry certification schemes. If the market demand for sustainably harvested timber can be increased, the supply side will respond. This demand however will not emerge unless there is awareness that forest practices can be independently certified to comply with forest sustainability criteria. Therefore, there is a need for mass information campaigns, directed at consumers, advising that it is possible to source sustainable timber.

In the United States, a study examined consumers' knowledge and understanding of forest certification logos, based upon 1,948 written survey responses.[132] Two experiments were conducted: consumers' reaction to different styles of eco-labels was tested as well as the factors that

---

131 Forest Stewardship Council (1996), *FSC International Standard: FSC Principles and Criteria for Forest Stewardship*, 2.

132 Teisl, Mario (2003), 'What We May Have is a Failure to Communicate: Labelling Environmentally Certified Forest Products', *Forest Science*, **49** (5), 668–680.

influence consumers when purchasing timber products. The eco-label section found that altering the amount of information on the label significantly altered the credibility of the certifying body.[133] The research confirmed earlier information that the credibility of an eco-seal could be greatly enhanced by the inclusion of contact information. The research used two current eco-labels (the FSC label and the Sierra Club Label). It also included the logotypes of the US Environmental Protection Agency and the Main Wood Producers Association. In the study, the Sierra Club was perceived to be the most credible label, and the FSC logo the least credible. It was suggested that the Council's label might have been found the least credible due to lack of consumer familiarity with the brand.[134] This suggests that the FSC should devote time and resources to promoting its label.

In relation to consumers and their timber-purchasing behaviour, the study found that presenting environmental information to respondents significantly influenced their choice of product. When no environmental information was presented, and the only distinguishing feature between products is price, respondents choose the lowest-priced product. It would, therefore, be in the Council's interest to explore options for mandatory disclosure requirements on timber products. The survey also found that when the respondents had to view multiple products with competing environmental seals, they assumed that the environmental characteristics of the higher-priced product were better. At least some of these individuals were willing to pay the price differential between the two certified products.[135] This finding is consistent with other findings that Northern consumers are prepared to pay a significant premium for certified timber products.[136] This again suggests that if consumers are aware of the availability of certified sustainable timber, the demand for such products will increase (which would allow the FSC to expand its efforts). Furthermore, it suggests that the FSC needs to actively promote why it is different from other forest certification bodies (such as industry-certified bodies).

The Council has undertaken some advertising and consumer outreach work in Europe. In the Netherlands, advertisements on television and in print inform customers that 'with FSC timber you save more than the forest'. Following on from these information campaigns in 2004, 63 per

---

133 Ibid. 673.
134 Ibid. 674.
135 Ibid. 676.
136 Leigh Taylor, above n 54, 142.

cent of the Dutch consumers recognized the FSC label (up from 12 per cent in 2001).[137] This suggests that a large component of future FSC action should be targeting consumers' awareness and understanding of the label. Research that compares the FSC's certification with fair trade initiatives suggests that:

> FSC could learn from Fair Trade how to create end-user demand and develop marketing channels for a Fair Trade approach to community-produced forest products. It could learn from Fair Trade how to establish more direct, personalised ties between producers and end consumers.[138]

The barriers discussed within this section are inter-related. The lack of price premiums for certified timber producers currently acts a barrier to wider participation by forest managers in the scheme. The lack of consumer knowledge about, and understanding of, forest certification schemes also means that there is little demand by consumers for certified timber products. Furthermore, even when they are available, this same lack of consumer understanding and knowledge means that consumers are not aware that a sustainably sourced item is available alongside items from non-certified sources. Therefore, the FSC needs to increase the incentives for participants in certification schemes, reduce fees, and increase access to 'green' markets by more effectively engaging with timber consumers.

C. Evolutions in Future Forest Certification Regulatory Regimes

In addition to expanding the existing area under FSC certification, other evolutions for the organization include: introducing a fair trade requirement as part of certification processes; expanding certification into non-timber forest products; and creating capacity within the FSC processes to measure and monitor carbon stored in forests. It is frequently noted in literature concerning the Council that it does not ensure that fair rates and wages are paid to timber harvesters.[139] A fair trade component would ensure that timber workers do obtain a fair wage. Potentially, it could be used to expand coverage in developing countries (because a fair wage requirement would act as an incentive for further involvement since this wage is likely to be higher than existing wages). Linked to the issue of fair trade is the generation of a market for community forest products. Timber

---

137 Klooster, above n 15, 413.

138 Leigh Taylor, above n 54, 143.

139 For example, see ibid.

is seen as the most important commercial community forest product and, as such, could be seen as an asset to be used to escape poverty.[140]

There has also been some research that assesses the potential of a certification mechanism for non-timber forest products.[141] The report found that certification would be a viable strategy for charismatic forest products with high profiles (such as brazil nuts, chicle, palm heart and rattan, and widely sold medicinal products such as cat's claws and yohimbe). It is currently possible to get FSC certification for both timber and non-timber forest products but the costs of doing so are prohibitive.[142] The FSC should also build capacity to measure and monitor the carbon stored in forests. This will provide the Council with a clear strategic advantage, as it would be able to offer its clients an eco-label and potential income during the growth period of the forest from forest carbon markets. Carbon storage accounting and management practices should be compatible with the international climate change regime requirements.

6. CONCLUSION

Forest certification schemes are a promising development in international forest policy and standards. Such schemes are able to complement a range of national-level forest management activities and might, in some cases, lead to increased enforcement of sustainable standards, improved land planning and use, expanded protected area networks, and agricultural policy reforms.[143] There are three ways to view certification processes[144]

- *Market-based mechanism* – a mechanism for changing consumer buying patterns;
- *Learning mechanism* – a mechanism for educating the public about sustainable harvesting practices;
- *Assurance mechanism* – a mechanism for verifying that timber products are from sustainably managed and harvested stocks.

---

[140] Macqueen, Duncan, (ed.) (2008), *Distinguishing Community Forest Products in the Market: Industrial Demand for a Mechanism that Brings Together Forest Certification and Fair Trade*, Edinburgh, UK: Russell Press.

[141] Shanley, above n 1.

[142] Ibid. 11.

[143] Ebeling and Yasue, above n 2, 1152.

[144] Overdevest, Christine and Mark Rickenbach (2006), 'Forest certification and institutional governance: An empirical study of forest stewardship council certificate holders in the United States', *Forest Policy and Economics*, **9** (1), 93–102.

All three roles are crucial in changing the way in which society views, consumes and manages forests. For such change to take place as a result of forest certification schemes, the percentage of certified timber sold on the international timber market must significantly increase. A number of barriers must be overcome to achieve this, including changing the cost structure of forest certification so that the consumer and not the producer bears the cost of implementing sustainable practices. This will provide an incentive for those in the timber industry to comply with sustainable standards. Just as important is the need to improve society's understanding of sustainable timber and eco-labelling processes. Such initiatives should aim to increase the demand for certified timber. This, and improving the incentives currently offered by the regime, is essential for ensuring that forest certification bodies maintain a role in the development of international forest policy and standards.

PART V

Conclusions

10. Evaluation and recommendations

1. OVERVIEW OF FOREST DOCTRINAL CONCEPTS

This book started its analysis with an overview in part two of the doctrines, principles, and rules that underpin global forest governance. The following key thematic aspects of forest regulation were identified: the rule of law; environmental justice; the rights and restrictions associated with sovereignty at the international level; the rights and restrictions associated with property at the domestic level; and consideration of forest governance arrangements. This book has argued that these concepts should and do influence the development of international forestry regulation.

The following components of this book examined different international regimes and their approach to forest regulation. Part three examined the role of public international institutions – defined as those bodies operating under the United Nations' umbrella. The institutions examined were the United Nations Forum on Forestry, the International Climate Change Regime and the World Bank. Part four focused on the role of non-state market regulation. 'Non-state' was defined to mean institutions that are neither strictly public international institutions, nor private international institutions. The common characteristics of non-state regulation involved being industry driven, using market-based instruments, and being voluntary in nature and transnational in character. The institutions examined in Part four involved analysis of forestry certification schemes (focusing on the operations of the Forest Stewardship Council) and the operation of forest market regulation. The analysis here will summarize the main findings from each chapter and comment upon their application in public and non-state regulation.

A. Rule of Law

Chapter 2 outlined the rule of law requirements as[1]

[1] International Development Law Organization (2012), 'Rule of Law Assistance Directory, available at http://www.idlo.int/english/Resources/ROL/Pages/default.aspx (accessed 14 March 2012).

- A set of legally binding rules that are known in advance;
- Rules enforced in practice;
- Mechanisms to ensure proper application of the rules by well-functioning institutions and mechanisms to allow for departure from the rules, as needed, according to established procedures;
- Conflicts in the application of these rules being resolved through binding decisions of an independent judicial or arbitral body, and;
- Having known procedures for amending these rules when they no longer serve their purpose.

It was established in Chapter 2 that global forest regulation does not accord with the rule of law criteria described above. After examining public and non-state international regulation it is now possible to draw some conclusions about the way in which these different types of regulation uphold rule of law criteria. Broadly speaking, it could be surmised that compliance with rule of law requirements is better observed in non-state institutions compared to public international institutions. Public international institutions do not have any legally binding forest rules, and compliance with, or practice of, existing standards is weak. Furthermore, the procedure for changing existing rules takes place through a process of negotiation. While this process is well established, it has failed to amend rules that are not achieving their desired objectives.

Non-state regulation appears to have been more successful in meeting the rule of law criteria above. It has been able to create legally binding obligations and ensure compliance by providing financial and other incentives. The processes for amending rules are simplified, as such bodies are able to respond directly to stakeholder feedback and amend their policy according to their own internal requirements without the need for protracted negotiations requiring consensus. Conflicts are also easier to resolve, as they usually involve two parties (the applicant and the regulatory body) as opposed to what might be any number of parties that might be present in public international institutional disputes. This is not to suggest that non-state regulation is at all times compliant with the rule of law. However, at least within current international forest regulation, it appears that non-state regulation is better than public international regulation at complying with the rule of law requirements created by the International Development Legal Organization.

B. Environmental Justice

One of the main recommendations of this book is a call for the theory of environmental justice to be embedded in international forest instruments.

Environmental justice can be considered from the perspectives of distribution, recognition, and participation. 'Distributional' analysis examines the allocation of forest resources and forest responsibilities. 'Recognition' requires that the conditions leading to inequitable distribution be taken into consideration, and 'participation' requires all relevant parties to be given an opportunity to participate and contribute. Recommendations for improving the distribution and recognition components of environmental justice are discussed later in this chapter. Participation opportunities are more widely available in public international forest institutions. This is because participation is not based upon a capacity to implement or comply but is, instead, premised upon all relevant stakeholders being given the opportunity to contribute. Conversely, participation in non-state institutions' activities is dependent upon a capacity to comply with the obligations of the institution. In many instances (as discussed in more detail in Part four), there are a number of barriers preventing wider participation, ranging from economic and property to governance and capacity issues.

C. Rights and Restrictions of Sovereignty

The rights associated with state sovereignty have had a significant impact upon global forest governance. Sovereignty-based considerations have shaped public international forest regulation more than non-state forest regulation. The rights associated with sovereignty have been used to block the creation of a legally binding forest instrument. States have been unwilling to negotiate a legally binding instrument that will have an impact on domestic timber economies. More success has been found when regimes linked sustainable forest management and another policy target (that is, poverty or carbon). The rights and restrictions associated with state sovereignty have not influenced non-state forest regulation to the same degree. This is because these processes are voluntary in nature where states agree to restrict their sovereign in order to comply with the scheme.

D. Rights and Restrictions Associated with Property

Rights and restrictions associated with property are of more relevance to non-state regulatory approaches. This is because non-state regulation often impacts upon forest use and management rights. Forest market mechanisms require the creation of property-based interests in forest values or services as a precursor to allowing trade in the product. In this regard, a domestic landholder's interests in forest areas are more directly influenced by the operation of such schemes. Public international forest regulation has the potential to impact significantly upon domestic forest landholders' interests. This would happen through the imposition of domestic legislation reflecting international requirements. The concept of sustainable forest management would operate to limit forest landholders' rights and responsibilities. Such an approach is yet to occur globally in any significant sense.

E. Governance Arrangements

The governance arrangements of public and non-state regulation are quite different. The nature of public international law requires that policy be generated as a result of negotiation. This influences the governance arrangements for the administration and processes of the institution. While this has the advantage of providing more equal opportunities for all forest stakeholders, compliance and implementation of policy created by such institutions in the forestry sectors has been weak. Governance arrangements within non-state bodies are generally simplified, as the institution sets its own policy standards and then approves membership based on compliance with its standards. In general, this produces high levels of compliance with these standards compared to public international forest policy. Its shortcoming is the lack of participation available to all relevant forest stakeholders.

F. Summary

It appears that public and non-state institutions have different strengths and challenges in complying with and implementing the themes of forest regulation identified by this book. In terms of meeting the rule of law requirements, as defined by the International Development Legal Organization, non-state forest regulatory processes meet more of the requirements when compared to public forest institutions. However, public forest institutions provide more inclusive operating modalities in the context of environmental justice from a participatory perspective. This is because they incorporate all perspectives in the negotiating process, and membership is not based upon capacity to comply with and implement forest management standards.

The concept of state sovereignty has a much larger impact upon public international forest institutions as compared with other public international institutions. This is because the concept of sovereignty in the forestry context has been interpreted to restrict the application of international law dealing with state interests in natural resources. The rights and responsibilities associated with property-based interests are of more relevance to regulation by non-state bodies as the rules of these bodies restrict the manner in which forest areas may be used. The governance arrangements underpinning public regulation and non-state regulation are quite different in nature. The nature of public international law makes the negotiation, implementation and administrative processes more complex, while the governance arrangements for non-state regulation involve less time spent on negotiation, rules and processes. It should be noted that such processes involve complex methodologies and are only responsible for regulating a small percentage of the global forest estate.

2. REFORMING GLOBAL FOREST INSTITUTIONS

The main global institution involved in the regulation of sustainable forest management is the United Nations Forum on Forests (UNFF). This body is charged with 'promoting the management, conservation and sustainable development of all types of forests and to strengthen long-term political commitment to this end'.[2] As discussed in Chapter 5, the work of the UNFF has been somewhat underwhelming to date. The body has not succeeded in harnessing a global commitment to improve forest management. Nor has it seriously attempted to coordinate the abundance of global forest initiatives. As a result, there is no global authority coordinating international forest initiatives. This deficiency has been recognized by the World Bank.[3]

There is a trend within international forest initiatives to focus upon a particular forest issue. This is understandable given the broad range of cross-cutting issues that forestry regulation involves. Such an approach, however, results in a number of institutions creating individual international standards for sustainable forest management. This book has explored the operations of five institutions creating international forest policy. Suggestions for reform of these institutions are detailed below.

A. Recommendation for Reform of the United Nations Forum on Forests

UNFF has achieved minimal gains in lifting the international profile of forests or generating concerted international cooperation in resolving forest issues. Domestic governments have, in turn, decided to deal with

---

[2] United Nations Forum on Forestry (2007), 'About UNFF', available at http://www.un.org/esa/forests/about.html (accessed 2 December 2008).

[3] See Chapter 7, which discusses the creation of a Global Forest Partnership – an international body proposed by the World Bank to reform global forest governance arrangements.

forest related issues in an *ad hoc* manner, addressing forest regulation from either political or economic related pressures. Fundamentally, the international forest regime lacks strategic direction and authority.

The UNFF was created to regulate all forest values. This includes productive forest values, protective forest values, and social, cultural, economic and ecological forest values. In order to regulate all forest values, a system delineating the relationship between forest values must be created. In certain instances this would result in one forest value being given precedence over another connected with the area. The UNFF has not explored the creation of such a methodology (for example, the creation of an international forest management methodology that specifies the creation of productive forest estates managed according to certain criteria, or the creation of protective forest estates managed to meet other specified criteria).

UNFF has avoided formulating specific international forestry obligations; and has instead created documents that recognize the existence of competing values and interests in forest areas.[4] This type of instrument provides no guidance for domestic governments on resolving disputes arising from the existence of competing forest values within a single forest area. In order for UNFF to contribute to international forest norm and rule-making processes, it should aim to create instruments recognizing all forest values. This would examine the relationships between these values and establish methods for determining the most appropriate forest value to be recognized in the circumstances.

As proposed in Chapter 5, UNFF needs to evolve in order for the body to gain political legitimacy. Such a reform of the institution is desired by many sectors of society – and is most likely required in order to attract future United Nations operational funding. The forum should be reformed to perform one of the roles described below

- The institution recognizes that it does not have the ability or influence to enforce binding legal commitments and forest targets. Instead, the forum could focus on acting as a coordinator for global international and intergovernmental programs on forests. In this

[4] For example, see *Non-legally Binding Instrument on All Types of Forests*, Economic and Social Council, Seventh Session, 2007, E/2007/42, e/cn.18/2007/8, principle 2 (c), which requires: 'Major groups as identified in Agenda 21, local communities, forest owners and other relevant stakeholders contribute to achieving sustainable forest management and should be involved in a transparent and participatory way in forest decision-making processes that affect them, as well as implementing sustainable forest management, in accordance with national legislation'.

new role, the institution could focus on aligning the research priorities of major global forest initiatives and implementing a streamlined single reporting framework that could be used to meet most international reporting requirements. In this coordination role, the body would identify thematic areas of regulation and concentrate on connecting relevant stakeholders with the forest governing body most useful to them.

- Alternatively, the institution might win back the respect of its members and stakeholders by taking a strict approach and creating a specific program of work with binding commitments and binding targets. This action would be controversial, and it is likely that a number of parties would withdraw from this type of process. The withdrawal of these members would allow the program to move forward and develop – and these parties might then be induced to return to the regime at a later date by market and non-market incentives. The forum could follow the lead of the climate change regime by creating binding commitments for developed countries in its first commitment period, with a view to increasing capacity within developing countries so that future commitment periods create obligations for these parties.

The second proposal aligns with the earlier suggestion of creating clear methodologies that outline the obligations for managing productive and protective forest estates. Domestic environmental policy is shaped and influenced by international environmental standards. The lack of a clear and authoritative international forest institution means that domestic forest regulation is influenced by a number of international forest standards. The fate of domestic forest regulation depends on the emergence of clear international forest methodologies, combined with political and financial support to assist in the implementation of them.

B. Recommendations for Reform of Forest Regulation under an International Climate Regime

Forests are regulated under the international climate change regime for both their sink and source qualities. Forest activities that produce greenhouse gas emissions include: forest harvesting; land clearing; illegal or legal deforestation; forest fires; and forest management practices (for example, the use of fertilizers). These activities are sources of greenhouse gas emissions. Other forest activities such as reforestation, afforestation, forest protection, and forests management (for example, the adoption of land use practices that reduce emissions) are activities that can act as sinks of greenhouse gas emissions. Forest sink and source regulation within the climate change regime is fragmented. There are separate guidelines for domestic forest activities (LULUCFG), CMD forest projects, and the REDD+ initiative. In the future, it would seem logical to have one policy that applied to all forest sink or source activities.

The biggest hurdle for forestry projects within the climate change regime relates to the non-permanent sequestration of carbon in forest projects and leakage related considerations. Permanence concerns arise due to uncertainty surrounding the duration for which forests are able to store carbon. Leakage is a concept used to refer to situations where improved forest practices in one region are offset by unsustainable forest practices transferred to another region. Permanence and leakage concerns have contributed to increasingly complicated accounting methodologies, which are less attractive to participants in carbon markets.[5] Reform of international climate change forestry guidelines is required to make accounting processes and methodologies easier to use and observe. This will, in turn, increase the use of such instruments, resulting in better forest management and use.

The international climate change regime has a stronger political platform than UNFF. Given this, it is suggested that reforms made in relation to forest use and management under the climate change regime are likely to achieve more than reforms made by UNFF. It could improve the condition of the world's forest estate if the climate change regime were to introduce new obligations for forest use and management. For example, a binding obligation could be introduced that requires parties to increase their forest cover by a certain percentage over time. Failure to comply with this new obligation might lead to an economic sanction. A similar obligation could require parties to increase the percentage of protected forest areas over a particular time span – and failure to comply might also attract an economic sanction.

C. Review of the World Bank's Forest Regulation

World Bank forest policy has two major goals: first, reducing poverty and improving forest conditions for forest dwellers; and, secondly, expanding the operation of forest certification schemes to promote the broader

---

[5] For a discussion of investment in land use, land use change, and forestry projects, see: Bosquet, B. (2005), 'Specific Features of Land Use, Land-Use Change and Forestry Transactions', in David Freeston and Charlotte Strek (eds), *Legal Aspects of Implementing the Kyoto Protocol Mechanisms: Making Kyoto Work*, New York, USA: Oxford University Press, pp. 282–283.

implementation of sustainable forest management principles. A review of the World Bank's forest strategies identifies a number of avenues for improvement in the implementation of its policies. The review suggests four avenues to incorporate the forest strategy more meaningfully

- Mainstreaming the forest strategy into key Bank instruments: poverty reduction strategies; country assistance strategies; development policy loans; and broader investment projects;
- Using the forest strategy to promote investment lending in forest projects;
- Enhancing the role of partnerships and linkages to implement the forest strategy;
- Building due diligence to ensure that the Bank's key operational policies trigger the application of the forest strategy and identify avenues for improvement in the operation of the forest strategy.

The World Bank has embraced a partnership approach when implementing forest projects. Such partnerships allow institutions to pool resources and expertise, usually leading to improved outcomes. Participation with forest partnerships has produced five benefits:

1. The use of a target-driven approach has increased the accountability and transparency of operations
2. The partnership has opened the doors of communication between a range of international institutional forest bodies and allowed for a pooling of resources and expertise
3. The partnerships have provided a number of valuable lessons that can inform future forest policy and practice
4. In a number of instances, these partnerships led to highly successful individual projects that evolved and developed, even after the initial partnership period ended
5. Success experienced within the partnerships led to the renewal of certain partnerships and, in some instances, of the adoption of a target-driven regulatory approach.

The World Bank has started the international dialogue about the need for a global body for forest regulation. The nature, form and role of this body is still up for negotiation. It appears that the World Bank is willing to provide the negotiating platform and it can be assumed that the Bank will also be able to provide a degree of administrative and technical capacity assistance for such an initiative. The World Bank should be encouraged to lead the creation of this partnership, given its experience in using strategic partnerships. The creation of a global forest partnership has the potential to be a system, which can bring together closely linked initiatives and create partnerships between overlapping programs. Such integration might just be the catalyst required to achieve progress in the implementation of sustainable forest management.

D. Reform of Forest Markets

There are two key issues which must be overcome in order for market mechanisms to be increasingly used in forest regulation. First, a demand for forest ecosystem services needs to be created. Market mechanisms work on the basis that the value being traded is desired by society. For market mechanisms to play a larger role in forest regulation, the demand for such forest services must be significantly increased. Secondly, there are a large number of methodological issues yet to be resolved in the forest-trading context. The complexity of forest ecosystems is now matched by the complexity of the system designed to regulate trade in forest products. The systems and processes need to be streamlined so as to be more practical and straightforward while maintaining an acceptable level of environmental integrity.

Demand and a willingness to pay for forest services can arise in two instances: under private voluntary investment; or, as a result of government intervention. Private voluntary investment is usually made on the premise that the item or service is of economic worth, and that this worth will not decrease in value. While there is some evidence of voluntary investment in reforestation and afforestation initiatives,[6] it is likely that implementation of REDD+ initiative will increase the level of voluntary investments made in forest areas. The second method of generating demand is through regulatory intervention. This approach increases the demand for forest ecosystem services by regulating and prescribing quantities and qualities for certain forest values.

To implement forest market mechanisms, property rights to forest land need to be clear and enforceable. Rights to forest land itself, and rights to particular forest products (for example, biodiversity or carbon), may also need to be created together with a system allowing for recognition and enforcement of those rights. Additionality, permanence, leakage and double-counting concerns are worthy of further consideration.

---

[6] See generally: Harris, Elizabeth, International Institute for Sustainable Development (2007), *The Voluntary Carbon Offsets Market: An Analysis of Market Characteristics and Opportunities for Sustainable Development.*

However, solutions to these issues must remain clear, straightforward and practicable.

E. Reform of Forest Certification Schemes

Forest certification schemes (where implemented appropriately) provide an example of an advanced forest methodology that recognizes and promotes a range of forest values. The biggest hurdle for forest certification schemes is the challenge of expanding their operations to those regions most in need of sustainable forest management reform. Expanding operations will also mean that certified forest products gain greater market power and recognition.

The Forest Stewardship Council (FSC) is deservedly proud of bringing 5 per cent of the world's productive forest estate under its regime. This accomplishment is significant in the light of forestry-related political, economic, social and environmental issues that have plagued progress in other global forest regimes. While 5 per cent is significant, there is clearly a need for greater expansion of areas under forest stewardship council certification. Developments needed to encourage the wider implementation of forest certification schemes include

- Fostering consumer demand and a willingness to pay a price premium for certified timber products;
- Improving consumer understanding of forest certification logos and processes;
- Increasing incentives for those bodies that comply with certification standards;
- Making forest certification more accessible to small-scale forest producers;
- Significantly increasing the capacity of tropical forest countries to participate in certification processes;
- Reforming certification guidelines to include fair trade conditions (this will allow certification schemes to leverage support from fair trade institutions).

There is need for this form of non-state governance to expand in countries where compliance with sustainable timber practices is weak or non-existent. Such non-government governance fills gaps in forest regulation in regions where domestic regulatory approaches indicate a lack of capacity, or an unwillingness, to implement sustainable timber practices. Non-state approaches are more likely to be implemented if there are appropriate incentives and support at the government and community level. The incentives offered by forest certification must increase so that timber-producer countries see value in participating in forest certification initiatives. Possible incentives include: increased market access; potential price premiums for sustainably harvested timber (something that forest certification schemes have been unable to offer their participants to date); or improved community engagement and control over forest estates.

3. MILESTONE FOREST INSTRUMENTS

UNFF has created two sets of non-legally binding forest principles, which are considered to be milestones in the development of international forestry. Such recognition is given despite the weak implementation and application of the principles.[7] Both the 1992 and 2007 principles represent a global consensus on issues surrounding forest management and use. While the principles outlined in the instruments are adequate, it is the lack of implementation that reduces their value. The major output from the principles has been the creation of National Forest Policy instruments in the majority of UNFF member countries. National forest policies, however, vary significantly from country to country and, as such, there is limited consistency and limited incorporation of these instruments.

The criteria and indicator processes produced instruments specifying the requirements of sustainable forest management, and these have been very influential in shaping domestic forestry regimes. Nearly all countries are members of a criteria and indicator process and have adopted the outcomes of these processes as the official means to measure sustainable forest management capacity at the domestic level. This, therefore, creates some consistency among nations in defining sustainable forest management at the national level. These processes, however, do not specify that certain standards be met; rather they serve as definitional tools only.

The REDD+ mechanism, currently evolving within the climate change regime, has the potential to change forest management fundamentally in developing countries. This is because it is the first international forest instrument that promises to deliver money and capacity building to protect forest estates. The success of the REDD+ initiative will be dependent upon its ability to deliver on environmental, social and economic

---

7 Humphreys, David (ed.) (2006), *Logjam: Deforestation and the Crisis of Global Governance*, London: Earthscan Publications.

factors. The rapid growth and interest in REDD+ policy demonstrates a global interest in forest conservation and in improving the livelihoods of forest dwelling communities. In order for REDD+ to receive continued global support and interest, it must deliver some substantial outcomes during the first phase of implementation.

4. EFFECTIVENESS OF THE INTERNATIONAL FOREST REGIME

The international forest regime is not operating effectively.[8] Such a conclusion can be drawn by applying the criteria for assessing effectiveness developed in Chapter 4. They are

- The existence of clear global forest objectives, with a preference for target-measured goals;
- Evidence of good global forest governance arrangements (integrated, practical and accountable, and equitable);
- Evidence of implementation of the concept (definitional processes, reporting on progress made, and future progress needed);
- Improvement in forest conditions – a tangible improvement in the quality and quantity of forest ecosystem services.

A. Clear Forest Objectives

The United Nations Forum on Forestry has created four global forest goals

- Reverse the loss of forest cover worldwide through sustainable forest management – including protection, restoration, afforestation and reforestation – and increased efforts to prevent forest degradation
- Enhance forest-based economic, social and environmental benefits (including improving the livelihoods of forest dependent people)
- Significantly increase the area of protected and sustainably managed forests worldwide (including the proportion of forest products from sustainably managed forests)
- Reverse the decline in official development assistance to sustainable forest management and mobilize new and additional

---

[8] This is widely acknowledged within existing literature on the international forest regime. Humphreys, above n 7.

financial resources for the implementation of sustainable forest management.[9]

The development of these global goals is a positive development for international forestry law because they create a focus for international forest policy – reversing deforestation, promoting economic and social forest values, protecting forest areas, and expanding forest finance. For these goals to be meaningful the following suggestions are made

1. The goals should be target-orientated. For example, the first goal could involve quantifying the rate of forest loss to be allowed within a specified period. This would require identifying the existing rate of forest loss, and then requiring a reduction in that rate. This would bind the institution to measure current rates of deforestation and report in the future upon progress made towards reducing these levels. Countries would be required to report upon levels of deforestation and policies undertaken to prevent it. The goal as it currently stands is not specific enough to require any action or follow up action on rates of deforestation.
2. These goals need to be incorporated across all global forest institutions. Presently, each international forest institution has a number of objectives, many of which would align with the four global goals above. International forest policy would be better integrated if all institutions dealing in forest policy identifying common goals.
3. The political authority and support attributed to the UNFF is weak compared to other international environmental institutions (for example, the climate change regime). Therefore, the status and significance of these goals may be questioned, especially in light of the non-legally binding status of the instrument from which they are derived. It is recommended that relevant international forest institutions meet to negotiate a number of goals using the existing ones as the starting point. Each institution should ensure that their forest objectives are compatible with the global ones, and that these become target driven. As outlined, this would assist in measuring and reporting the progress made towards achieving the goals. In addition to specifying a number of global goals, a common definition of sustainable forest management should also be created using the definition from criteria and indicator processes as the basis for negotiations.

---

9 United Nations Forum on Forestry, *Non-legally Binding Instrument on All Types of Forests*, 2nd Comm, 62nd sess, Agenda Item 54, UN Doc A/C.2/62/L.5 (22 October 2007) annex.

B. Good Governance Considerations

As already suggested, extreme fragmentation of the regime has lead to duplication, overlapping, and confusion about the global goals for forest use and management. It is a difficult task to identify all the bodies involved in global forest governance, let alone piece together the individual forest policies and relationships between the institutions. In combination with this extreme fragmentation, there is evidence of policy and implementation failure with a lack of compliance with, and implementation of, principles and standards, which also operates to undermine the effectiveness of the regime.

International forest policy is not clear, consistent or practical. This book has attempted to examine some of the international forest institutions and instruments and, during this process, it has been found that policies are often convoluted, unnecessarily complex, and often fail to explain how the policy is to be implemented (taking in to consideration budgetary and capacity issues). Policy fails when it does not achieve its stated purpose – and if this definition were strictly applied to the forest policy examined within this book, every single international forest policy would be a failure. Alternatively, it could be argued that it is not the fault of the forest policy itself, but its lack of implementation that is the problem. It is suggested, however, that designing forest policy with unrealistic expectations, and without specifying the mechanisms necessary for its implementation, is a failure of policy design.

The current global forest regime cannot be considered equitable. The concept of environmental justice should be incorporated into the global forest regime. This would ensure that those with the capacity – and, arguably, the responsibility, to act – are given significant obligations and are required to assist those countries unable to act. The design of the Reduced Emissions from Deforestation and Degradation (REDD) instrument is the first global instrument to embrace the ideas behind environmental justice in the forest context.

C. Evidence of Implementation

Implementation of international forest law is hindered by the regime's inability to address the longstanding issues that have plagued it since its inception. The inherent political issues associated with forests can be reduced to money and capacity. Forest products make significant contributions to many national economies and as such, international policies that detrimentally affect national financial interests will remain controversial unless some form of incentive is provided to promote compliance. The forest regime needs to develop forest policy that provides dual incentives, for example, improved forest outcomes and improved livelihood outcomes – such as the forest policy designed by World Bank. International forest policy also needs to work towards creating policy that aims to improve capacity within countries to implement sustainable forest management. In designing this type of policy, developed countries will need to undertake obligations to positively assist in such activities.

In terms of definitional processes, some positive developments have taken place. The definition and requirements of sustainable forest management (SFM) were established by criteria and indicator processes. These processes provide a definition of SFM and establish thematic areas for states to measure progress towards SFM. Reporting on the criteria and indicators of SFM is however voluntary (like all other international forest reporting requirements). Presently, there are no mandatory reporting obligations concerning SFM at the international level. The Collaborative Partnership on Forestry, a body formed under the United Nations, has been developing a streamlined approach for international forest reporting, designed for use by all international bodies requiring reports on SFM. Any mandatory SFM reporting obligations developed in the future should make use of this tool.

D. Improvement in Forest Conditions

The 2007 *State of the World's Forests* report, compiled by the Food and Agricultural Organization,[10] provides a number of relevant statistics concerning the improvement of forest conditions. They include[11]

- Over the 15 years from 1990 to 2005, the world lost 3 per cent of its total forest area.
- From 1990 to 2005, forests with conservation status increased by 32 per cent – this is an increase of 96 million hectares of conservation forest estates.
- 6 million hectares of primary[12] forest are lost or modified each year
- The area of productive forest plantations increased by 2.5 million hectares between 2000 and 2005, indicating that a larger proportion

---

[10] Food and Agricultural Organization of the United Nations (2007), *State of the World's Forests 2007*.

[11] Ibid. 68–70.

[12] 'Primary forest' is defined as forests of native species where there are no clearly visible indications of human activity and ecological processes are not significantly disturbed.

of wood removal may come from forest plantations in the future rather than primary forests.

In summary:

> [A] legal, policy and institutional framework is perhaps the most important factor in setting the stage for sustainable forest management. Positive change is evident in all regions. There are signs of political commitment towards sustainable forest management in the vast majority of countries. In the 15 years since the United Nations Conference on Environment and Development (UNCED), most countries have enacted new, more progressive forest laws and policies. Over 100 countries have established national forest programmes in an attempt to manage forests more holistically.[13]

This statement highlights some of the positive developments that have occurred in forest management since 1992. However, any praise of progress should be tempered with the reality that the progress made to date is quite small, especially when one considers that these improvements spanned 15 years. Current issues of concern include: the ongoing loss of primary forest, and the anticipated growth in demand for forest products expected in coming years. The *2009 State of the Worlds Forests*[14] report predicts a steep increase in the demand for wood products. The demand is the result of increases expected in the consumption of wood products, as well as an increase in the use of renewable energy technologies (biofuel). This suggests that it is crucial to increase plantation forest globally, in order to respond to this anticipated demand for wood products.

5. IMPROVING INTERNATIONAL FOREST REGULATION AND DOMESTIC POLICY

A. Global Forest Governance Reform

International forest regulation needs an authoritative body to plan strategically and to oversee the implementation of international sustainable forest management initiatives. The current global international forestry governance regime needs a body to coordinate and steer the global forest agenda. If such a body existed, it could align similar forest initiatives that

---

13 Food and Agricultural Organization of the United Nations (2007), *State of the World's Forest 2007*, 70.

14 Food and Agricultural Organization of the United Nations (2009), *State of the World's Forest 2009*.

would assist in removing duplication and overlap and allow for processes to pool skills and expertise and learn from each other's experiences. This would also promote improved knowledge transfer among these institutions, leading to improved forest policy that could take into account practical lessons from trial implementation programs.

UNFF is desperately in need of reform. This body has the potential to become the ultimate authority on international forest regulation because it is the one body that focuses on all forest values. As outlined earlier, reform of this institution can take place in two ways. One is to focus on being a coordinator for global international and intergovernmental programs on forests. This is an administrative-driven reform; it would require the UNFF to collate information about relevant forest initiatives, to hold meetings and negotiations with bodies working on similar tasks, and create reports on how each individual forest value is regulated. Such reform would also require strategic planning, which would go some way to improving monitoring, reporting and implementation of sustainable forest management principles and practice.

The UNFF may not be the most appropriate body for such a role, because it may not be able to generate the requisite political support and authority required for such reform. The World Bank has proposed the creation of a Global Forest Partnership and this may, ultimately, be more successful in creating a peak international forest institution due to the Bank's political platform and status. In any event, the UNFF should actively participate in the development and administration of the global forest body. This will go some way to avoiding further overlap duplication in the international forest regime.

A second way would be to reform the Forum to increase its political standing through improving the authority of its standards and rules. A target-driven approach should be adopted so that progress can be measured and reported. Such an approach may not be reached by consensus – however, similar to the development of the Kyoto Protocol, a progressive membership may take place. This means that countries with a strong interest in sustainable forest management would sign a legally binding document in the hope that other countries will join in the future. For such affirmative action, the political significance of forest issues will need to be raised in many countries.

Regardless of the nature of the reform undertaken (administrative versus legally binding forest commitments), the essence of the reforms should also address the following

- The adoption of a common concept of sustainable forest management, and a system for defining, implementing, and reporting on it. This would involve an expansion of the criteria and indicator approach to defining sustainable forest management;

- The creation of international management standards for protective forest areas. This standard should include information about different forest ecosystems (for example, boreal, tropical), and should be compatible with the requirements for forest market mechanisms (carbon, biodiversity);
- The creation of international management standards for productive forest areas. This should be aligned with forest certification guidelines as much as possible;
- The development of more advanced forest management methodologies that provide guidance on how forest areas can be managed to meet a number of forest values;
- The development of a timber policy to prevent trade in illegally sourced timber.

To address the lack of implementation of the international forest regimes instruments, three barriers must be overcome

1. Financial support must be provided to developing countries to support their implementation of sustainable forest management principles and practices[15]
2. The capacity of developing countries to implement and enforce sustainable forest management principles and practices must be addressed. For forest programs to be successful, there must be an underlying political structure that is secure, transparent and accountable. In the absence of such a system, the implementation of sustainable forest management will continue to be weak[16]
3. There needs to be policy integration of forest issues into broader sustainable development and development strategies.

B. Incorporating Environmental Justice into Global Forest Regulation

The lack of capacity in developing countries to implement sustainable forest management, coupled with the consumption of forest products by developed countries, should be used as a basis for incorporating

---

[15] The REDD+ initiative is a positive and welcome development that should assist in raising the funds available for SFM.

[16] FLEGT initiative and REDD+ investment should assist in improving capacity for forest governance.

environmental justice considerations into global forest regulation.[17] This requires a re-examination of the relationship between countries that are producers and those that are consumers of forest products, and a re-examination of the ability of developed countries (and, arguably, their global duty) to assist the implementation of sustainable forest management in developing countries. It is important that the wide range of different political, economic and social conditions present in developing countries be accommodated and recognized by policy initiatives in this area. Presently, there is no global consensus on who, ultimately, should be responsible for implementing sustainable forest management. Existing principles of international environmental law, such as state sovereignty, suggest that sovereign states should be responsible for managing their natural resources.

Incorporating justice-related considerations into international forest regulation would assist in mending the relationship between developing and developed countries, as well as timber producing and consumer countries. Examining justice-related issues goes to the heart of three contentious issues within international forest regulation

- Is it fair and reasonable for some countries to bear a heavier burden in relation to the implementation of sustainable forest management?
- How should liability for the implementation of sustainable forest management be calculated?
- Is it fair for stewards of large forest areas to bear global forest responsibilities when they lack the capacity to address forest issues?

Environmental justice, like many other environmental concepts, is capable of being defined in a number of ways. Particularly relevant to international forest regulation and justice are considerations linked to distributive justice, and justice as recognition. A distributive justice approach would require the development of:

- two sets of sustainable forest management standards (reflecting the capacity of developed and developing countries to implement them);
- a system of liability (which acknowledges that the consumption of timber in developed countries contributes significantly to unsustainable timber practices in developing countries); and

---

[17] In support of this, see: Greber, Thomas, Melinda Janki, Morcos Orellana, Annalisa Savaresi and Dinah Shelton (2009), 'Conservation with Justice: A Rights-based Approach', Policy Paper No 71, International Union for the Conservation of Nature, Ch. 4.

- a system that shares liability for the implementation of sustainable forest management practices among the global community (reflecting the benefits that the global community obtains from sustainable forest management practices).

Justice as recognition, in the international forestry context, could be used as a basis to argue for:

- consumption patterns of forest products to bear some relation to the liability imposed for the implementation of sustainable forest management practices;
- recognition of the circumstances that have pushed certain forest dwellers into their current position (and the need for a system to assist them with the implementation of sustainable forest management practices); and
- for the law to recognize the legitimate claims of indigenous or customary owners of forest areas (leading to tenure and land rights being secured for them).

C. Domestic Forest Governance Reform

From the national governance perspective, improvements in the legal regulation of forest resources might be made possible by re-examining property and tenure rights, and their relationship to sustainable forest management. This would involve examining the effects of public and private landholding patterns, and the effect these have on the implementation of sustainable forest management. Further work is needed to identify the gaps in the law concerning property rights and forest rights. These gaps will differ from country to country, but issues in these areas can be generalized into developed country forest property issues, and developing country forest property issues.

Developed countries usually have regulatory frameworks that provide citizens with clearly defined and enforceable tenure rights to land. However, laws still need to be developed so that rights and interests in forest property can be precisely defined. This would be of particular importance in the event of market-based regulation becoming a dominant form of environmental regulation. Tenure rights in forest property that require recognition and definition include:

- rights in forest property (such as interests in carbon);
- clarification of forest property rights in leasehold land; and
- the ability of land registries to recognize multiple forest interests in a single allotment.

Developing countries face different challenges in relation to land tenure, as there are issues concerning the implementation of the rule of law in forestry practices. In many instances, tenure rights to forest property are not clear, certain, stable or enforceable. Stable tenure rights provide citizens with an incentive to manage and use land to achieve environmental, social and economic benefits, by providing security to the holder of the rights. Developing countries require capacity building initiatives aimed at creating and implementing an accountable land tenure regulatory framework. The introduction of such a system would serve as the initial building block of a series of reforms aimed at improving forest management and restricting the use of forest resources in unsustainable processes.

6. CONCLUDING THOUGHTS

The development of appropriate legal rules and processes is critical to implementing sustainable forest management at both the international and national levels. The concept of environmental justice has been referred to through out this book to support a call for a change in liability for global forest management. Ultimately, it is proposed that those countries with the capacity to implement sustainable forest management, along with those who are high consumers of forest products, should bear more responsibility for implementing SFM. Such an approach is more practical and ethical and will lead to improvements in protecting and managing the global forest estate. The FLEGT and REDD+ initiatives both recognize that a system of liability based on sovereignty is not appropriate for distributing obligations for SFM. These two initiatives both apply environmental justice considerations by distributing liability to finance SFM to developing and consumer countries. These two mechanisms will significantly influence global forest governance in the coming decades and it is hoped that these mechanisms can assist in transforming the use and management of the world's forest estate.

References

Abbot, Carolyn (2006), 'Environmental Command Regulation' in Benjamin J. Richardson and Stepan Wood (eds), *Environmental Law for Sustainability*, Oxford; Portland, Oregon: Hart Publishing, pp. 61–73

ABC News Online (2009), 'Asset Sell-Off Only Decided Last Week: Bligh', available at http://www.abc.net.au/news/stories/2009/06/03/2587961.htm (accessed 11 June 2009)

Africa Forest Law Enforcement and Governance, Ministerial Conference, Yaounde, Cameroon, 16 October 2003, Ministerial Declaration

Alchian, Armen, and Harold Demsetz (1973), 'The Property Right Paradigm', *The Journal of Economic History*, **33** (16), 16–27

Ali, Asghar (2006), 'A Conceptual Framework for Environmental Justice based on Shared but Differentiated Responsibilities' in Tony Shallcross and John Robinson (eds), *Global Citizenship and Environmental Justice*, Amsterdam; New York: Rodopi, pp. 41–50

Ali, Paul, and Kanako Yano (eds) (2004), *Eco-Finance: The Legal Design and Regulation of Market-based Environmental Instruments*, The Hague: Kluwer Law International

Al Jurf, Saladin (1999), 'Participatory Development and NGOs: A Look at the World Bank', *Transnational Law and Contemporary Problems*, **9** (1), 175–185

Allen, Craig D., Alison K. Macalady, Haroun Chenchouni, Dominique Bachelet, Nate McDowell, Michel Vennetier, Thomas Kitzberger, Andreas Rigling, David D. Breshears, E.H. (Ted) Hogg, Patrick Gonzalez, Rod Fensham, Zhen Zhang, Jorge Castro, Natalia Demidova, Jong-Hwan Lim, Gillian Allard, Steven W. Running, Akkin Semerci, Neil Cobb (2009), 'A Global Overview of Drought and Heat-induced Tree Mortality Reveals Emerging Climate Change Risks for Forests', *Forest Ecology and Management*, **259** (4), 660–684

Alley, Patrick, (2011), 'Corruption A Root Cause of Deforestation and Degradation' in Transparency International (eds), *Global Corruption Report: Climate Change*, London; Washington DC: Earthscan, pp. 299–311

Anand, Ruchi (1980), 'Development and the Environment: these Case of Developing Countries', *Indian Journal of International Law*, **20**, 1–19

Ananda, Jayanath, and Gamini Herath (2003), 'Incorporating Stakeholder Values into Regional Forest Planning: A Value Function Approach', *Ecological Economics*, **45** (1), 75–90

Angelsen, Arild and Sven Wunder (eds) (2003), *Exploring the Forest-Poverty Link: Key Concepts, Issues and Research Implications*, Bogor, Indonesia: Centre for International Forestry Research

Angelstam, Per (2003), 'Forest Biodiversity Management–the Swedish model' in David Lindenmayer and Jerry F. Franklin (eds), *Towards Forest Sustainability*, Collingwood, Victoria: CSIRO Publishing, pp. 143–159

Arnold, Craig Anthony (2002), 'The Reconstitution of Property: Property as a Web of Interests', *Harvard Environmental Law Review*, **26** (2), 281–295

Ashcroft, Ross (2008), 'Carbon Capture and Storage: A Need for Re-conceiving Property Interests and Resource Management in the Australian Legal System', *Journal of the Law Association for Asia and the Pacific*, **1** (1), 48–61

Ashford, Nicholas and Charles Caldart (eds) (2008), *Environmental Law, Policy and Economics: Reclaim the Environmental Agenda*, Cambridge, Mass: MIT Press

Asquity, Nigel, Maria Vargas and Sven Wunder (2008), 'Selling Two Environmental Services: In Kind Payments for Bird Habitat and Watershed Protection in Los Negros, Bolivia', *Ecological Economics*, **65** (4), 675–684

Australian Department of Agriculture, Fisheries and Forestry (2000), *Summary of Proposals for Action: A Tool to Assist Countries to Measure Progress and Establish Priorities for Sustainable Forest Management*

Australian Government: Department of Agriculture, Fisheries and Forestry (2007), 'The RFA Process: The Commonwealth Position', available at http://www.daff.gov.au/rfa/about/process/cwth-position (accessed 6 February 2009)

Australian Government: Department of Climate Change (2008), 'Carbon Pollution Reduction Scheme Green Paper', available at http://www.climatechange.gov.au/greenpaper/index.html (accessed 28 November 2008)

Australian Greenhouse Office, (2005), *Planning Forest Sink Projects: A Guide to Legal, Taxation and Contractual Issues*

Barbier, E., J. Burgess and C. Folke (eds) (1994), *Paradise Lost? The Ecological Economics of Biodiversity*, London: Earthscan Publications

Bass, Stephen, 'Global Forest Governance: Emerging Impacts of the Forest Stewardship Council' (paper presented at the International SUSTRA Workshop 'Architecture of the Global System of Governance of Trade and Sustainable', Berlin, 9–10 December 2002)

Bates, Gerry (ed.) (2006), *Environmental Law in Australia*, (6th edition), Chatswood, NSW: LexisNexis Butterworths

Bates, Gerry (ed.) (2010) *Environmental Law in Australia*, (7th edition), Chatswood, NSW: LexisNexis Butterworths

Bayon, Ricardo, Forest Trends (2004), *Making Environmental Markets Work: Lessons from Early Experience with Sulphur, Carbon, Wetlands, and Other Related Markets*

Beder, Sharon, (ed.) (2006) *Environmental Principles and Policies: An Interdisciplinary Approach*, Sydney: UNSW Press

Bell, Justine (2008), 'Can the Torrens System Adapt to Ecologically Sustainable Development?', *Queensland Environmental Practice Reporter*, **13** (62), 218–222

Bell, Stephen and Hindmoor, Andrew (2011), 'Governance with Government? The Case of the Forestry Stewardship Council', *Public Administration*, 1–16

Bellassen, Valentin, and Vincent Gitz (2008), 'Reducing Emissions from Deforestation and Degradation in Cameroon – Assessing Costs and Benefits', *Ecological Economics*, **68** (1–2), 336–341

Benndorf, R., S. Federicib, C. Fornerc, N. Penad, E. Rametsteinere, M.J. Sanzf, Z. Somogyi (2007), 'Including Land Use, Land-Use Change and Forestry In Future Climate Change Agreements: Thinking Outside The Box', *Environmental Science and Policy,* **10** (4), 283, 284

Bernstein, Lenny, Peter Bosch, Osvaldo Canziani, Zhenlin Chen, Renate Christ, Ogunlade Davidson, William Hare, Saleemul Huq, David Karoly, Vladimir Kattsov, Zbigniew Kundzewicz, Jian Liu, Ulrike Lohmann, Martin Manning, Taroh Matsuno, Bettina Menne, Bert Metz, Monirul Mirza, Neville Nicholls, Leonard Nurse, Rajendra Pachauri, Jean Palutikof, Martin Parry, Dahe Qin, Nijavalli Ravindranath, Andy Reisinger, Jiawen Ren, Keywan Riahi, Cynthia Rosenzweig, Matilde Rusticucci, Stephen Schneider, Youba Sokona, Susan Solomon, Peter Stott, Ronald Stouffer, Taishi Sugiyama, Rob Swart, Dennis Tirpak, Coleen Vogel, Gary Yo, Intergovernmental Panel on Climate Change (2007), *Climate Change 2007: Synthesis Report: Summary for Policy Makers*

Bernstein, Steven and Benjamin Cashore (2000), 'Globalization, Four Paths of Internationalization and Domestic Policy Change: The Case Of Eco-Forestry Policy Change in British Columbia, Canada', *Canadian Journal of Political Science*, **33** (1), 67–99

Bhattari, Madhusudan, and Micheal Hammig (2001), 'Institutions and the Environmental Kuznets Curve for Deforestation: A Cross-Country Analysis for Latin America, Africa and Asia', *World Development*, **29** (6), 995–1010

Binkley, C. and D. Brand (2002), 'Carbon Sink by Forest Sector – Options and Needs for Implementation', *Forest Policy and Economics*, **4** (1), 65–77

Blom, Benjamin, Terry Sunderland and Daniel Murdiyarso (2010), 'Getting REDD to Work Locally: Lessons Learned from Integrated Conservation and Development Projects', *Environmental Science and Policy*, **13** (2), 164–172

Blomley, Tom, and Acacia Consulting, International Institute for Environment and Development (2009), *Evaluation of the Work of the Forest Governance Learning Group 2005-2009*

Bluemel, Erik (2005), 'Overcoming NGO Accountability Concerns in International Governance', *Brooklyn Journal of International Law*, **31** (1), 139–206

Bodansky, Daniel and Diringer, Elliot (2010), 'The Evolution of Multilateral Regimes: Implications for Climate Change', available at http://www.environmentportal.in/files/evolution-multilateral-regimes-implications-climate-change.pdf (accessed 26 January 2012)

Bosquet, B. (2005), 'Specific Features of Land Use, Land-Use Change and Forestry Transactions', in David Freeston and Charlotte Strek (eds), *Legal Aspects of Implementing the Kyoto Protocol Mechanisms: Making Kyoto Work*, New York, USA: Oxford University Press, pp. 282–283

Bosselmann, Klaus (1999), 'Justice and the Environment: Building Blocks for a Theory on Ecological Justice' in Klaus Bosselmann and Benjamin Richardson (eds), *Environmental Justice and Market Mechanisms: Key Challenges for Environmental Law and Policy*, London: Kluwer Law International, pp. 30–31

Bosselmann, Klaus (2004), 'Environmental Governance: A new approach to Territorial Sovereignty' in Robert J. Goldstein (ed.), *Environmental Ethics and Law*, Aldershot: Ashgate, pp. 293–313

Bosselmann, Klaus (ed.) (2008), *The Principle of Sustainability: Transforming Law and Governance,* Aldershot, England; Burlington, VT: Ashgate

Bosselmann, Klaus and Benjamin J. Richardson (1999), 'Introduction: New Challenges for Environmental Law and Policy', in Klaus Bosselmann and Benjamin J. Richardson (eds) *Environmental Justice and Market Mechanisms: Key Challenges for Environmental Law and Policy*, London: Kluwer Law International, pp. 1–3

Bowers, John (2005), 'Instrument Choice for Sustainable Development: An Application to the Forestry Sector', *Forest Policy and Economics*, **7** (1), 97–107.

Boyd, E. and M. Guiterrez (2007), 'Small-scale Forest Carbon Projects: Adapting CDM to Low-income Communities', *Global Environmental Change*, **17** (2), 250–259

Boyd, Emily, Esteve Corbera and Manuel Estrada (2008), 'UNFCCC negotiations (pre Kyoto to COP 9): What the Process Says About the

Politics of CDM-sinks', *International Environmental Agreements*, **8** (2), 95–108

Brand, David, 'Forest Investment and Emerging Environmental Markets' (paper presented at the second Annual Green Trading Summit, New York, 7–8 April 2003)

Brown, Chris, Patrick Durst and Thomas Enters (2005), 'Perception of Excellence: Ingredients of Good Forest Management', in Patrick Durst, Chris Brown, Henrylito D. Tacio and Miyuki Ishikawa (eds) *In Search of Excellence: Exemplary Forest Management in Asia and the Pacific*, Regional Community Forestry Training Centre for Asia and the Pacific, Bangkok: Food and Agriculture Organization of the United Nations, pp. 7–28

Brown, K. and E. Corbera (2003), 'Exploring Equity and Sustainable Development in the New Carbon Economy', *Climate Policy Special Supplement on Climate Change and Sustainable Development*, **3** (1), 41–56

Bryant, Dirk, Daniel Nielsen and Laura Tangley (eds) (1997), *The Last Frontier Forests: Ecosystems and Economies on the Edge*, Washington DC: World Resources Institute, Forest Frontiers Initiative

Buchy M., Shover and C. Averill, (eds) (1999), *Understanding Public Participation in Forest Planning in Australia: How Can We Learn from Each Other?* Canberra, ACT: Department of Forestry; and Jack Siry, F.W. Cubbage and M.R. Ahmed (2005), 'Sustainable Forest Management: Global Trends and Opportunities', *Forest Policy and Economics*, **7** (4), 551–561

Burgin, Shelley (2008), 'Biobanking: An Environmental Scientist's View of the Role of Biodiversity Banking Offsets in Conservation', *Biodiversity Conservation*, **17** (4), 807–816

Button, Jillian (2008), 'Carbon: Commodity or Currency? The Case for an International Carbon Market based on the Currency Model', *Harvard Environmental Law Review*, **32** (2), 571–596

Cashore, Benjamin, 'Governing Through Markets' (paper presented at International Regime, Avoided Deforestation and the Evolution of Public and Private Forest Policies in the South, Paris, France, 21–23 April 2007)

Cashore, Benjamin, G.C. van Kooten, Ilan Vertinsky, Graeme Auld, Julia Affolderbach (2005), 'Private or Self Regulation? A Comparative Study of Forest Certification Choices in Canada, the United States and Germany', *Forest Policy and Economics*, **7** (1), 53–69

Cashore, Benjamin, Graeme Auld and Deánna Newsom (2003), 'Forest Certification (Eco-Labelling) Programs and their Policy-Making Authority: Explaining Divergence among North American and European Case Studies', *Forest Policy and Economics*, **5** (3), 225–247

Cashore, Benjamin, Graeme Auld, Steven Bernstein and Constance McDermott (2007), 'Can Non-state Governance "Ratchet up" Global Environmental Standards? Lessons from the Forest Sector', *Review of European Community and International Environmental Law*, **16** (2), 158–172

Cavigila-Harris, Jill, Dustin Chambers and James Kahn (2009), 'Taking the "U" out of Kuznets: A Comprehensive Analysis of the EKC and Environmental Degradation', *Ecological Economics*, **68** (4), 1149–1159

Centre for International Economics Canberra and Sydney (2009), *Proposed New Policy on Illegally Logged Timber: Issues Paper*

Chan, Sander, and Philipp Pattberg (2008), 'Private Rule-Making and the Politics of Accountability: Analysing Global Forest Governance', *Global Environmental Politics,* **8** (3), 103–121

Chandani, Achala (2007), 'Distributive Justice and Sustainability as a Viable Foundation for the Future Climate Regime', *Carbon and Climate Law Review*, (2), 152–163

Chokkalingam, Unna and Vanniarachchy, Anvradha (2011), 'CDM AR did not Fail and is not Dead', available at http://www.forestcarbonasia.org/articles/cdm-ar-did-not-fail-and-is-not-dead/ (accessed 8 February 2012)

Chomitz, Kenneth (2007), *At Loggerheads? Agricultural Expansion, Poverty Reduction, and Environment in the Tropical Forests*, Washington DC: The International Bank for Reconstruction and Development/ The World Bank

Chomitz, Kenneth C., Esteban Brenes and Luis Constantion, (1998), 'Financing Environmental Services: The Costa Rica Experience and its Implications', available at http://www.elaw.org/system/files/Financing.Environmental.Services.pdf (accessed 20 February 2009)

Clemencon, Raymond (2006), 'What Future for the Global Environmental Facility?', *The Journal of Environment and Development*, **15** (1), 50–74

Clemencon, Raymond (2007), 'Funding for Global Environmental Facility Continues to Decline', *The Journal of Environment and Development*, **16** (1), 3–7

Climate Funds Update (2012), 'Forest Carbon Partnership Facility', available at http://www.climatefundsupdate.org/listing/forest-carbon-partnership-facility (accessed 27 February 2012)

Coase, R.H. (1960), 'The Problem of Social Cost', *The Journal of Law and Economics*, **3** (1), 1–44

Colchester, Marcus, Tejaswini Apte, Michel Laforge, Alois Mandondo and Neema Pathak 2003, 'Bridging the Gap: Communities, Forests and International Networks' (Occasional Paper No 41, Centre for International Forestry Research)

Colfer, C. and R. Wadley, Centre for International Forestry Research (1996), *Assessing Participation in Forest Management*

Collaborative Partnership on Forestry (1998), 'International Forestry Directory', available at http://www.fao.org/forestry/site/1662/en (accessed 16 June 2007)

Collaborative Partnership on Forestry (2003), *Towards a Common Information Framework for Forest-Related National Reporting to International Processes*

Collaborative Partnership on Forestry (2006), 'Streamlining Reporting', available at www.fao.org/forestry/site/6308/en (accessed 6 June 2007)

Collaborative Partnership on Forestry (2007), 'Sourcebook on Funding for Sustainable Forest Management', available at www.fao.org/forestry/site/7148/en (accessed 6 June 2007)

Combes Motel, P., R. Pirard, and J.L. Combes (2008), 'A Methodology to Estimate Impacts of Domestic Policies on Deforestation: Compensated Successful Efforts for "Avoided Deforestation" (REDD)', *Ecological Economics*, **68** (3), 680–691

Commonwealth Government (1992), 'National Forest Policy Statement: A New Focus for Australia's Forests', available at http://www.daffa.gov.au/__data/assets/pdf_file/0019/37612/nat_nfps.pdf (accessed 4 February 2009)

Conference of the Parties (2011), 'Draft Decision -/CP.17, Guidance on Systems For Providing Information on how Safeguards are Addressed and Respected Modalities Relating to Forest Reference Emission Levels and Forest Reference Levels as Referred to in Decision 1/CP.16', available at http://unfccc.int/files/meetings/durban_nov_2011/decisions/application/pdf/cop17_safeguards.pdf (accessed 21 March 2012)

Conference of the Parties (2011), 'Outcome of the Work of the ad hoc Working Group on Further Commitments for Annex I Parties under the Kyoto Protocol at its Sixteenth Session', available at http://unfccc.int/files/meetings/durban_nov_2011/decisions/application/pdf/awgkp_outcome.pdf (accessed 21 March 2012)

Convention on Biological Diversity, opened for signature 5 June 1992, UNTS 30619 (entered into force 29 December 1995)

Corbera, Esteve, Katrina Brown and W. Neil Adger (2007), 'The Equity and Legitimacy of Markets for Ecosystem Service', *Development and Change*, **38** (4), 587–613

Couillard, V., J. Gilbert, J. Kenrick & C. Kidd, Forest Peoples Program (2009), 'Land Rights and the Forest Peoples of Africa, Historical, Legal And Anthropological Perspectives – Overview: Analysis And Context', available at http://www.forestpeoples.org/sites/fpp/files/publication/2010/05/overviewlandrightsstudy09eng.pdf (accessed 27 January 2012)

Counsell, Simon, and Kim Terje Lorras (eds) (2002), *Trading in Credibility: the Myth and the Reality of the Forest Stewardship Council*, London: Rainforest Foundation

Crittenden, J. and Wilder, M. (2008), 'Bringing the Forest to Market: Structuring Avoided Deforestation Projects', *The Carbon & Climate Law Review*, **2** (3), 273–280

Cronkleton, Peter, Peter L. Taylor, Deborah Barry, Samantha Stone-Jovicich, Marianne Schmink (2008), 'Environmental Governance and the Emergence of Forest-Based Social Movements' (Occasional Paper No 49, Centre for International Forestry Research)

Crook, Carolyn, and Roger Clapp (1998), 'Is Market-Oriented Forest Conservation a Contradiction in terms?', *Environmental Conservation*, **25** (2), 131–142

Cubbage, Frederick, and David Newman (2006), 'Forest Policy Reformed: A United States Perspective', *Forest Policy and Economics*, **9** (3), 261–270

Cubbage, Frederick, Patrice Harou and Erin Sills (2007), 'Policy Instruments to Enhance Multi-Functional Forest Management', *Forest Policy and Economics*, **9** (7), 833–851

Culas, R.J. (2007), 'Deforestation and the Environmental Kuznets Curve: An Institutional Perspective', *Ecological Economics*, **61** (2–3), 429–437

Curnow, Paul, and Louisa Fitz-Gerald (2006), 'Biobanking in New South Wales: Legal Issues in the Design and Implementation of a Biodiversity Offsets and Banking Scheme', *Environmental and Planning Law Journal*, **23** (4), 298–308

Cutter, Susan (1995), 'Race, Class and Environmental Justice', *Progress in Human Geography*, **19** (1), 111–122; Manuel Pastor, Jim Sadd and John Hipp (2001), 'Which Came First? Toxic Facilities, Minority Move-In, and Environmental Justice', *Journal of Urban Affairs*, **23** (1), 1–21

Datta-Chaudhuri, Mrinal (1990), 'Market Failure and Government Failure', *The Journal of Economic Perspectives*, **4** (3), 25–39

Davenport, Deborah, and Peter Wood (2006), 'Finding the Way Forward for the International Arrangement on Forests: UNFF-5,-6 and-7', *Review of the European Community and International Environmental Law*, **15** (3), 316–326

Department of Agriculture, Fisheries and Forestry – Australia (AFFA) Program on Forests (PROFOR) at the World Bank (2003), 'Implementing the Proposals for Action of the Intergovernmental Panel on Forests and the Intergovernmental Forum on Forests: A Tool to Assist National-level Assessment of Progress and Priorities for Action toward Sustainable Forest Management, Developed in Support of the United Nations Forum on Forests', available at http://www.profor.info/profor/Documents/pdf/IPF_IFFtext.pdf (accessed 26 March 2012)

Department of the Economic Secretariat of the United Nations Forum on Forestry, *Review of the Effectiveness of the International Arrangement on Forests*, 5th sess, Agenda Item 5, UN Doc CN.18/2005/6 (11 March 2005)

Development Law Service: FAO Legal Office (2000) 'Legal Framework Analysis for Rural and Agriculture Investment Projects' available at http://www.fao.org/legal/prs-ol/lpo12.pdf (accessed 20 January 2012)

Dimitrov, Radoslave (2005), 'Hostage to Norms: States, Institutions and Global Forest Politics', *Global Environmental Politics*, **5** (4), 1–24

Dingwerth, Klaus (2008), 'North-South Parity in Global Governance: The Affirmative Procedures of the Forest Stewardship Council', *Global Governance*, **14** (1), 53–71

Doelle, Meinhard (2004), 'Climate Change and Human Rights: The Role of International Human Rights in Motivating States to take Climate Change Seriously', *Macquarie Journal of International and Comparative Environmental Law*, **1** (2), 179–216

Doherty, Emma and Heike Schroeder (2011), 'Forest Tenure and Multilateral Governance in Avoiding Deforestation under REDD+', *Global Environmental Politics*, **11** (4), 66–88

Dovers, Stephen (2003), 'Are Forests Different as a Policy Challenge' in David Lindenmayer and Jerry Franklin (eds), *Towards Forest Sustainability*, Collingwood, Victoria: CSIRO Publishing, pp. 15–30

Dovers, Stephen and Robin Connor (2006), 'Institutional and Policy Change for Sustainability' in Benjamin J. Richardson and Stepan Wood (ed.), *Environmental Law for Sustainability*, Oxford; Portland, Oregon: Hart Publishing, pp. 21–29

Dovers, Stephen and Su Wild River (eds) (2003), *Managing Australia's Environment,* Annandale, NSW: Federation Press

Dresner, Simon, Paul Etkins, Kate McGeevor and Julia Tomei (2007), 'Forests and Climate Change: Global Understanding and Possible Responses', in Freer-Smith, P., Broadmeadow M. & J. Lynch (eds), *Forestry and Climate Change*, Wallingford, Oxon: CABI Publishing, pp. 33–48

Duraiappah, Anantha Kumar, International Institute for Sustainable Development (2006), *Markets for Ecosystem Services: A Potential Tool for Multilateral Environmental Agreements*

Durst, Patrick, Chris Brown, Henrylito D. Tacio and Miyuki Ishikawa (eds) (2005), *In Search of Excellence: Exemplary Forest Management in Asia and the Pacific*, Bangkok: Food and Agricultural Organization of the United Nations and Regional Community Forestry Training Centre for Asia and the Pacific

Ebeling, Johannes, and Mai Yasue (2009), 'The Effectiveness of

Market-based Conservation in the Tropics: Forest Certification in Ecuador and Bolivia', *Journal of Environmental Management*, **90** (2), 1145–1153

Eckersley, Robyn (1995), 'Markets, the State and the Environment: An Overview' in Robyn Eckersley (ed.) *Markets, the State and the Environment: Towards Integration*, South Melbourne: Macmillan Education Australia, pp. 8–9

Eckert, Sandra and Richard McKellar (2008), 'Securing Rights to Carbon Sequestration: The Western Australian Experience', *Sustainable Development Law and Policy*, **8** (2), 30–33

Economic and Social Council of the United Nations, *United Nations Forum on Forestry: Report of the Ninth Session (24 January – 4 February 2011)*, UN ESCOR, 9th sess, Supp No 22, UN Doc E/2011/42, e/cn.18/2011/20 (1 May 2009 and 24 January to 4 February 2011)

Economic and Social Council of the United Nations, *United Nations Forum on Forestry: Report of the Seventh Session*, UN ESCOR, 7th sess, Supp No 22 (24 February 2006 and 16–27 April 2007)

Eden, Sally (2009), 'The Work of Environmental Governance Networks: Traceability, Credibility and Certification by the Forest Stewardship Council, *Geoforum*, **40** (3), 383–394

El Lakany, Hosny, Michael Jenkins, and Michael Richards (2007), PROFOR, *Background Paper on Means of Implementation (Contribution by PROFOR to discussions at UNFF-7)*

Ellis, Jay, and Stepan Wood (2006), 'International Environmental Law' in Benjamin J. Richardson and Stepan Wood (ed.), *Environmental Law for Sustainability*, Oxford; Portland, Oregon: Hart Publishing, pp. 343–344

Engel, Stefanie and Charles Palmer (2008), 'Payments for Environmental Services as an Alternative to Logging under Weak Property Rights: The Case of Indonesia', *Ecological Economics*, **65** (4), 799–809

Engel, Stefanie, Stefano Pagiola and Sven Wunder (2008), 'Designing Payments of Environmental Services in Sheory and Practice: An Overview of the Issues', *Ecological Economics*, **65** (4), 663–674

Epstein, Richard (ed.) (1985), *Takings: Private Property and the Power of Eminent Domain*, Cambridge, Mass: Harvard University Press

Eshbach, R. (1990), 'A Global Approach to the Protection of the Environment: Balancing State Sovereignty and Global Interests', *Temple International and Comparative Law Journal*, **4** (2), 271–305

European Commission Directorate-General for Environment (1994), *Potential Benefits of Integration of Environmental and Economic Policies: An Incentive-based Approach to Policy Integration*

European Commission (2006), *Public Consultation on Additional Options to Combat Illegal Logging Analysis and Report*

European Union (2007), 'Forest Law Governance', available at http://ec.europa.eu/environment/forests/flegt.htm (accessed 24 September 2008)

Evans, Michael (ed.) (2000), *Principles of Environmental and Heritage Law*, St Leonards, NSW: Prospect Media

Falkner, Robert (2003), 'Private Environmental Governance and International Relations: Exploring the Links', *Global Environmental Politics*, **3** (2), 72–87

Farrier, David, Andrew Kelly and Angela Langdon (2007), 'Biodiversity offsets and Native Vegetation Clearance in New South Wales: The Rural/Urban Divide in the Pursuit of Ecologically Sustainable Development', *Environmental and Planning Law Journal*, **24** (6), 427–449

FCPF (2011), 'Forest Carbon Partnership Facility 2011 Annual Report', available at http://www.forestcarbonpartnership.org/fcp/sites/forestcarbonpartnership.org/files/Documents/PDF/Oct2011/FCPF_Carbon_AR_FINAL_10_3.pdf (accessed 8 February 2012)

Feje, Lesniewska (2005), 'Laws for Forests: An Introductory Guide to International Forest and Forest Related Materials that Shape Forest Ethics and Practice', available at http://www.policy-powertools.org/Tools/Engaging/docs/law%20for%20forests2.pdf (accessed 1 February 2012)

Fisher, Bob, Cor Veer and Sango Mahanty (2008), *Poverty Reduction and Forests: Tenure, Market, and Policy Reforms,* Rights and Resource Initiative

Fisher, Brendan, R. Kerry Turner and Paul Morling (2009), 'Defining and Classifying Ecosystem Services for Decision Making', *Ecological Economics*, **68** (3), 643–653

Fisher, Douglas (ed.) (2003), *Australian Environmental Law*, Pyrmont, NSW: Lawbook Co.

Fisher, Douglas (ed.) (2006), *Implementing the National Water Initiative: A Generic Set of Arrangements for Managing Interests in Water*, Kelvin Grove, Qld: Queensland University of Technology

Fisher, Douglas (2006), 'Markets, Water Rights and Sustainable Development', *Environmental and Planning Law Journal*, **23** (2), 100–112

Fisher, Douglas (2008), 'Legal and Paralegal Rules for Biodiversity Conservation: A Sequence of Conceptual, Linguistic and Legal Challenges', in Jeffrey, Michael, Jeremy Firestone and Karen Bubna-Litic (eds), *Biodiversity, Conservation, Law and Livelihoods: Bridging the North-South Divide*, New York: Cambridge University Press, pp. 94–131

Fisher, Douglas (ed.) (2009), *The Law and Governance of Water Resources:*

The Challenge of Sustainability, Cheltenham, UK and Northampton, MA, USA: Edward Elgar Publishing
Fisher, Douglas (ed.) (2010), *Australian Environmental Law: Norms, Principles and Rules,* second ed., Rozelle, NSW: Thomson Reuters
Fisher, D. E., 'Rights of Property in Water: Confusion or Clarity' (2004) 21 *Environmental and Planning Law Journal* **21**(3), 200–226
FLEGT Expert Group of the European Commission, European Commission (2007), *FLEGT Briefing Note Number 01 – What is FLEGT?*
FLEGT Expert Group of the European Commission, European Commission (2007), *FLEGT Briefing Note Number 02 – What is Legal Timber?*
FLEGT Expert Group of the European Commission, European Commission (2007), *FLEGT Briefing Note Number 03 – A Timber Legality Assurance System*
FLEGT Expert Group of the European Commission, European Commission (2007), *FLEGT Briefing Note Number 05 – Legality Assurance Systems: Requirements for Verification*
FLEGT Expert Group of the European Commission, European Commission (2007), *FLEGT Briefing Note Number 06 – Voluntary Partnership Agreement*
Food and Agriculture Organization (2010), 'FAO Global Forest Assessment 2010', available at http://www.fao.org/forestry/fra/fra2010/en/ (accessed 20 January 2012)
Food and Agricultural Organization of the United Nations (2007), *State of the World's Forests 2007*
Food and Agricultural Organization of the United Nations (2009), *State of the World's Forests 2009*
Food and Agriculture Organization (2011), 'Reforming Forest Tenure: Issues, Principles and Process', available at http://www.fao.org/docrep/014/i2185e/i2185e00.pdf (accessed 27 January 2012).
Forest Carbon Partnership Facility (2011), 'FCFP Readiness Fund, Readiness Package Content and Assessment Approval' available at http://www.forestcarbonpartnership.org/fcp/sites/forestcarbonpartnership.org/files/FMT%20Note%202011-14%20Readiness%20Package%20Concept.pdf (accessed 14 February 2012)
Forest Law Enforcement and Governance, East Asia Ministerial Conference, Bali, Indonesia, 11–13 September 2001, Ministerial Declaration
Forest Stewardship Council (1994), *Forest Stewardship Council A.C. By-Laws* (Ratified September 1994)
Forest Stewardship Council (1996), *FSC International Standard: FSC Principles and Criteria for Forest Stewardship*

Forest Stewardship Council (2004), *FSC Standard: SLIMF Eligibility Criteria*

Forest Stewardship Council (2008), 'History', available at http://www.fsc.org/history.html (accessed 4 September 2009)

Forest Stewardship Council (2009), '5 Steps towards FSC certification', available at http://www.fsc.org/106.html (accessed 11 September 2009)

Forest Stewardship Council (2009), 'Global FSC certificates: type and distribution', available at http://www.fsc.org/fileadmin/web-data/public/document_center/powerpoints_graphs/facts_figures/09-06-15_Global_FSC_certificates_-_type_and_distribution_-_FINAL.pdf (accessed 16 September 2009)

Forest Stewardship Council (2009), 'FSC facts and figures', available at http://www.fsc.org/facts-figures.html (accessed 16 September 2009)

Forest Stewardship Council (2009), *Statute* (Revised August 2000, November 2002, June 2005, February 2009)

Forest Stewardship Council International Centre, Forest Stewardship Council (2003), *FSC Social Strategy: Building and Implementing a Social Agenda*

Fraser, Nancy (ed.) (1997), *Justice Interruptus: Critical Reflections on the "Postsocialist" condition,* New York: Routledge

Fry, Ian (2002), 'Twists and Turns in the Jungle: Exploring the Evolution of Land-use, Land-Use Change and Forestry Decisions within the Kyoto Protocol', *Review of European Community and International Environmental Law*, **11** (2), 159–168

Fry, Ian (2007), 'More Twists, Turns and Stumbles in the Jungle: A Further Exploration of Land Use, Land-Use Change and Forestry Decision within the Kyoto Protocol', *Review of the European Community and International Environmental Law*, **16** (3), 341–355

FSC Board of Directors and Staff, Forest Stewardship Council (2007), *Strengthening Forest Conservation, Communities and Markets: The Global Strategy of the Forest Stewardship Council*

Fuller, Lon (ed.) (1969), *The Morality of Law*, New Haven, CT: Yale University

Gardner, Royal (1997), 'Taking the Principle of Just Compensation Abroad: Private Property Rights, National Sovereignty, and the Cost of Environmental Protection', *University of Cincinnati Law Review,* **65** (2), 539–593

Garnaut, Ross (2008), *The Garnaut Climate Change Review*

General Assembly, Report of the United Nations Conference on Environment and Development, UN Doc A/CONF.151/26 (Vol. III) (14 August 1992) annex III, available at http://www.un.org/documents/ga/

conf151/aconf15126-3annex3.htm (accessed 26 March 2012), principle 1.

'General Guidelines for the Sustainable Management of Forests in Europe', available at http://www.mcpfe.org/resolutions/helsinki/resolution_h1.pdf (accessed 30 November 2006)

Gerdung, Anja (2004), 'Global Environmental Governance and the Role of Civil Society Groups', *New Zealand Journal of Environmental Law*, **8**, 55–98

Gianaris, William N. (1991), 'Weighted Voting in the International Monetary Fund and the World Bank', *Fordham International Law Journal*, **14** (4), 910–945

Global Environment Facility (1995), *Operational Strategy of the Global Environmental Facility – Revised Draft GEF Operational Strategy*

Global Witness (2011), 'Forest Carbon, Cash and Crime' available at http://www.globalwitness.org/sites/default/files/library/Forest%20Carbon,%20Cash%20and%20Crime.pdf (accessed 20 January 2012)

Global Witness (2011), 'Making the Forest Sector Transparent: Annual Transparency Report 2010', available at http://www.foresttransparency.info/ (accessed 20 January 2012)

Gluck, P., R. Tarasofsky, N. Byron and I. Tikkanen, European Commission, European Forest Institute, The World Conservation Union and the Centre for International Forestry Research (1997), *Options for Strengthening the International Legal Regime for Forests,* 26

Goldsmith, Arthur (2007), 'Is Governance Reform a Catalyst for Development?', *Governance: An International Journal of Policy, Administration, and Institutions*, **20** (2), 165–186

Graichen, Patrick (2005), 'Can Forestry Gain from Emissions Trading? Rules Governing Sinks Projects Under the UNFCCC and the EU Emissions Trading System', *Review of the European Community and International Environmental Law*, **14** (1), 11–18

Greber, Thomas, Melinda Janki, Morcos Orellana, Annalisa Savaresi and Dinah Shelton, 'Conservation with Justice: A Rights-based Approach' (Policy Paper No 71, International Union for the Conservation of Nature, 2009)

Gregg, R. Juge, and Amelia Porges (2008), 'Amendment to the US Lacey Act: Implications for Chinese Forest Products Exporters', available at http://www.forest-trends.org/documents/publications/Lacey%20Act%20Brief_China_EN.pdf (accessed 11 June 2009)

Grêt-Ragamey, Ardrienne, Eugene Hendrick, Sebastian Hetsch, Kim Pingoud, Sebastian Rüter, *Challenges and Opportunities of Accounting for Harvested Wood Products*, Swiss Federal Office for the Environment, UNECE/FAO and MCPFE (2008)

Griffiths, Tom, Forest People Program (2008), *Seeing 'REDD'? Forests, Climate Change Mitigation and the Rights of Indigenous Peoples and Local Communities*

Gulbrandsen, Lars (2004), 'Overlapping Public and Private Governance: Can Forest Certification Fill the Gaps in the Global Forest Regime?', *Global Environmental Politics*, **4** (2), 75–99

Gunningham, Neil and Peter Grabosky (eds) (1998), *Smart Regulation: Designing Environmental Policy*, Oxford: Clarendon

Hahn, Robert, and Robert Stavins (1991), 'Incentive-Based Environmental Regulation: A New Era from an Old Idea', *Ecology Law Quarterly*, **18** (1), 1–42

Hardin, Garrett (1968), 'The Tragedy of the Commons', *Science*, **162**, 1243–1248

Harris, Elizabeth, International Institute for Sustainable Development (2007), *The Voluntary Carbon Offsets Market: An Analysis of Market Characteristics and Opportunities for Sustainable Development*

Harris, Paul G. (1999), 'Common but Differentiated Responsibility: The Kyoto Protocol and United States Policy', *New York University Environmental Law Journal*, **7** (1), 27–48

Hatcher, Jeffrey, and Luke Bailey, Rights and Resource Initiative (2009), *Tropical Forest Tenure Assessment: Trends, Challenges and Opportunities*

Hayes, Tanya, and Elinor Ostrom (2005), 'Conserving the World's Forests: Are Protected Areas the Only Way?', *Indiana Law Review*, **38** (3), 595–617

Hegel, Georg Wilhelm Friedrich (1821), *Hegel's Philosophy of Right* (T.M. Knox trans., Oxford University Press, 1967) [trans. of: *Grundlinien der Philosophie des Rechts* (first published 1821)]

Hemmati, Minu (ed.) (2002), *Multi-Stakeholder Processes for Governance and Sustainability: Beyond Deadlock and Conflict*, London; Sterling, VA: Earthscan Publications

Hermosilla, Arnoldo Contreras, and Markku Simula, The World Bank (2007), *The World Bank Forest Strategy: Review of Implementation*

Hickey, Gordon (2008), 'Evaluating Sustainable Forest Management', *Ecological Indicators*, **8** (2), 109–114

Hiraldo, Rocío and Thomas Tanner (2011), 'Forest Voices: Competing Narratives over REDD+', *IDS Bulletin*, **42** (3), 42–51

Hodge, Graeme, and Ken Coghill (2007), 'Accountability in the Privatized State', *Governance: An International Journal of Policy, Administration, and Institutions*, **20** (4), 675–702

Humphreys, David (ed.) (1996), *Forest Politics: The Evolution of International Cooperation*, London: Earthscan Publications

Humphreys, David (2003), 'The United Nations Forum on Forests: anatomy of a stalled international process', *Global Environmental Change*, **13** (4) 319–323

Humphreys, David (ed.) (2006), *Logjam: Deforestation and the Crisis of Global Governance*, London: Earthscan Publications

Hunt, Colin (2007), 'Economy and Ecology of Emerging Markets and Credits for Bio-Sequestered Carbon on Private Land in Tropical Australia', *Ecological Economics*, **66** (2–3), 309–318

Hunt, Geoffrey (1992), 'Is There a Conflict between Environmental Protection and the Development of the Third World?' in Robin Attfield and Barry Wilkins (eds), *International Justice and the Third World*, London; New York: Routledge, pp. 117–151

Intergovernmental Panel on Climate Change (1996), 'Revised IPCC guidelines for National Greenhouse Gas Inventories', available at http://www.ipcc-nggip.iges.or.jp/public/gl/invs1.html (accessed 7 February 2012)

Intergovernmental Panel on Climate Change (2003), 'Good Practice Guidelines for Land Use, Land-Use Change and Forestry', available at http://www.ipcc-nggip.iges.or.jp/public/gpglulucf/gpglulucf_files/0_Task1_Cover/Cover_TOC.pdf (accessed 7 February 2012)

Intergovernmental Panel on Climate Change (2004), 'About the IPCC', available at http://www.ipcc.ch/about/index.htm (accessed 7 January 2008)

Intergovernmental Panel on Climate Change (2006), 'IPCC Guidelines of National Greenhouse Gas Inventories Online', available at http://www.ipcc-nggip.iges.or.jp/public/2006gl/index.html (accessed 14 December 2011)

International Bank for Reconstruction and Development (ed.) (2004), *Sustaining Forests: A Development Strategy*, Washington D.C.: The World Bank

International Bank for Reconstruction and Development (2008), *Charter Establishing the Forest Carbon Partnership Facility*

International Centre for Settlement of Investment Disputes (2006), 'Convention on the Settlement of Investment Disputes between States and Nationals of Other States', available at http://icsid.worldbank.org/ICSID/StaticFiles/basicdoc/CRR_English-final.pdf (accessed 26 March 2012)

International Centre for Settlement of Investment Disputes (2006), 'ICSID Convention, Regulations and Rules', available at http://icsid.worldbank.org/ICSID/StaticFiles/basicdoc/CRR_English-final.pdf (accessed 26 March 2012)

International Development Law Organization (2012), 'Rule of Law

Assistance Directory, available at http://www.idlo.int/english/Resources/ROL/Pages/default.aspx (accessed 14 March 2012)
International Institute for Environment and Development (2008), *Towards a Global Forest Partnership: Consultation, Assessment and Recommendations*
International Network for Environmental Compliance and Enforcement (2008), 'Recent Amendments to U.S. Lacey Act Should Help Protect Forests', available at http://www.inece.org/climate/ClimateComplianceAlert_LaceyAct.pdf (accessed 16 December 2009)
International Tropical Timber Organization (1992), 'Criteria for the Measurement of Sustainable Tropical Forest Management', available at http://www.itto.or.jp/live/PageDisplayHandler?pageId=201 (accessed 30 November 2006)
International Tropical Timber Organization (2004–2009), 'Market Access', available at http://www.itto.int/en/feature11/ (accessed 11 June 2009)
International Tropical Timber Organization (2008), *Annual Review and Assessment of the World Timber Situation*
ITTO Working Group, International Tropical Timber Organisation (2008), *ITTO Action Plan 2008-2011: ITTO Policy Development Series No 18*
ITTO Working Group, International Tropical Timber Organisation, *ITTO Yokohama Action Plan 2002-2006, ITTO Policy Development Series No 11* (2002)
Jack, Kelsey, Carolyn Kousky, and Katharine Sims (2008), 'Designing Payments for Ecosystem Services: Lessons from Previous Experience with Incentive-based Mechanisms', *Proceedings of the Nations Academy of Sciences*, **105** (28), 9465–9470
Jacquemont, Frederic, Alejandro Caparros (2002), 'The Convention on Biological Diversity and Climate Change Convention 10 Years after Rio: Towards a Synergy of the Two Regimes?', *Review of the European Community and International Environmental Law*, **11** (2), 169–180
Jung, Martin (2005), 'The Role of Forestry Projects in the Clean Development Mechanism', *Environmental Science and Policy*, **8** (2), 87–104
Kaldor, Mary (2003), 'The Idea of Global Civil Society', *International Affairs*, **79** (3), 583–593
Kant, Immanuel (1797), *The Metaphysical Elements of Justice – Part 1 of the Metaphysical Morals* (John Ladd trans., Hackett Publishing, 1965) [trans. of *Die Metaphysiche Anfangsgründe der Rechtslehre*, Part 1 of the *Metaphysik der Sitten* (first published 1797)]
Kant, Promode and Wu Shuirong (2008), 'Reducing Deforestation and Degradation through Post-Colonial Settlement of Land Rights: A Case Study in India', *Carbon and Climate Law Review*, (3), 300–305

Kant, Shashi (2004), 'Economics of Sustainable Forest Management', *Forest Policy and Economics,* **6** (3–4), 197–203

Katere, Yemi (2009), 'REDD in Copenhagen', available at http://www.un-redd.org/Newsletter5_Introduction/tabid/2890/language/en-US/Default.aspx (accessed 2 February 2010)

Kehoe, Jo (2009), 'Environmental Law Making in Queensland: The Vegetation Management Act 1999 (Qld)', *Environmental and Planning Law Journal*, **26** (6), 392–410

Keohane, Nathaniel, Richard Revesz and Robert Stavins (1998), 'The Choice of Regulatory Instruments in Environmental Policy', *Harvard Environmental Law Review*, **22** (2), 313–367

Kimber, Richard (1983), 'Black Lightning Aborigines and Fire in Central Australia and the Western Desert', *Archaeology in Oceania*, **18** (1), 38–45

Kinrade, Peter (1995), 'Towards Ecologically Sustainable Development: The Role and Shortcomings of Markets' in Robyn Eckersley (ed.) *Markets, the State and the Environment: Towards Integration*, South Melbourne: Macmillan Education Australia, pp. 86–99

Klooster, Dan (2005), 'Environmental Certification of Forests: The Evolution of Environmental Governance in a Commodity Network', *Journal of Rural Studies*, **21** (4), 403–417

Ku, Charlotte, and Paul Diehl (2006), 'Filling in the Gaps: Extra Systemic Mechanisms for Addressing Imbalances between the International Legal Operating System and the Normative system', *Global Governance*, **12** (1), 161–183

Kuehls, T (1998), 'Between Sovereignty and Environment: An Exploration of the Discourse of Government' in K.Liftin (ed), *The Greening of Sovereignty in World Politics,* Cambridge, Mass: MIT Press, pp. 31–53

Kuznets, S. (1955), 'Economic Growth and Income Inequality', *American Economic Review*, **45** (1), 1–28

Laaksonen-Craig, Susanna (2004), 'Foreign Direct Investment in the Forest Sector: Implications for Sustainable Forest Management in Developed and Developing Countries', *Forest Policy and Economics*, **6** (3–4), 359–370

Lambin, Eric F., B.L. Turner, Helmut J. Geist, Samuel B. Agbola, Arild Angelsen, John W. Bruce, Oliver T. Coomes, Rodolfo Dirzo, Gunther Fischer, Carl Folke, P.S. George, Katherine Homewood, Jacques Imbernon, Rik Leemans, Xiubin Li, Emilio F. Moran, Michael Mortimore, P.S. Ramakrishnan, John F. Richards, Helle Skanes, Will Steffen, Glenn D. Stone, Uno Svedin, Tom A. Veldkamp, Coleen Vogel and Jianchu Xu (2001), 'The Causes of Land-Use and Land-Cover Change: Moving beyond the Myths', *Global Environmental Change*, **11** (4), 261–269

Landell-Mills, Natasha and Ina Porras, International Institute for Environment and Development (2002), *Silver Bullet or Fools' Gold? A Global Review of Markets for Forest Environmental Services and their Impacts on the Poor*

Lane, M.B. (1999), 'Regional Forest Agreements: Resolving Resource Conflicts or Managing Resource Politics?', *Australian Geographical Studies*, **37** (2), 142–153

Lane, Marcus (2003), 'Decentralization or Privatization of Environmental Governance? Forest Conflict and Bioregional Assessment in Australia', *Journal of Rural Studies*, **19** (3), 283–294

Larson, Anne, (2011), 'Forest Tenure Reform in the Age of Climate Change: Lessons for REDD+', *Global Environmental Change*, **21** (2), 540–549

Larson, Anne, Peter Cronkleton, Deborah Barry, Pablo Pacheco, Centre for International Forestry Research (2008), *Tenure Rights and Beyond: Community Access to Forest Resources in Latin America*

Law Book Company, *Laws of Australia*, vol 28 (at 24 July 2009) 28 Real Property, '28.1.73 Leasehold Interests'

Law Book Company, *Laws of Australia*, vol 28 (at 24 July 2009) 28 Real Property, '28.10.550 Profit à prendre'

Law Book Company, *Laws of Australia*, vol 28 (at 24 July 2009) 28 Real Property, '28.13.1 Easements'

Law Book Company, *Laws of Australia*, vol 28 (at 24 July 2009) 28 Real Property, '28.14.1 Restrictive Covenants'

Law Book Company, *Property Law / Land Titles Law and Practice (Qld)*, Division 4A, 'Covenants'

Law Book Company, *Property Law / Land Titles Law and Practice (Qld)*, Division 4B, 'Profits a Prendre'

Leape, James and Paul Wolfowitz (2005), *WWF/World Bank Global Forest Alliance: Alliance Annual Report 2005*

Leohane, Nathaniel, Richard Revesz and Robert Stavins (1998), 'The Choice of Regulatory Instruments in Environmental Policy', *Harvard Environmental Law Review*, **22** (2), 313–367

Leopold, Aldo (ed.) (1966), *A Sand Country Almanac*, New York: Oxford University Press

Lewis, Melea, Charles Sampford and Ramesh Thakur (2008), 'Introduction' in Jacobsen, Trudy Charles Sampford and Ramesh Thakur (eds), *Re-envisioning Sovereignty: The End of Westphalia,* UK: Ashgate Publishers, pp. 1–8

Liftin, K. (1998), 'The Greening of Sovereignty: An Introduction' in K Liftin (ed.) *The Greening of Sovereignty in World Politics*, Cambridge, Mass: MIT Press, pp. 1–2

Lindsay Jon, Ali Mekour and Lawrence Christy, FAO Legal Papers Online #27, FAO Development Law Service (2002), 'Why Law Matters: Design Principles for Strengthening the Role of Forestry Legislation in Reducing Illegal Activities and Corrupt Practices' available at http://www.fao.org/legal/prs-ol/lpo27.pdf (accessed 21 March 2012)

Lipschutz, Ronnie D. (2001), 'Why is There No International Forestry Law? An Examination of International Forestry Regulation, both Public and Private', *Journal of Environmental Law and Policy*, **19** (1), 153–179

Lough, Peter B., and Alastair D. Cameron (2008), 'Forestry in the New Zealand Emission Trading Scheme: Design and Prospects for Success', *Carbon and Climate Law Review*, (3), 281–291

Lyster, Rosemary (2002), '(De)regulating the Rural Environment', *Environmental and Planning Law Journal,* **19** (1), 445–468

Lyster, Rosemary (2009), 'The New Frontier of Climate Law: Reducing Emissions from Deforestation and Degradation' *Environmental and Planning Law Journal*, **26** (6), 417–456

Macqueen, Duncan, (ed.) (2008), *Distinguishing Community Forest Products in the Market: Industrial Demand for a Mechanism that Brings Together Forest Certification and Fair Trade*, Edinburgh, UK: Russell Press

Magraw, D. and D. Hawke (2007), 'Sustainable Development' in Bodansky, Daniel, Ellen Brunée, Ellen Hay (eds), *Oxford Handbook of International Environmental Law*, Oxford: Oxford University Press, pp. 613–638

Maguire, Rowena (2008), 'Environmental Offsets in Queensland: New Mechanisms for Managing Natural Resources?', *Queensland Environmental Practice Reporter*, **13** (61), 160–167

Maguire, Rowena (2008), 'The Compatibility of Leasehold and Freehold Tenure Arrangements with Environmental Offset and Trading Initiatives', *Queensland Environmental Practice Reporter*, **14** (63), 27–38

Maguire, Rowena (2011), 'Opportunities for Forest Finance: Compliance and Voluntary Markets', *Carbon and Climate Law Review*, (1), 100–112

Maguire, Rowena and Angela Springer (2011), 'The Role of Property Law in Environmental Management: An Examination of Environmental Markets', *Environmental and Planning Law Journal,* **28** (4), 215–242

Manguiat, Maria Socorro, Z. Verheyen, Roda Mackensen and Gerald Jens Scholz (eds) (2005), *Legal Aspects in the Implementation of CDM Forestry Projects*, Gland, Switzerland: IUCN in collaboration with GTZ and UNEP

Mayers, James, and Stephen Bass (eds) (1999), *Policy that Works for Forests and People*, London: International Institute for Environment and Development

Mayers, J., S. Bass and D. Macqueen, International Institute for Environment and Development (2005), *The Pyramid: a Diagnostic and Planning Tool for Good Forest Governance*

McDermott, Michael (2011), 'International Environmental Disputes and the Need for Court-Commissioned Independent Experts', *Boston College International and Comparative Law Review*, **34**, 67–80

McDonald, Jan (1999), 'Regional Forest (DIS) Agreements: The RFA Process and Sustainable Forest Management', *Bond Law Review*, **11** (2), 295–335

McGrath, Chris (ed.) (2010), *Does Environmental Law Work? How to Evaluate the Effectiveness of an Environmental Legal System*, Saarbrücken, Germany: Lambert Academic Publishing

McGregor, Joan L. (1999), 'Property Rights and Environmental Protection: Is this Land Made for You and Me?', *Arizona State Law Journal*, **31** (2), 391–437

Melkas, Erika (2002), 'Sovereignty and Equity within the Framework of the Climate Regime', *Review of European Community and International Environmental Law*, **11** (2), 115–128

Mendoza, Guillermo and Ravi Prabhu (2000), 'Development of a Methodology for Criteria and Indicators of Sustainable Forest Management: A Case Study on Participatory Assessment', *Environmental Management*, **26** (6), 659–673

Metz, B, O. Davidson, P. Bosh, R. Dave, and L. Meyer (eds) (2007), *Summary for Policymakers in Climate Change 2007: Mitigation, Contribution of Working Group III to the Fourth Assessment Report of the Intergovernmental Panel on Climate Change*, Cambridge, United Kingdom; New York, NY, USA: Cambridge University Press

Michaelowa, Axel (2007), 'Unilateral CDM – Can Developing Countries Finance Generation of Greenhouse Gas Emission Credits on their Own?', *International Environmental Agreements*, **7** (1), 17–34

Michaelowa, Axel, Hamburgisches WeltWirstschafts Institut (2005), *CDM: Current Status and Possibilities for Reform*

Mickelson, Karen (1998), 'Rhetoric and Rage: Third World Voices in International Legal Discourse', *Wisconsin International Law Journal*, **16** (2), 353–419

Mickelson, Karin (1996), 'Seeing the Forest, the Trees and the People: Coming To Terms With Developing Perspectives On The Proposed Global Forest Regime', in S. Johnson (ed.), *Global Forests and International Environmental Law*, London; Boston: Kluwer Law International, pp. 239–264

Miller, Alan (2000), 'The Global Environmental Facility and the Search for Financial Strategies to Foster Sustainable Development', *Vermont Law Review*, **24** (4), 1229–1244

Milner-White, Greg (2007), 'The Legal Implications of Climate Change in New Zealand for the Forestry Industry', *New Zealand Journal of Environmental Law*, **11**, 141, 148–180

Minany, Peter Akong, Stefan Jungcurt, Vanessa Meadu, Deborah Murphy (2009), 'The REDD Negotiations: Moving into Copenhagen, International Institute for Sustainable Development, available at http://www.iisd.org/pdf/2009/redd_negotiations.pdf (accessed 14 December 2011)

Minany, Peter, Hans Bressers, Margaret Skutsh and Michael McCall (2007), 'National Forest Policy as a Platform for Biosphere Carbon Management: The Case of Community Forestry in Cameroon', *Environmental Science and Policy*, **10** (3), 204–218

Mitchell, Andrew and Jennifer Beard (eds) (2009), *International Law in Principle*, Pyrmont, NSW: Lawbook Co.

Molnar, A., Forest Trends (2003), *Forest Certification and Communities: Looking Forward to the Next Decade*

Moran, Alan (1995), 'Tools of Environmental Policy: Market Instruments versus Command-and Control' in Robyn Eckersley (ed.) *Markets, the State and the Environment: Towards Integration*, South Melbourne: Macmillan Education Australia, pp. 73–76

Morgan, G. (1993), 'The Dominion of Nature: Can Law Embody a New Attitude?', *Bulletin of the Australian Society of Legal Philosophy*, **18** (60) 43–76

Murphy, Craig (2009), 'Privatizing Environmental Governance', *Global Environmental Politics*, **9** (3), 134–138

Musselwhite, Gary, and Gamini Hearth (2005), 'Australia's Regional Forest Agreement Process: Analysis of the Potential and Problems', *Forest Policy and Economics*, **7** (4), 579–588

Myers, Norman (1997), 'Consumption in Relation to Population, Environment and Development', *The Environmentalist*, **17** (1), 33–44

New South Wales Government: Department of Environment and Climate Change (2007), 'Biobanking: Biodiversity Banking and Offsets Scheme, Scheme Overview', available at http://www.environment.nsw.gov.au/resources/biobanking/biobankingoverview07528.pdf (accessed 1 July 2008)

New South Wales Government: Department of Environment and Climate Change (2007), 'Questions and Answers about the Biodiversity Banking and Offset Scheme' available at http://www.environment.nsw.gov.au/threatspec/biobankingqa.htm (accessed 7 November 2007)

Nitschke, Craig R. and John L. Innes (2008), 'Integrating Climate Change into Forest Management in South-Central British Columbia: An Assessment of Landscape Vulnerability and Development of a

Climate-Smart Framework', *Forest Ecology and Management*, **256** (3), 313–327
Nozick, Robert (2000), 'Distributive Justice', in Richard Epstein (ed.), *Modern Understanding of Liberty and Property*, New York: Garland Pub, pp. 317–495
O'Sullivan, Robert, and Charlotte Streck (2008), 'Conservation Carbon: A New Voluntary Market Mechanism to Protect Forests', *Carbon and Climate Law Review*, (3), 312–321
Overdevest, Christine and Mark Rickenbach (2006), 'Forest Certification and Institutional Governance: An Empirical Study of Forest Stewardship Council Certificate Holders in the United States', *Forest Policy and Economics*, **9** (1), 93–102
Pagiola, Stefano (2004), 'Paying for Water Services in Central America: Learning from Costa Rica', in Stefano Pagiola, Joshua Bishop and Natasha Landell-Mills (eds), *Selling Forest Environmental Services: Market-based Mechanisms for Conservation and Development*, New York: Routledge, pp. 37–56
Pagiola, Stefano (2008), 'Payments for Environmental Services in Costa Rica', *Ecological Economics*, **65** (4), 712–724
Pagiola, Stefano, Natasha Landell-Mills and Joshua Bishop (2004), 'Market-based Mechanisms for Forest Conservation and Development', in Stefano Pagiola, Natasha Landell-Mills and Joshua Bishop (eds), *Selling Forest Environmental Services: Market-based Mechanisms for Conservation and Development*, New York: Routledge, pp. 20–32, 261–274
Pagiola, Stefano, Joshua Bishop and Natasha Landell-Mills (eds) (2004), *Selling Forest Environmental Services: Market-based Mechanism for Conservation and Development*, London: Earthscan Publications
Palm, Matilda, Madelene Ostwald and John Reilly (2008), 'Land Use and Forestry Based CDM in Scientific Peer-Reviewed Literature pre and post COP 9 in Milan', *International Environmental Agreements*, **8** (3), 249–275
Panayotou, Theodore (ed.) (1994), *Economic Instruments for Environmental Management and Sustainable Development*, Nairobi: United Nations Environment Program
Parker, C., A. Mitchell, M. Trivedi, N. Mardas and K. Sosis (2009), 'The Little REDD+ Book: An Updated Guide to Governmental and Non-Governmental Proposals for Reducing Emission from Deforestation and Degradation', available at http://unfccc.int/files/methods_science/redd/application/pdf/the_little_redd_book_dec_08.pdf (accessed 14 December 2011)
Passero, Michelle (2008), 'The Nature of the Right of Interest Created by a Market for Forest Carbon', *Carbon and Climate Law Review*, (3), 248–253

Passmore, John (ed.) (1974), *Man's Responsibility for Nature: Ecological Problems and Western Traditions*, London: Duckworth

Pattberg, Philipp (2005), 'The Forest Stewardship Council: Risk and Potential of Private Forest Governance', *The Journal of Environment and Development,* **14** (3), 356–374

Pearce, David W. (ed.) (1993), *Economic Values and the Natural World*, London: Earthscan Publications

Peck, T.J. and E.G. Richards (2005), 'Fifty Years of International Co-operation in Forestry', available at http://www.unece.org/timber/docs/publications-other/JCHistory.pdf (accessed 11 February 2010)

Penner, J.E. (1996), 'The Bundle of Rights Picture of Property', *UCLA Law Review*, **43** (3), 711–820

Pennington, Mark (2001), 'Environmental Markets vs. Environmental Deliberation: A Hayekian Critique of Green Political Economy', *New Political Economy*, **6** (2), 171–190

Pfaff, Alexander, S. Kerr, R. F. Hughes, S. G. Liu, G. A. Sanchez-Azofeifa, D. Schimel, J. Tosi, V. Watson (2000), 'The Kyoto Protocol and Payments for Tropical Forest: An Interdisciplinary Method for Estimating Carbon-Offset Supply and Increasing the Feasibility of a Carbon Market under the CDM', *Ecological Economics*, **35** (2), 203–221

Pigou, A.C. (ed.) (1932), *The Economics of Welfare*, London: Macmillan and Co. Ltd.

Potvin, Catherine and A. Bovarnick (2008), 'Reducing Emissions from Deforestation and Forest Degradation in Developing Countries: Key Actors, Negotiations and Actions', *Carbon and Climate Law Review*, (3), 264–273

Prior, S. and C. Streck (2006), Centre for International Sustainable Development Law, *Submission to the COP UNFCCC in Response to the Call for Views on the Issue of Avoided Deforestation*

Program on Forests (2007), 'About PROFOR', available at http://www.profor.info/about.html (accessed 29 September 2008)

Program on Forests (2007), *Forest Poverty Toolkit*

Program on Forests, United Nations Development Program (1999), *Forest Management to Support Sustainable Livelihoods: Framework for the UNDP Programme on Forests*

Queensland Department of Natural Resources and Water (2007), 'Statutory Covenants' available at http://www.nrw.qld.gov.au/factsheets/pdf/land/198.pdf (accessed 8 March 2012)

Queensland Government Department of Natural Resources and Mines, Statutory Covenants Working Group (2003), *Statutory Covenants: Guidelines for their Use in Queensland*

Rametsteiner, Ewald (2002), 'The Role of Governments in Forest

Certification – A Normative Analysis based on New Institutional Economics Theories', *Forest Policy and Economics*, **4** (3), 163–173

Rametsteiner, Ewald and Markku Simula (2003), 'Forest Certification – An Instrument to Promote Sustainable Forest Management?', *Journal of Environmental Management*, **67** (1), 87–98

Rangan, Haripriya, and Marcus Lane (2001), 'Indigenous People and Forest Management: Comparative Analysis of Institutional Approaches in Australia and India', *Society and Natural Resources*, **14** (2), 145–160

Raunikar, Ronald, Joseph Buongiorno, James A. Turner and Shushuai Zhu (2010), 'Global Outlook for Wood and Forests with the Bioenergy Demand Implied by Scenarios of the Intergovernmental Panel on Climate Change', *Forest Policy and Economics*, **12** (1), 48–56

Rawls, John (ed.) (1972), *A Theory of Justice*, Oxford: Clarendon Press

Redwood, Jill (2001), 'Sweet RFA', *Briefs*, **26** (5), 255–256

Reed, David (1993), 'The Global Environmental Facility and Non-Governmental Organizations', *American University Journal of International Law and Policy*, **9** (1), 191–213

Reitze Jr, A.W. (1984), 'Environmental Policy – It is Time for a New Beginning', *Columbia Journal of Environmental Law*, **14** (1), 111–156

Resolution H1 – General Guidelines for the Sustainable Management of Forests in Europe (1993), 'The European Process on Criteria and Indicators for Sustainable Forest Management', available at http://www.foresteurope.org/filestore/foresteurope/Conferences/Helsinki/helsinki_resolution_h1.pdf (accessed 26 March 2012)

Resolutions and Decisions of the Economic and Social Council (2000), '2000/35 Report on the Fourth Session of the Intergovernmental Forum on Forests', available at http://www.un.org/esa/forests/pdf/2000_35_E.pdf (accessed 26 March 2012)

Richards, Michael (2000), 'Can Sustainable Tropical Forestry be made Profitable? The Potential and Limitations of Innovative Incentive Mechanisms', *World Development*, **28** (6), 1001–1016

Richardson, Benjamin J. (ed.) (2009), *Climate Law and Developing Countries: Legal and Policy Challenges for the World Economy*, Cheltenham, UK and Northampton, MA, USA: Edward Elgar Publishing

Richardson, Benjamin, and Donna Craig (2006), 'Indigenous Peoples, Law and the Environment' in Benjamin J. Richardson and Stepan Wood (eds), *Environmental Law for Sustainability*, Oxford; Portland, Oregon: Hart Publishing, pp. 210–224

Richardson, Benjamin, and Jona Razzaque (2006), 'Public Participation in Environmental Decision-Making' in Benjamin J. Richardson and Stepan

Wood (eds), *Environmental Law for Sustainability*, Oxford; Portland, Oregon: Hart Publishing, pp. 166–177

Robeyns, Ingrid (2003), 'Is Nancy Fraser's Critique of Theories of Distributive Justice Justified', *Constellations*, **10** (4), 538–554

Rose, Carol (1986), 'The Comedy of the Commons: Custom, Commerce and Inherently Public Property', *The University of Chicago Law Review*, **53** (3), 711–781

Rosenbaum, Kenneth, Dieter Schone, Ali Mekouar, Food and Agriculture Organization of the United Nations (2004), *Climate Change and the Forest Sector: Possible National and Subnational Legislation, FAO Forestry Paper No 144*

Rosendale, Kristin (2001), 'Overlapping International Regimes: The Case of the Intergovernmental Forum on Forests (IFF) between Climate Change and Biodiversity', *International Environmental Agreements: Politics, Law and Economics*, **1** (4), 447–468

Rothman, Dale (1998), 'Environmental Kuznets curves – real progress or passing the Buck? A Case for Consumption-based approaches', *Ecological Economics*, **25** (2), 177–194

Sagemuller, Imke (2006), 'Forest Sinks under the United Nations Framework Convention on Climate Change and the Kyoto Protocol: Opportunity or Risk for Biodiversity?', *Columbia Journal of Environmental Law*, **31** (2), 189–242

Sandberg, Audun (2007), 'Property Rights and Ecosystem Properties', *Land Use Policy*, **24** (4), 613–623

Sands, Philippe (2003), *Principles of International Environmental Law*, second ed., UK: Cambridge University Press

Saunders, Jade, Hohannes Ebeling and Ruth Nussbaum, ProForest (2008), *Reduced Emissions from Deforestation and Degradation: Lessons from a Forest Governance Perspective*

Saunders, Phillip (1996), *Global Forests & International Environmental Law*, Boston: Kluwer Law International

Sayer, Jeffrey, Unna Chokkalingam and John Poulsen (2004), 'The Restoration of Forest biodiversity and Ecological Values', *Forest Ecology and Management*, **201** (1), 3–11

Sayer, Jeffrey, Jeffrey McNeely, Stewart Maginnis, Into Boedhihartono, Gill Shepherd and Bob Fisher, Rights and Resource Initiative (2008), *Local Rights and Tenure for Forests: Opportunity or Threat for Conservation?*

Scanlon, Jane (2007), 'An Appraisal of the NSW Biobanking Scheme to Promote the Goal of Sustainable Development in NSW', *Macquarie Journal of International and Environment Law*, **4** (1), 71–133

Schlamadinger, Bernhard, N. Bird, T. Johns, S. Brown, J. Canadell, L.

Ciccarese, M. Dutschke, J. Fiedler, A. Fischlin, P. Fearnside, C. Forner, A. Freibauer, P. Frumhoff, N. Hoehne, M.U.F. Kirschbaum, A. Labat, G. Marland, A. Michaelowa, L. Montanarella, P. Moutinho, D. Murdiyarso, N. Pena, K. Ingoud, Z. Rakonczay, E. Rametsteiner, J. Rock, M.J. Sanz, U.A. Schneider, A. Shvidenko, M. Skutsch, P. Smith, Z. Somogyi, E. Trines, M. Ward, Y. Yamagata (2007), 'A Synopsis of Land Use, Land-Use Change and Forestry (LULUCF) under the Kyoto Protocol and Marrakech Accords', *Environmental Science and Policy*, **10** (4), 271–282

Schlamadinger, Bernhard, T. Johns, L. Ciccarese, M. Braun, A. Sato, A. Senyaz, P. Stephens, M. Takahashi, X. Zhang (2007), 'Options for Including Land Use in a Climate Agreement Post 2012: Improving The Kyoto Protocol Approach', *Environmental Science and Policy*, **10** (4), 295–305

Schlosberg, David (ed.) (2007), *Defining Environmental Justice: Theories, Movements, and Nature*, Oxford; New York: Oxford University Press

Schrijver, Nico (ed.) (1997), *Sovereignty Over Natural Resources: Balancing Rights and Duties*, Cambridge; New York: Cambridge University Press

Schulp, Catharina, Gert-Jan Nabuurs and Peter Verbug (2008), 'Future Carbon Sequestration in Europe – Effects of Land Use Change', *Agriculture, Ecosystems and Environment*, **127** (3–4), 251–264

Scott, Anthony (ed.) (2008), *The Evolution of Resource Property Rights*, Oxford; New York: Oxford University Press

Scott, Megan, Department of the Environment and Heritage, Australian Greenhouse Office, Australian Government (2005), *Planning Forest Sink Projects: A Guide to Legal, Taxation and Contractual Issues*

Secretariat of the Convention on Biological Diversity, UNEP (2001), *The Value of Forest Ecosystems*, Convention on Biological Diversity Technical Series No 4

Secretary General, Economic and Social Council, *Assessment of Progress Made on the Implementation of the Non-Legally Binding Instruments on all Types of Forests and towards the Achievement of the Four Global Objectives on Forests*, 9th sess, Agenda Item 3, UN Doc E/CN/18/2011/2 (15 November 2010)

Secretary General, Economic and Social Council, *Preparations for International Year of the Forest 2011*, 9th sess, Agenda Item 6, UN Doc E/CN.18/2011/7 (8 November 2010)

Sedjo, Roger, Resources for the Future Washington (2006), *Forest and Biological Carbon Sinks after Kyoto*

Shanley, Patricia, Alan Pierce, Sarah Laird and Dawn Robinson, Rights and Resources Initiatives (2008), *Beyond Timber: Certification and Management of Non-Timber Forest Products*

Sheahan, Mark, The Winston Churchill Memorial Trust of Australia

(2001), *Credit for Conservation: A Report on Conservation Banking and Mitigation Banking in the USA and its Applicability to New South Wales*
Sheehan, John, and Garrick Small (2005), 'Biota and the Problem of Property', *Environmental and Planning Law Journal*, **22** (2), 158–163
Sheehan, John, 'Conceiving Property Rights in Carbon' (paper presented to the Natural Resources and Water Climate Change Conference, Brisbane Convention and Exhibition Centre, Brisbane, 23 May 2007)
Shelton, Dinah (1999), 'Environmental Justice in the Postmodern World' in Klaus Bosselmann and Benjamin Richardson (eds), *Environmental Justice and Market Mechanisms: Key Challenges for Environmental Law and Policy*, London: Kluwer Law International, pp. 21–26
Sherbinin, Alex de, David Carr, Susan Cassels and Leiwen Jiang (2007), 'The Annual Review of Environmental Resources: Population and Environment', available at http://arjournals.annualreviews.org/doi/pdf/10.1146/annurev.energy.32.041306.100243?cookieSet=1 (accessed 6 May 2008)
Sherbinin, Alex de, David Carr, Susan Cassels and Leiwen Jiang (2007), 'Population and Environment', *The Annual Review of Environmental Resources*, **32** (1), 345–373
Simula, Markku, Advisory Group on Finance of the Collaborative Partnership on Forests and Program on Forests of the World Bank, *Financing Flows and Needs to Implement the Non-Legally Binding Instrument on all Types of Forests* (2008)
Siry, Jacek, Frederick W. Cubbage and Miyan Rukunuddin Ahmed (2005), 'Sustainable Forest Management: Global Trends and Opportunities' *Forest Politics and Economics*, **7** (4), 551–561
Sivayoganathan, Shasi (2007), 'Forestry Tasmania v Brown: Biodiversity Protection – An Empty Promise?', *National Environmental Law Review*, **3**, 21–29
Skutsch, M, N. Bird, E. Trines, M. Dutschke, P. Frumhoff, B.H.J. de Jong, P. van Laake, O. Masera and D. Murdiyarso (2007), 'Clearing the Way for Reducing Emissions from Tropical Deforestation', *Environmental Science and Policy*, **10** (4), 322–334
Slee, Bill (2001), 'Resolving Production-Environment Conflicts: The Case of the Regional Forest Agreement Process in Australia', *Forest Policy and Economics*, **3** (1–2), 17–30
Sterk, Wolfgang, and Bettina Wittneben (2006), 'Enhancing the Clean Development Mechanism through Sectoral Approaches: Definitions, Applications and Ways Forward', *International Environmental Agreements*, **6** (3), 271–287
Stern, David, Michael Common and Edward Barbier (1996), 'Economic Growth and Environmental Degradation: The Environmental Kuznets

Curve and Sustainable Development', *World Development*, **24** (7), 1151–1160

Stern, David (1998), 'Progress on the Environmental Kuznets Curve', *Environmental and Development Economics*, **3** (2), 173–196

Stern, David (2004), 'The Rise and Fall of the Environmental Kuznets Curve', *World Development*, **32** (8), 1419–1439

Stern, N. (2006), 'Stern Review on the Economics of Climate Change', available at http://www.occ.gov.uk/activities/stern.htm (accessed 26 November 2008)

Stern, Nicholas (2008), 'The Economics of Climate Change', *American Economic Review: Papers and Proceedings*, **98** (2), 1–37

Stewart, R.B. (1992), 'Models for Environmental Regulation: Central Planning versus Market-based Approaches', *Boston College Environmental Affairs Law Review,* **19** (3), 547–562

St Petersburg Declaration, Forest Law Enforcement and Governance, Ministerial Conference 22–25 November 2005, St Petersburg, Russia, Ministerial Declaration

Streck, Charlotte (2008), 'Forests, Carbon Markets, and Avoided Deforestation: Legal Implications', *Carbon and Climate Law Review*, (3), 239–247

Streck, Charlotte, and Jolene Lin (2008), 'Making Markets Work: a Review of CDM Performance and the Need for Reform', *European Journal of International Law*, **19** (2), 409–442

Stuart, Jill (2009), 'Unbundling Sovereignty, Territory and the State in Outer Space: Two Approaches' in Bormann, Natalie and Michael Sheehan (eds), *Securing Outer Space: International Relations Theory and the Politics Of Space*, Oxford, UK: Routledge, pp. 8–23

Subsidiary Body for Scientific and Technological Advice, *Report on The Workshop on Methodological Issues Relating to Reducing Emissions from Deforestation and Forest Degradation in Developing Countries*, 29th sess, Agenda item 5, FCCC/SBSTA/2008/11 (8 September 2008)

Subsidiary Body for Scientific and Technological Advice, *Reducing Emissions from Deforestation in Developing Countries: Approaches to Stimulate Action*, 11th sess, Agenda item 6, FCCC/SBSTA/2009/L.9 (11 November 2005)

Subsidiary Body for Scientific and Technological Advice, UNFCCC, *Intergovernmental Panel on Climate Change Guidelines for National Greenhouse Gas Inventories*, 26th Session, Agenda Item 7(a) – Methodical Issues under the Convention, UN Doc FCCC/SBSTA/2007/L.5, GE.07-70214 (15 May 2007)

Suchman, M.C. (1995), 'Managing Legitimacy: Strategic and Institutional Approaches', *Academic Management Review*, **20** (3), 571–610

Sunderlin, William D., Sonya Dewi and Atie Puntodewo, (2007) 'Poverty and Forests: Multi-Country Analysis of Spatial Association and Proposed Policy Solutions' (Occasional Paper No 47, Centre for International Forestry Research)

Sunderlin, William, Jeffrey Hatcher and Megan Liddle, Rights and Resource Initiative (2008), *From Exclusion to Ownership? Challenges and Opportunities in Advancing Forest Tenure Reform*

Szaro, Robert, David Langor and Atse Yapi (2000), 'Sustainable Forest Management in the Developing World: Science Challenges and Contributions', *Landscape and Urban Planning*, **47** (3–4), 135–142

Takacs, David (2009), 'Forest Carbon Law and Property Rights' (Conservation International), available at http://www.conservation.org/Documents/CI_Climate_Forest-Carbon_Law-Property-Rights_Takacs_Nov09.pdf (accessed 27 January 2012)

Tarasofsky, Richard (1999), 'Assessing the International Forest Regime: Gaps, Overlaps, Uncertainties and Opportunities' in Richard Tarasofsky (ed.), *Assessing the International Forest Regime*, Gland, Switzerland: IUCN – The World Conservation Union, pp. 1–3

Tarasofsky, Richard, ICUN – The World Conservation Union (1999), *Assessing the International Forest Regime* 4

Tarlock, Dan (1992), 'The Role of Non-Governmental Organizations in the Development of International Environmental Law', *Chicago Kent Law Review*, **68** (1), 61–76

Tarlock, Dan (1997), 'Exclusive Sovereignty versus Sustainable Development of a Shared Resource: The Dilemma of Latin American Rainforest Management', *Texas International Law Journal*, **32** (1), 37–66

Tarrant, Michael, and Ken Cordell (2002), 'Amenity Values of Public and Private Forests: Examining the Value-Attitude Relationship', *Environmental Management*, **30** (5), 692–703

Taylor, Dorceta (2000), 'The Rise of the Environmental Justice Paradigm: Injustice Framing and the Social Construction of the Environmental Discourses', *American Behavioral Scientist*, **43** (4), 508–580

Taylor, Peter Leigh (2005), 'In the Market but not of it: Fair Trade Coffee and Forest Stewardship Council Certification as Market-Based Social Change', *World Development*, **33** (1), 129–147

Teisl, Mario (2003), 'What We May Have Is a Failure to Communicate: Labelling Environmentally Certified Forest Products', *Forest Science*, **49** (5), 668–680

Teixeira, Marcos, Marcus Luke Murray and M.G. Carvalho (2006), 'Assessment of Land Use and Land Use Change and Forestry (LULUCF) as CDM projects in Brazil', *Ecological Economics*, **60** (1), 260–270

The World Justice Project (2011), 'Dimensions on the Rule of Law', available at http://worldjusticeproject.org/rule-of-law-index/dimensions (accessed 20 January 2012)

Thorpe, Amelia, and Kristy Graham, 'A Payment for Ecosystem Services Scheme in Papua New Guinea: Incorporating Lessons from Costa Rica's PSA' (paper presented at the New Zealand Centre for Environmental Law Conference 2009: Property Rights and Sustainability, Auckland, New Zealand, 16–18 April 2009)

UNEP, *Convention on Biological Diversity: Summary of the Second Global Biodiversity Outlook*, 8th mtg, Agenda Items 13 and 20, UN Doc UNEP/CBD/COP/8/12 (15 February 2006) annex

UNEP, *Declaration of the United Nations Conference on the Human Environment*, 21st plen mtg, UN Doc A/Conf.48/14/Rev. 1 (16 June 1972)

Unerman, Jeffery (2007), 'Stakeholder Engagement and Dialogue', in Jeffrey Unerman, Jan Bebbington and Brendan O'Dwyer (eds), *Sustainability Accounting and Accountability*, London; New York: Routledge, pp. 86–89

UNFCCC, *Land Use, Land-Use Change and Forestry*, Decision 11/CP.7, UN Doc FCCC/CP/2001/13/Add.1 (10 November 2001), available at http://maindb.unfccc.int/library (accessed 16 January 2008)

UNFCCC, *Report of the Conference of the Parties on its Seventh Session, held at Marrakesh from 29 October to 10 November 2001*, 7th sess, UN Doc FCCC/CP/2001/13/Add.1 (21 January 2002)

UNFF Secretariat (2006), 'Composite Draft Text for a Non-Legally Binding Instrument on all Types of Forests', available at http://www.un.org/esa/forests/pdf/aheg/nlbi/composite_text_NLBI.pdf (accessed 21 March 2012)

United Nations, Brundtland Report (1972), *Our Common Future*, available at http://www.un-documents.net/wced-ocf.htm (accessed 26 March 2012)

United Nations, Commission on Human Rights, *Report of the Special Rapporteur on the Situation of Human Rights and Fundamental Freedoms of Indigenous People, Rodolfo Stavenhagen*, 61st sess, Agenda Item 15, UN Doc A/HRC/4/32 (27 February 2007)

United Nations Conference on Environment and Development, *Non-Legally Binding Authoritative Statement of Principles for a Global Consensus on the Management, Conservation and Sustainable Development of all Types of Forests*, UN Doc A/Conf 151/126 (14 August 1992)

United Nations Convention on Climate Change (2007), 'Intergovernmental Organisations and the Climate Change Process', available at http://unfccc.int/parties_and_observers/igo/items/3720.php (accessed 9 January 2008)

United Nations Convention on Climate Change (2007), 'List of Admitted

International Organisations', available at http://maindb.unfccc.int/public/igo.pl?mode=wim (accessed 9 January 2008)

United Nations Declaration on the Rights of Indigenous People, GA Res 61/295, 107th plen mtg, UN Doc A/61/L.67/ (13 September 2007)

United Nations Forum on Forestry (2004), *Ad hoc Expert Group on Consideration with a View to Recommending the Parameters of a Mandate for Developing a Legal Framework on all Types of Forests*

United Nations Forum on Forestry (2007), 'About UNFF', available at http://www.un.org/esa/forests/about.html (accessed 2 December 2008)

United Nations Forum on Forestry (2007), 'Member States and Bureau', available at http://www.un.org/esa/forests/about.html (accessed 2 December 2008)

United Nations Forum on Forestry (2007), 'Multi-year Programme of Work 2007-2015', available at http://www.un.org/esa/forests/multi-year-work.html (accessed 1 December 2009)

United Nations Forum on Forestry, *Non-Legally Binding Instrument on all Types of Forests*, 2nd Comm, 62nd sess, Agenda Item 54, UN Doc A/C.2/62/L.5 (22 October 2007) annex

United Nations Forum on Forestry (2007), 'UNFF Secretariat', available at http://www.un.org/esa/forests/secretariat.html (accessed 2 December 2008)

United Nations Framework Convention on Climate Change, *Decision 2/CMP.6, The Cancun Agreements: Land Use, Land Use Change and Forestry*, UN Doc FCCC/KP/CMP2010/12/Add.1 (15 March 2011)

United Nations Framework Convention on Climate Change, *Decision 2/CP.13 Reducing Emissions from Deforestation in Developing Countries: Approaches to Stimulate Action*, UN Doc FCCC/CP/2007/6/Add.1 (14 March 2008)

United Nations Framework Convention on Climate Change, *Decision 13/CP.10 Incorporation of the Modalities and Procedures for Afforestation and Reforestation Project Activities Under the Clean Development Mechanism Guidelines Under Article 7 and 8 of the Kyoto Protocol,* UN Doc FCCC/CP/2004/10/Add.2 (19 April 2005)

United Nations Framework Convention on Climate Change, Decision CMP.7, Land Use, Land Use Change and Forestry (2011), available at http://unfccc.int/files/meetings/durban_nov_2011/decisions/application/pdf/awgkp_lulucf.pdf (accessed 14 March 2012)

United Nations Framework Convention on Climate Change: Clean Development Mechanism (2006), 'Facilitating Reforestation for Guangxi Watershed Management in Pearl River Basin', available at http://cdm.unfccc.int/Projects/DB/TUEV-SUED1154534875.41/ (accessed 13 February 2008)

United Nations Framework Convention on Climate Change (2005), 'Input to UNFCCC on Double Counting and Methodological Issues', available at http://cdm.unfccc.int/public_inputs/meth_double_counting/DC4_Inputs_DoubleCounting_MA_SS.pdf (accessed 27 June 2008)
United Nations Framework Convention on Climate Change (2007), 'Meetings', available at http://unfccc.int/meetings/items/2654.php (accessed 7 January 2008)
United Nations Framework Convention on Climate Change, opened for signature 4 June 1992, 1771 UNTS 107 (entered into force 21 March 1994)
United Nations Framework Convention on Climate Change (2008), 'Reducing Emissions from Deforestation in Developing Countries: Approaches to Stimulate Action', available at http://unfccc.int/resource/docs/2008/sbsta/eng/misc04.pdf (accessed 27 June 2008)
United Nations Framework Convention on Climate Change (2007), 'The Secretariat', available at http://unfccc.int/secretariat/items/1629.php (accessed 7 January 2008)
United Nations Institute for Training and Research (UNITAR), *Progressive Development of the Principles and Norms of International Law Relation to the New International Economic Order*, UNITAR/DS/5 (15 August 1982)
United Nations, *International Tropical Timber Agreement 2006*, 4th pt, Agenda Item 7, UN Doc TD/TIMBER.3/12 (1 February 2006)
UN REDD Programme (2009), 'Supporting Countries to Get Ready for REDD', available at http://www.un-redd.org/Home/tabid/565/language/en-US/Default.aspx (accessed 2 February 2010)
UN REDD Programme (2012), 'About the UN REDD Programme', available at http://www.un-redd.org/AboutUNREDDProgramme/tabid/583/Default.aspx (accessed 8 February 2012)
Vahenen, Tina, 'Introduction to REDD', (paper presented at the World Forestry Congress, Argentina, 18–23 October, 2009)
Van der Vyver, Johan (2009), 'The Environment: State Sovereignty, Human Rights and Armed Conflict', *Emory International Law Review*, **23** (1), 85–112
van Kooten, G. Cornelis, Harry Nelson and Ilan Vertinsky (2005), 'Certification of Sustainable Forest Management Practices: A Global Perspective on Why Countries Certify', *Forest Policy and Economics*, **7** (6) , 857–867
Vanderzwagg, D. and D. Mackinlay (1996), 'Towards a Global Forest Convention: Getting out of the Woods and Barking up the Right Tree' in Johnson S. (ed.), *Global Forests and International Environmental Law*, Boston: Kluwer Law International, pp. 1–4
Vedeld, Paul, Arild Angelsen, Jan Bojo, Espens Sjaastad, Gertrude

Kobugabe Berg (2007), 'Forest Environmental Incomes and the Rural Poor', *Forest Policy and Economics*, **9** (7), 869–879

Verheyen, Roda (2005), *Climate Change Damage and International Law: Prevention, Duty and State Responsibility*, Leiden; Boston: Martinus Nijhoff Publishers

Victor, David, Kal Raustialia and Eugen Sknolnikoff (eds) (1998), *The Implementation and Effectiveness of International Environmental Commitments: Theory and Practice*, Laxenburg, Austria: International Institute for Applied Systems Analysis; Cambridge, Mass: MIT Press

Visseren-Hamakers, Ingrid, and Peter Glasbergen (2007), 'Partnerships in Forest Governance', *Global Environmental Change*, **17** (3–4), 408–417

Vogel, David (ed.) (1995), *Trading Up: Consumer and Environmental Regulation in a Global Economy*, Cambridge, Mass: Harvard University Press

Wang, Sen (2004), 'One Hundred Faces of Sustainable Forest Management', *Forest Policy and Economics* **6** (3–4), 205–213

Watson, Robert, Ian R. Noble, Bert Bolin, N. H. Ravindranath, David J. Verardo and David J. Dokken, Intergovernmental Panel on Climate Change (2000), *IPCC Special Report Land Use, Land-Use Change and Forestry: Summary for Policymakers*

Weale, A. (ed.) (1992), *The New Politics of Pollution*, Manchester; New York: St Martin's Press

Webster Dictionary (2008), 'Definition of Permanence', available at http://www.webster dictionary.net/definition/permanence (accessed 8 May 2008)

Weiss, Edith Brown (1984), 'The Planetary Trust: Conservation and Intergenerational Equity', *Ecology Law Quarterly*, **11** (4), 495–582

Weiss, Edith Brown (ed.) (1989), *In Fairness to Future Generations: International Law, Common Patrimony and Intergenerational Equity*, Tokyo, Japan: United Nations University; Dobbs Ferry N.Y.: Transnational Publishers

Weiss, Edith Brown (1993), 'International Environmental Law: Contemporary Issues and the Emergence of a New World Order', *Georgetown Law Journal*, **81** (3), 675–710

Weiss, Edith Brown (1995), 'Environmental Equity: The Imperative for the Twenty-First Century' in Winfried Lang (ed.), *Sustainable Development and International Law*, London; Boston: Graham & Trotman/M. Nijhoff, pp. 7–23

Weiss, Edith Brown (1999), 'Understanding Compliance with International Environmental Agreements: The Bakers Dozen Myth', *University of Richmond Law Review*, **32** (5), 1555–1589

White, Andy, and Alejandra Martin, Forest Trends, *Who Owns the World's Forests? Forest Tenure and Public Forests in Transition* (2002)

Wilkinson, Jessica and Jared Thompson, Environmental Law Institute (2006), *2005 Status Report on Compensatory Mitigation in the United States*
Woods, Ngaire (2000), 'The Challenge of Good Governance for the IMF and the World Bank Themselves', *World Development*, **28** (5), 823–841
World Bank (2004), *Sustaining Forests: A Development Strategy*
World Bank (2007), *A Guide to the World Bank*, second ed.
World Bank (2002), 'Bank Policy 4.36 – Forests', available at http://web.worldbank.org/WBSITE/EXTERNAL/PROJECTS/EXTPOLICIES/EXTOPMANUAL/0,,contentMDK:20141282~menuPK:4564187~pagePK:64709096~piPK:64709108~theSitePK:502184,00.html (accessed 26 March 2012)
World Bank (2002), 'Operational Policy 4.36 – Forests', available at http://web.worldbank.org/WBSITE/EXTERNAL/PROJECTS/EXTPOLICIES/EXTOPMANUAL/0,,contentMDK:20064668~menuPK:64701637~pagePK:64709096~piPK:64709108~theSitePK:502184~isCURL:Y,00.html (accessed 26 March 2012)
World Bank (2009), 'Exploring a New Approach to Partnerships in Forestry', available at http://web.worldbank.org/WBSITE/EXTERNAL/TOPICS/EXTARD/EXTFORESTS/0,,contentMDK:21840644~pagePK:210058~piPK:210062~theSitePK:985785,00.html (accessed 26 March 2012)
World Bank Group (1989), 'International Bank for Reconstruction and Development Articles of Agreement, as amended effective February 16', available at http://web.worldbank.org/WBSITE/EXTERNAL/EXTABOUTUS/0,,contentMDK:20049557~menuPK:58863~pagePK:43912~piPK:44037~theSitePK:29708,00.html (accessed 1 September 2008)
World Commission on Environmental and Development (1990), *Our Common Future*
Wulder, Michael, Stephanie M. Ortlepp, Joanne C. White, Nicholas C. Coops and Sam B. Coggins (2009), 'Monitoring the Impacts of Mountain Pine Beetle Mitigation', *Forest Ecology and Management*, **258** (7), 1181–1187
Wunder, Sven, 'Payments for Environmental Services: Some Nuts and Bolts' (Occasional Paper No 42, CIFOR, 2005)
Yasmin, F. and J. Depledge (eds) (2004), *The International Climate Change Regime: A Guide to Rules, Institutions and Procedures*, Cambridge, UK; New York: Cambridge University Press
Young, Iris (ed.) (1990), *Justice and the Politics of Difference*, Princeton, N.J.: Princeton University Press
Zellmer, Sandra and Jessica Harder (2008), 'Unbundling Property in Water', *Alabama Law Review*, **59** (3), 679–745

Zomer, Robert, Antonio Trabucco, Deborah A. Bossio and Louis V. Verchot (2008), 'Climate Change Mitigation: A Spatial Analysis of Global Land Suitability for Clean Development Mechanism Afforestation and Reforestation', *Agriculture, Ecosystems and Environment*, **126** (1–2), 67–80

Index